T0238003

Lecture Notes in Civil Engineering

Volume 235

Lecture Notes in Civil Engineering (LNCE) publishes the latest developments in Civil Engineering—quickly, informally and in top quality. Though original research reported in proceedings and post-proceedings represents the core of LNCE, edited volumes of exceptionally high quality and interest may also be considered for publication. Volumes published in LNCE embrace all aspects and subfields of, as well as new challenges in, Civil Engineering. Topics in the series include:

- Construction and Structural Mechanics
- Building Materials
- Concrete, Steel and Timber Structures
- Geotechnical Engineering
- Earthquake Engineering
- Coastal Engineering
- Ocean and Offshore Engineering; Ships and Floating Structures
- Hydraulics, Hydrology and Water Resources Engineering
- Environmental Engineering and Sustainability
- Structural Health and Monitoring
- Surveying and Geographical Information Systems
- Indoor Environments
- Transportation and Traffic
- Risk Analysis
- Safety and Security

To submit a proposal or request further information, please contact the appropriate Springer Editor:

- Pierpaolo Riva at pierpaolo.riva@springer.com (Europe and Americas);
- Swati Meherishi at swati.meherishi@springer.com (Asia—except China, Australia, and New Zealand);
- Wayne Hu at wayne.hu@springer.com (China).

All books in the series now indexed by Scopus and EI Compendex database!

Shuren Wang · Jingan Li · Kui Hu · Xingxian Bao
Editors

Proceedings of the 2nd International Conference on Innovative Solutions in Hydropower Engineering and Civil Engineering

 Springer

Editors
Shuren Wang
Henan Polytechnic University
Jiaozuo, China

Jingan Li
Zhengzhou University
Zhengzhou, China

Kui Hu
Henan University of Technology
Zhengzhou, China

Xingxian Bao
China University of Petroleum, East China
Qingdao, China

ISSN 2366-2557   ISSN 2366-2565 (electronic)
Lecture Notes in Civil Engineering
ISBN 978-981-99-1825-6   ISBN 978-981-99-1748-8 (eBook)
https://doi.org/10.1007/978-981-99-1748-8

This Springer imprint is published by the registered company Springer Nature Singapore Pte Ltd.
The registered company address is: 152 Beach Road, #21-01/04 Gateway East, Singapore 189721, Singapore

Foreword

With the success of the past 2020 HECE conference, it is with this pride and honor that invite you to attend the 2022 2nd International Conference on Innovative Solutions in Hydropower Engineering and Civil Engineering, which was held online in Hangzhou, China, during December 25–27, 2022.

Despite being considered by many to be a mature field, Hydropower Engineering and Civil Engineering research continues to be a vibrant topic of research throughout the world, with the number of publications continually increasing over time all around the world. The continued interest in the application of these areas cries out for different new ways for sharing cutting-edge research in the field. It is important not only from a fundamental point of view but also for its engineering and industrial applications.

HECE 2022 brings you an immense opportunity to be a part of scientific acceleration to world-class personalities, young scholars, scientific delegates and young scientists to join in this conference to utilize the expertise and novelties that brings a new era for innovations in the field of which brings well-versed scrutinizers. It also provides a special channel to have open discussions, knowledge sharing and interactive sessions with field experts.

This volume includes rudimentary and challenging studies concerning the basic and advanced technology, theoretical developments, new empirical data and innovative method of civil engineering and other related research, such as structural engineering, construction materials and technology, bridge health monitoring, numerical simulation of structural vibration, reliability and durability of structures, tunnel construction, geotechnical and geological engineering, surveying engineering and hydraulic engineering, et al.

HECE 2022 is organized by China University of Geosciences (CUG) and Ryerson University Toronto, technically supported by CCCC, Beijing Jiaotong University, China, Xi'an Jiaotong University, China (XJTU), University of Missouri, University of Ferrara, Lincoln University, Texas A&M University-Kingsville, University of Texas, X-wave Innovations, Inc., China Agricultural University, etc.

On behalf of the HECE 2022 organizing committee, we sincerely appreciate the keynote speakers Prof. Guangliang Feng, Institute of Rock and Soil Mechanics,

Chinese Academy of Science, Prof. Lei, Junqing, Beijing Jiaotong University, China, Prof. Hany El Naggar, Dalhousie University, Canada and Prof. Yupeng Wang, Xi'an Jiaotong University, China (XJTU). We also extend our thanks to the international reviewers and the members of the program committee for their contribution and commitment to putting together a program of presentations. Meanwhile, thanks to all the participants to contribute your latest research in this proceeding.

Said Easa
Ryerson University
Toronto, Canada

Committee Members

Conference Chair

Prof. Said Easa, Ryerson University Toronto, Canada

Co-chairs

Prof. Wanqing Lu, China Construction Seventh Engineering Division Corp. Ltd., China
Prof. Ripendra Awal, Prairie View A&M University, USA
Prof. Alaeddin Bobat, University of Kocaeli, Turkey
Prof. İlgi Karapınar, Dokuz Eylul University, Turkey

Program Chairs

Prof. Chamil Abeykoon, Northwest Composites Centre, UK
Prof. Zhengji Li, China Construction Seventh Engineering Division Corp. Ltd., China

Editors

Prof. Shuren Wang, Henan Polytechnic University, China
Prof. Kui Hu, Henan University of Technology, China

Program Committee Members

Prof. S. K. Singh, Academy of Scientific Innovative Research (AcSIR)
Prof. Rohitashw Kumar, National Institute of Technology, India
Dr. Rui Xue, China Academy of Railway Sciences Co. Ltd., China
Dr. Samuel Bimenyimana, Huaqiao University, China
Dr. Ahmed Abed Gatea, University of Wasit, Kut, Iraq
Dr. Richard Grünwald, Yunnan University, China
Dr. Abadou Yacine, University of Djelfa, Algeria
Prof. Sherif Ali Younis, Egyptian Petroleum Research Institute, Egypt
Dr. Hongyang Chu, University of Science and Technology Beijing, China
Prof. Abdul Muntaqim Naji, Balochistan University of IT, Pakistan
Prof. Ramtin Moeini, University of Isfahan, Isfahan
Dr. Ghrieb Abderrahmane, University Ziane Achour, Algeria
Prof. Sadek Deboucha, University of Bourdj Bouarreridj, Algeria
Dr. Amin Atarodi, Semnan University, Iran
Prof. Firas Alrawi, University of Baghdad, Iraq

Contents

Effects of Web Stiffeners Locations on Flexural Capacities of SupaCee Sections About the Weak Axis

Ngoc Hieu Pham

Abstract The paper investigates the influence of variation of web stiffener locations on the sectional capacities of SupaCee sections under bending about the weak
axis. The SupaCee is the new section made on the basis of the traditional channel
section by adding several stiffeners in the web of the channel section to increase
stability. The variation of stiffener locations has specific impacts on the flexural
capacities of SupaCee sections about the weak axis. With the asymmetrical character
of the SupaCee section about the weak axis, the behaviour of this section is analysed
when the moment direction is changed. The flexural capacities of cold-formed steel
SupaCee sections are determined according to the Australian/New Zealand Standard
AS/NZS 4600:2018. Based on the investigated results, it is found that the behaviour
of SupaCee sections depends on the moment directions. Also, the web stiffeners
should be kept far from the flanges which will be more beneficial for the flexural
capacities of SupaCee sections about the weak-axis.

Keywords Web Stiffener locations · Flexural capacities · SupaCee · Weak axis

1 Introduction

Cold-formed steel structures have been progressively used in buildings due to their
convenience in manufacturing, fabrication, transportation and assembly. Their applications are seen in industrial or commercial buildings. More details of the applications
can be found in Yu et al. [1]. Cold-formed channel section is a popular product that
has been available in the worldwide market for decades. This section has a wide and
flat web that is very sensitive to local buckling. This drawback subsequently has been
solved by adding stiffeners in this web to create a new section called SupaCee. This
SupaCee section will be considered in this investigation.

In terms of design for cold-formed steel structures, Direct Strength Method (DSM)
can be seen as a new design method that has attracted the mentions of researchers and

N. H. Pham (✉)
Faculty of Civil Engineering, Hanoi Architectural University, Hanoi, Vietnam
e-mail: hieupn@hau.edu.vn

© The Author(s) 2023
S. Wang et al. (eds.), *Proceedings of the 2nd International Conference on Innovative
Solutions in Hydropower Engineering and Civil Engineering*, Lecture Notes in Civil
Engineering 235, https://doi.org/10.1007/978-981-99-1748-8_1

1

designers due to its advantages compared to the traditional effective width method (EWM) [2–4]. The new method allows the designers to easily determine the capacities of complex section shapes such as the SupaCee section. This method has been regulated in the Australian/New Zealand Standard AS/NZS 4600:2018 [5]. Elastic buckling analyses are compulsory for the application of this DSM method that can be carried out by using commercial software packages such as CUFSM [6] or THIN-WALL-2 [7, 8]. The design procedure was reported by Pham and Vu [9]. The DSM method will be used for the investigation in this paper.

In literature, a huge number of studies on channel or SupaCee sections under flexure have been available. Almost of the studies are flexure about the strong-axis [4, 10–15] whereas these on flexure about the weak-axis remained limited [16, 17]. Impacts of web stiffeners on the capacities of SupaCee sections under bending about the strong axis were studied in previous works [18] whereas their impacts on the weak-axis were not reported. Flexure about the weak axis of SupaCee sections can be found in numerous cases and this has a specific influence on the flexural capacities of SupaCee sections. The strength and behaviour of SupaCee sections under bending about the weak-axis are investigated in this paper. The distances between web stiffeners are subsequently varied in order to study their distance impacts on the flexural capacities about the weak-axis.

2 Sectional Capacities of Cold-Formed Steel Sections Under Bending Using the Direct Strength Method

In this paper, global buckling failures are prevented by using bracing systems as illustrated in Fig. 1 and discussed in Pham and Vu [9]. The member capacities can be determined as the sectional capacities with the replacement of the global buckling moment by the yield moment in the design formulae using the DSM method regulated in the Australian/New Zealand Standard AS/NZS 4600:2018 [5]. The flexural capacity is the smaller of the local buckling moment (M_{bl}) and the distortional buckling moment (M_{bd}) as follows:

$$M_s = \text{Min}\,(M_{bl},\,M_{bd}) \tag{1}$$

$$M_{bl} = \begin{cases} M_y & \text{for } \lambda_l \le 0.776 \\ \left[1 - 0.15\left(\frac{M_{ol}}{M_y}\right)^{0.4}\right]\left(\frac{M_{ol}}{M_y}\right)^{0.4} M_y & \text{for } \lambda_l > 0.776 \end{cases} \tag{2}$$

$$M_{bd} = \begin{cases} M_y & \text{for } \lambda_d \le 0.673 \\ \left[1 - 0.22\left(\frac{M_{od}}{M_y}\right)^{0.5}\right]\left(\frac{M_{od}}{M_y}\right)^{0.5} M_y & \text{for } \lambda_d > 0.673 \end{cases} \tag{3}$$

Fig. 1 The arrangement of
lateral bracing systems

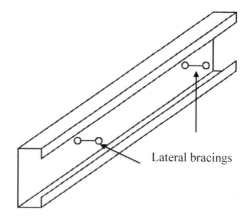

Lateral bracings

where $\lambda_l = \sqrt{M_y/M_{ol}}$; $\lambda_d = \sqrt{M_y/M_{od}}$; M_y is the yield moment; M_{ol} and M_{od} are the elastic local and distortional buckling moments respectively and are determined as presented in Sect. 3.

3 Behaviour and Strength of SupaCee Sections Under Flexure About the Weak Axis

SupaCee sections for the investigation are selected from the commercial sections provided by BlueScope Lysaght [19] as listed in Table 1, and their dimensions are illustrated in Fig. 2. Each section is labelled on the basis of its depth and thickness. For instance, "SC25015" means that: "SC" stands for SupaCee; "250" shows the nominal depth of 250 mm; "15" indicates the thickness of 1.5 mm. The grade G450 is used for the investigation as regulated in the Australian Standard AS1397 [20] including yield stress $f_y = 450$ MPa; Young's modulus E = 200 GPa.

Due to the asymmetry of SupaCee sections about the weak axis, the bending directions are considered in the investigation with the directions regulated in Fig. 3. Elastic buckling analyses are conducted by using the software program THIN-WALL-2 [7, 8]. The elastic buckling stresses are obtained and given in Table 2 for both two bending directions. For the negative direction, two lips are in tension whereas web

Table 1 The nominal dimensions of SupaCee sections

Sections	t	D	B	L_1	L_2	GS	S	α_1	α_2
SC25015	1.5	254	76	11	11	166	42	5	35
SC25019	1.9	254	76	11	11	166	42	5	35
SC25024	2.4	254	76	11	11	166	42	5	35

Note The inner radius $r_1 = r_2 = 5$ mm; t, D, B, L1, L2, GS, S (mm); α_1, α_2 $(^0)$

Fig. 2 Nomenclature for
SupaCee sections

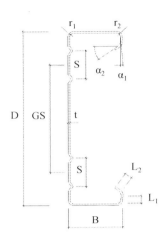

areas are in compression. This leads to the observation of local buckling in the web as illustrated in Fig. 4a, and distortional buckling does not occur in this situation. For the positive direction, tension is found in the webs and compression is observed in the lips and flange areas adjacent to the lips. Local buckling is seen in the flange areas and distortional buckling is also observed, as illustrated in Fig. 4b, c. Based on the obtained elastic buckling stresses, the flexural capacities of the investigated SupaCee sections are determined according to the DSM method as presented in Sect. 2. The results are given in Table 3. It is found that the local buckling occurs for the negative moment direction and this is distortional buckling for the positive direction.

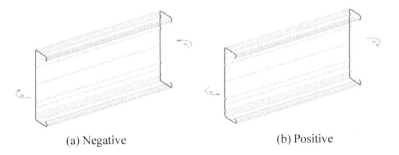

(a) Negative   (b) Positive

Fig. 3 The conventions of bending directions

Table 2 Sectional buckling stresses of SupaCee sections under flexure about the weak-axis

Sections	$M^{(-)}$	$M^{(+)}$	
	f_{ol} (MPa)	f_{ol} (MPa)	f_{od} (MPa)
SC25015	51.35	1114.47	479.47
SC25019	74.87	1536.89	604.03
SC25024	110.91	2245.01	807.68

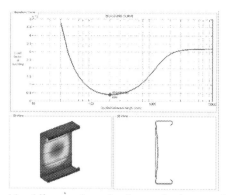

(a) The local buckling mode under flexure in the negative direction

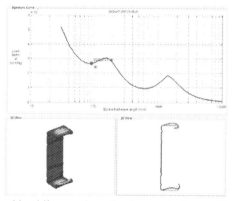

(b) The local buckling mode under flexure in the positive direction

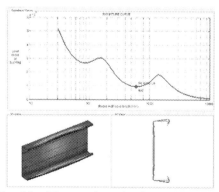

(c) The distortional buckling mode under flexure in the positive direction

Fig. 4 The behaviours of the investigated sections with the variation of bending directions

Table 3 Flexural capacities of SupaCee sections about the weak-axis (Unit: kNm)

Sections	$M^{(-)}$	$M^{(+)}$	$M^{(+)}/M^{(-)}$
SC25015	1.148	2.328	2.029
SC25019	1.743	3.327	1.909
SC25024	2.483	4.493	1.809

Table 3 shows that the capacities in the positive direction ($M^{(+)}$) are significantly higher than those in the negative direction ($M^{(-)}$). Also, it is found that the ratios $M^{(+)}/M^{(-)}$ are seen as a downward trend when the section thicknesses increase. This is explained that the increase in the thickness allows the webs to become more stable which has significant impacts on the negative moments $M^{(-)}$.

The negative moment is the detrimental case for the flexural capacities of SupaCee sections about the weak-axis, and this moment is also significantly affected by stiffeners. Therefore, the negative direction is used to investigate the impacts of web stiffener locations on the flexural capacities of SupaCee sections about the weak-axis, as presented in Sect. 4.

4 Effects of Web Stiffener Locations on the Flexural Capacities of SupaCee Sections About the Weak-Axis

As seen in Fig. 2, SupaCee has two couples of stiffeners, and these couples are close to the flanges. The distance between two couples is termed "GS". In this paper, two couples of stiffeners will reach the centroid of the webs; it means that the distance "GS" will decrease, as given in Table 4.

In terms of elastic buckling analyses, elastic buckling stresses are also determined by using the commercial software program THIN-WALL-2 for the negative bending direction, as listed in Table 4. These buckling stresses are utilised for the determination of flexural capacities of SupaCee sections as presented in Table 4.

Table 4 shows the increase in the negative moments if the distances "GS" decrease. It means that the flexural capacities of SupaCee sections become more beneficial if two couples of stiffeners are towards the centroid of the webs. The moment improvements are more significant for sections with small thicknesses. This improvement is about 6% for SC25015, but this is only about 3.6% for SC25024.

5 Conclusions

The paper investigates the behaviour of SupaCee sections under flexure about the weak-axis and the impacts of the distances between web stiffeners of these SupaCee sections on their flexural capacities. Based on the investigated results, several remarks are given as follows:

Table 4 Flexural capacities of SupaCee sections under flexure about the weak-axis

Sections	GS	f_{ol} (MPa)	M_{bl} (kNm)	Δ %
SC25015	166	51.35	1.15	–
	146	54.19	1.17	1.74%
	126	56.44	1.19	3.48%
	106	58.58	1.21	5.22%
	86	60.87	1.22	6.09%
SC25019	166	74.87	1.74	–
	146	77.45	1.77	1.72%
	126	80.27	1.79	2.87%
	106	83.3	1.81	4.02%
	86	85.38	1.83	5.17%
SC25024	166	110.91	2.48	–
	146	113.91	2.51	1.21%
	126	116.3	2.53	2.02%
	106	118.91	2.55	2.82%
	86	121.65	2.57	3.63%

Note Δ % is the flexural capacity deviation of SupaCee sections in comparison with the original SupaCee sections (GS = 166 mm)

The behaviour of SupaCee sections under bending about the weak-axis depends on the moment directions due to the asymmetry of these sections.

The failures of SupaCee sections are governed by the local buckling for the negative moment direction, and they are governed by distortional buckling for the positive moment direction. In general, the flexural capacities of SupaCee sections about the weak-axis are taken as the negative direction moments.

As the couples of stiffeners reach toward the centroid of the webs, the negative moments become more beneficial. These stiffener couples should be kept close to the centroid of the webs.

References

1. Yu WW, Laboube RA, Chen H (2020) Cold-formed steel design. Wiley
2. Schafer BW, Peköz T (1998) Direct strength prediction of cold-formed members using numerical elastic buckling solutions. In: Fourteenth international specialty conference on cold-formed steel structures
3. Schafer BW (2002) Local, distortional, and euler buckling of thin-walled columns. J Struct Eng 128(3):289–299
4. Schafer BW (2008) Review: the direct strength method of cold-formed steel member design. J Constr Steel Res 64:766–778

5. AS/NZS 4600-2018 (2018) Australian/New Zealand standard TM cold-formed steel structures. The Council of Standards Australia
6. Li Z, Schafer BW (2010) Buckling analysis of cold-formed steel members with general boundary conditions using CUFSM: conventional and constrained finite strip methods. Saint Louis, Missouri, USA
7. Nguyen VV, Hancock GJ, Pham CH (2015) Development of the thin-wall-2 for buckling analysis of thin-walled sections under generalised loading. In: Proceeding of 8th international conference on advances in steel structures
8. Nguyen VV, Hancock GJ, Pham CH (2017) New developments in the direct strength method (DSM) for the design of cold-formed steel sections under localised loading. Steel Constr 10(3):227–233
9. Pham NH, Vu QA (2021) Effects of stiffeners on the capacities of cold-formed steel channel members. Steel Constr 14(4):270–278
10. Rasmussen KJR, Hancock GJ (1993) Design of cold-formed stainless steel tubular members. II: Beams. J Struct Eng 119(8):2368–2386
11. Niu S (2014) Interaction buckling of cold-formed stainless steel beams. PhD Thesis, University of Sydney, Sydney, Australia
12. Niu S, Rasmussen KJR (2013) Experimental investigation of the local-global interaction buckling of stainless steel i-section beams. Research Report—University of Sydney, Department of Civil Engineering, no 944, pp 1–81
13. Niu S, Rasmussen KJR, Fan F (2015) Local—global interaction buckling of stainless steel i-beams. II: numerical study and design. J Struct Eng 141(8)
14. Wang L, Young B (2014) Design of cold-formed steel channels with stiffened webs subjected to bending. Thin-Walled Struct 85:81–92
15. Yu C, Schafer BW (2003) Local buckling tests on cold-formed steel beams. J Struct Eng 129(12):1596–1606
16. Oey O, Papangelis J (2020) Nonlinear analysis of cold-formed channels bent about the minor axis. In: Proceedings of the cold-formed steel research consortium colloquium
17. Oey O, Papangelis J (2021) Behaviour of cold-formed steel channels bent about the minor axis. Thin-Walled Struct 164:107781
18. Pham NH (2022) Impacts of web stiffener locations on capacities of cold-formed steel SupaCee Sections. In: 3rd international conference on building science, technology and sustainability (ICBSTS2022)
19. BlueScope Lysaght: Supapurlins Supazeds & Supacees. Blue Scope Lysaghts (2014)
20. AS1397:2011 (2011) Continuous hot-dip metalic coated steel sheet and strip-coating of zinc and zinc alloyed with aluminium and magnesium. Standards Australia

A Hand Method for Assessment of Maximum IDR and Displacement of RC Buildings

Kanat Burak Bozdogan⬤ and **Duygu Ozturk**⬤

Abstract Maximum displacement and the maximum interstorey drift ratio are the important factors for the measurement of the vulnerability of multistorey buildings. For this reason, in this paper a method was proposed to calculate the maximum displacement and maximum interstorey drift ratio (IDR) values. In this model, reinforced concrete multistorey structure was modeled as an equivalent flexural-shear frame. Maximum displacement and the maximum IDR were calculated according to the Equivalent Static Loads Method and The Response Spectrum Method using the continuum model and the results were tabulated. With the help of the obtained tables by this study, the maximum displacement and the maximum IDR of the regular multistorey structures can be calculated quickly and practically. The axial deformation of the vertical elements (columns and shear walls) were approximately considered in the study. The convergence of the presented method to the Finite Elements Method was investigated by two examples in the last part of the study.

Keywords Hand method · Maximum displacement · Maximum interstorey drift ratio (IDR) · Response spectrum method · Equivalent static loads method

1 Introduction

One of the effective and practical methods used in the analysis of shear wall-frame systems is the continuum method. There have been many studies dealing with the method which was first used by Chitty in 1940 [1]. These studies include various situations such as static analysis and dynamic analysis [2–26]. Stafford Smith et al. proposed an approximate method for calculating the displacement of the high-rise buildings under triangular distributed loads, taking account of the continuum model

K. B. Bozdogan (✉)
Canakkale Onsekiz Mart University, Canakkale, Turkey
e-mail: kbbozdogan@comu.edu.tr

D. Ozturk
Ege University, Izmir, Turkey

© The Author(s) 2023
S. Wang et al. (eds.), *Proceedings of the 2nd International Conference on Innovative Solutions in Hydropower Engineering and Civil Engineering*, Lecture Notes in Civil Engineering 235, https://doi.org/10.1007/978-981-99-1748-8_2

[27]. Heidebrecht ve Rutenberg proposed an approach to determine the performance of frame systems by utilizing the intersorey drift ratios. In the study, the frame system was modeled as an equivalent shear beam [28]. Gülkan and Akkar proposed a method for determining IDR of the structures of which the bearing system consisting of frames. In the study, the multistorey frame system was idealized as an equivalent shear beam to achieve the maximum IDR [29]. In Miranda and Akkar's work, they proposed a method of using graphs for response spectrum analysis by using the flexural-shear beam to obtain the IDR [16]. Xie and Wen used the Timoshenko beam model to calculate the IDRs of multistorey structures [30]. Khaloo and Khosravi investigated the effect of modes in the structures under near fault pulse like ground motions. In the study, the structure was idealized as an equivalent flexural shear beam according to the continuum method [31].

Yang, Pan and Li used the flexural-shear beam model to determine the maximum IDR of the structures under near-fault ground motion [32].

Fardipour et al. proposed a practical method of determining the maximum IDR for Australia, depending on the height of the building, the type of bearing system, the change in the stiffness and the mass [33]. Shodja and Rofooei developed a method based on the discrete mass model for determining the drift spectrum. In the method, the variation of the structure stiffness along the height of the structure is also taken into consideration [34]. Tekeli, Atımtay and Turkmen proposed a method for determining the lateral displacement of the reinforced concrete frame structures under the static loads. In the method, the frame system was considered as an equivalent shear frame [35]. In this study, a method has been proposed in which the maximum displacement and the maximum IDR are quickly and practically determined by means of tables for both Response Spectrum Analysis and Equivalent Static Loads Method. Subsequent paragraphs, however, are indented.

In creating the tables for the method, it is assumed that

- The properties of the building is constant and the mass is uniformly distributed up to the height of the building.
- The shear deformations of the columns, shear walls and the beams are neglected.
- The axial deformations of the beams are neglected.
- The torsion is neglected.
- The material is considered as linear elastic and the nonlinear geometric effects are ignored.
- Slabs are assumed infinitely rigid in their own planes.

2   Response Spectrum Analysis by the Presented Method

The reinforced concrete multistorey building can be modeled as an equivalent flexural-shear beam as seen in Fig. 1.

In accordance with this model, the 4th order parabolic partial differential equation of the system with undamped free vibration is written as follows.

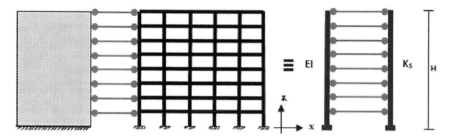

Fig. 1 The equivalent flexural-shear beam model of the structure

$$(EI)\frac{\partial^4 U}{\partial z^4} - K_s\frac{\partial^2 U}{\partial z^2} + \frac{m}{h}\frac{\partial^2 U}{\partial t^2} = 0 \tag{1}$$

where z is the vertical axis, and m/h is the distributed mass up to the height of the building. Ks is the equivalent shear stiffness and it is calculated by the following equation [4, 12].

$$K_s = \frac{12E}{h\left(\frac{1}{s} + \frac{1}{r}\right)} \tag{2}$$

where h is the storey height, E is the modulus of elasticity, r and s are the contributions of columns and beams to the shear stiffness which are calculated by the following equations.

$$s = \sum_{j=1}^{p} \frac{I_{cj}}{h} \tag{3}$$

$$r = \sum_{j=1}^{g} \frac{I_{bj}}{l_i} \tag{4}$$

In the combined shear wall- frame system in Fig. 2, the shear stiffness is calculated approximately by the following equation.

$$K_s = \frac{1.1Er}{h} \tag{5}$$

Here 1.1 is the correction factor.

The total bending stiffness of the shear walls and columns is shown as (EI) and it is calculated by the following equation [4, 12].

$$(EI) = \sum_{j=1}^{u} EI_{wj} + \sum_{j=1}^{p} EI_{cj} \tag{6}$$

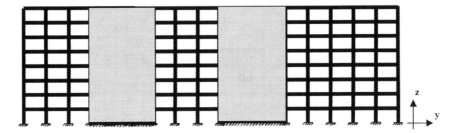

Fig. 2 The combined shear wall-frame system

If the partial differential Eq. (1) is separated into the variables using the Eq. (7), the ordinary differential equations in (8) and (9) are obtained.

$$U(z, t) = Y(z)T(t) \tag{7}$$

$$(EI)\frac{d^4y}{dz^4} - K_s\frac{d^2y}{dz^2} - \frac{m\omega^2}{h}\frac{d^2y}{dt^2} = 0 \tag{8}$$

$$\frac{d^2y}{dt^2} + \omega^2y = 0 \tag{9}$$

Here, y shows the modal shape and ω denotes the angular frequency. The boundary conditions of the 4th order constant coefficient homogeneous ordinary differential Eq. (8) are that, the displacement and rotation at the base of the structure are zero and the bending moment and the shear force at the top point are zero. These boundary conditions are given below.

$$y(0) = 0 \tag{10}$$

$$\frac{dy(0)}{dz} = 0 \tag{11}$$

$$(EI)\frac{d^2y(H)}{dz^2} \tag{12}$$

$$(EI)\frac{d^3y(H)}{dz^3} - K_s\frac{dy(H)}{dz} = 0 \tag{13}$$

To make differential Eq. (8) dimensionless, the following transformation can be used.

$$\varepsilon = \frac{z}{H} \tag{14}$$

Equation (15) is obtained when the transformation in Eq. (14) is applied to the Eq. (8).

$$\frac{d^4y}{d\varepsilon^4} - k^2 \frac{d^2y}{d\varepsilon^2} - \alpha y = 0 \tag{15}$$

With the aim of the reduction, the parameters in Eq. (15) are defined as follows

$$k = H\sqrt{\frac{K_s}{(EI)}} \tag{16}$$

$$\alpha = \frac{mH^4\omega^2}{(EI)h} \tag{17}$$

The solution of the differential Eq. (15) is obtained as follows.

$$y(\varepsilon) = c_1 \cosh(a_1\varepsilon) + c_2\sinh(a_1\varepsilon) + c_3 \cos(a_2\varepsilon) + c_4\sin(a_2\varepsilon) \tag{18}$$

Here, a_1 and a_2 are shown below.

$$a_1 = \sqrt{\frac{k^2 + \sqrt{k^4 + 4k^2\alpha}}{2}} \tag{19}$$

$$a_2 = \sqrt{\frac{-k^2 + \sqrt{k^4 + 4k^2\alpha}}{2}} \tag{10}$$

The boundary conditions given in the Eqs. (10), (11), (12) and (13) are written in dimensionless form as follows.

$$y(0) = 0 \tag{21}$$

$$\frac{1}{H}\frac{dy(0)}{d\varepsilon} = 0 \tag{22}$$

$$\frac{(EI)}{H^2}\frac{d^2y(1)}{d\varepsilon^2} \tag{23}$$

$$\frac{d^3y(1)}{d\varepsilon^3} - k^2\frac{dy(1)}{d\varepsilon} = 0 \tag{24}$$

If the non-dimensional boundary conditions are written in Eq. (18), the following frequency equation is obtained.

$$f = \left[a_2^4 + k^2a_2^2 + a_1^4 - k^2a_1^2\right] + \left[2a_1^2a_2^2 + k^2a_1^2 - k^2a_2^2\right]\cosh(a_1)\cos(a_2)$$

$$- \left[-a_2^3 a_1 + a_1^3 a_2 - k^2 a_1 a_2 \right] \sinh(a_1) \sin(a_2)$$
$$= 0 \tag{25}$$

If the necessary arrangements are made in this frequency equation, the following equation is obtained.

$$f = 2\alpha + \left[2\alpha + k^4 \right] \cosh(a_1) \cos(a_2) + (k^2 \sqrt{\alpha}) \sinh(a_1) \sin(a_2) = 0 \tag{26}$$

After calculating α values which make the Eq. (26) zero, with the help of Eq. (17), the natural vibration periods are found as follows.

$$T_i = Z_i H^2 \sqrt{\frac{m}{h(EI)}} \tag{27}$$

The Z_i period parameters in Eq. (27) are given in Table 1 depending on the k parameter for the first three modes.

Using the frequency equation and boundary conditions, the mode shapes are found by the following equation.

$$y = [\cosh(a_1 \varepsilon) - \cos(a_2 \varepsilon)] + E \left[\sinh(a_1 \varepsilon) - \frac{a_1}{a_2} \sin(a_2 \varepsilon) \right] \tag{28}$$

E can be calculated by the following equation.

$$E = -\frac{\left[a_1^2 \cosh(a_1) + a_2^2 \cos(a_2) \right]}{\left[a_1^2 \sinh(a_1) + a_1 a_2 \sin(a_2) \right]} \tag{29}$$

Table 1 Z values for the first three modes

k	Z_1	Z_2	Z_3	k	Z_1	Z_2	Z_3
0.0	1.788	0.285	0.102	11	0.328	0.103	0.056
1.0	1.529	0.276	0.101	12	0.304	0.096	0.053
2.0	1.160	0.254	0.098	13	0.282	0.090	0.050
3.0	0.908	0.227	0.094	14	0.264	0.085	0.047
4.0	0.744	0.200	0.089	15	0.248	0.080	0.045
5.0	0.631	0.178	0.083	16	0.234	0.075	0.043
6.0	0.547	0.160	0.078	17	0.221	0.072	0.041
7.0	0.483	0.144	0.073	18	0.209	0.068	0.039
8.0	0.432	0.132	0.068	19	0.199	0.065	0.037
9.0	0.391	0.121	0.064	20	0.190	0.062	0.036
10.0	0.357	0.111	0.060	30	0.129	0.042	0.025

The modal participation factor required for the Response Spectrum Analysis is calculated by the following equation as known from the literature [36].

$$\Gamma_i = \frac{\frac{m}{h}\int_0^1 y_i}{\frac{m}{h}\int_0^1 y_i^2} \tag{30}$$

For the i-th mode shape, the maximum displacement is calculated by the Eq. (31).

$$dep_i = \Gamma_i y_i(1) S_{di} = \mu_i S_{di} \tag{31}$$

where S_{di} is the displacement spectral ordinate for the i-th mode and it can be obtained either from the design spectrum, or from the displacement spectrum created for a particular earthquake record.

The μ_i values given here are calculated for the first three modes and they are given in Table 2.

By using the maximum displacements obtained for the three modes and with the help of the SRSS rule, the maximum displacements is calculated by Eq. (32).

$$d = \sqrt{\left(dep_1^2 + dep_2^2 + dep_3^2\right)} \tag{32}$$

For the dimensionless IDR, if the derivative of Eq. (28) is taken the following equation is obtained.

$$\frac{1}{H}\frac{dy(\varepsilon)}{d\varepsilon} = \frac{1}{H}[a_1 \sinh(a_1\varepsilon) + a_2 \sin(a_2\varepsilon)] + E[a_1 \cosh(a_1\varepsilon) - a_1 \cos(a_2\varepsilon)] \tag{33}$$

To find the maximum value of the Eq. (33), the derivation must be taken and equalized to zero.

Table 2 μ displacement coefficients for the first three modes

k	μ_1	μ_2	μ_3	k	μ_1	μ_2	μ_3
0.0	1.57	−0.87	0.51	11	1.31	−0.51	0.37
1.0	1.55	−0.85	0.5	12	1.30	−0.50	0.36
2.0	1.52	−0.82	0.50	13	1.30	−0.49	0.35
3.0	1.47	−0.77	0.48	14	1.30	−0.48	0.35
4.0	1.43	−0.70	0.47	15	1.29	−0.47	0.34
5.0	1.39	−0.66	0.47	16	1.29	−0.47	0.33
6.0	1.37	−0.62	0.44	17	1.29	−0.46	0.32
7.0	1.35	−0.59	0.42	18	1.29	−0.45	0.32
8.0	1.33	−0.58	0.41	19	1.29	−0.45	0.31
9.0	1.32	−0.56	0.38	20	1.29	−0.44	0.30
10.0	1.31	−0.52	0.37				

$$\frac{1}{H}\frac{d^2y_i}{d\varepsilon^2} = \frac{1}{H}\left[a_1^2\cosh(a_1\varepsilon) + a_2^2\cos(a_2\varepsilon)\right] + E\left[a_1^2\sinh(a_1\varepsilon) + a_1a_2\sin(a_2\varepsilon)\right] = 0 \tag{34}$$

ε_{imax} which makes the Eq. (34) equal to zero is different for three modes. For the ε values, which make the IDR maximum in related mode, IDRs for the other modes are calculated by the equation below

$$drift_{ij} = \Gamma_i\frac{1}{H}\left(\frac{dy_i(\varepsilon_{j\,max})}{d\varepsilon}\right)S_{di} = \beta_{ij}S_{di} \quad i = 1, 2, 3 \quad j = 1, 2, 3 \tag{35}$$

The β values in Eq. (35) are given in Tables 3, 4 and 5 considering the location of the maximum IDR in the first three modes. Also in Tables 6, 7 and 8, locations of maximum IDRs are given for the first three modes.

In this case, IDR values where the maximum IDR of the related mode is located are calculated according to the SRSS rule for the first three modes by the following equations.

$$dr_1 = \sqrt{\left(drift_{11}^2 + drift_{21}^2 + drift_{31}^2\right)} \tag{36}$$

$$dr_2 = \sqrt{\left(drift_{12}^2 + drift_{22}^2 + drift_{32}^2\right)} \tag{37}$$

$$dr_3 = \sqrt{\left(drift_{13}^2 + drift_{23}^2 + drift_{33}^2\right)} \tag{38}$$

And eventually, the maximum IDR is obtained as the largest of these three values.

Table 3 β values for the first mode

k	β_{11}	β_{21}	β_{31}	k	β_{11}	β_{21}	β_{31}
0.0	2.161	4.155	4.025	11	1.988	1.154	0.313
1.0	2.041	3.761	2.920	12	1.982	1.217	0.155
2.0	1.919	1.964	1.523	13	1.997	1.285	0.009
3.0	1.856	0.562	2.705	14	2.002	1.328	0.106
4.0	2.065	0.083	2.359	15	1.992	1.362	0.253
5.0	1.897	0.416	1.954	16	1.995	1.415	0.368
6.0	1.932	0.634	1.507	17	2.001	1.441	0.469
7.0	1.950	0.773	1.175	18	2.010	1.452	0.555
8.0	1.956	0.938	0.913	19	2.010	1.489	0.599
9.0	1.9820	1.036	0.653	20	2.010	1.492	0.694
10.0	1.971	1.071	0.456				

Table 4 β values for the second mode

k	β_{12}	β_{22}	β_{32}	k	β_{12}	β_{22}	β_{32}
0.0	2.161	4.155	4.025	11	1.082	2.221	1.188
1.0	2.032	3.984	3.919	12	1.077	2.208	1.126
2.0	1.757	3.678	3.781	13	1.069	2.188	1.129
3.0	1.469	3.279	3.361	14	1.061	2.159	1.251
4.0	1.291	2.845	2.810	15	1.056	2.133	1.143
5.0	1.184	2.645	2.385	16	1.071	2.161	1.007
6.0	1.134	2.495	1.939	17	1.047	2.110	1.097
7.0	1.115	2.398	1.591	18	1.055	2.083	1.086
8.0	1.085	2.402	1.500	19	1.051	2.087	1.161
9.0	1.077	2.360	1.292	20	1.051	1.931	1.004
10.0	1.081	2.230	1.164				

Table 5 β values for the third mode

k	β_{13}	β_{23}	β_{33}	k	β_{13}	β_{23}	β_{33}
0.0	2.161	4.155	4.025	11	0.636	1.763	2.432
1.0	2.032	3.984	3.999	12	0.626	1.731	2.398
2.0	1.746	3.670	3.828	13	0.625	1.729	2.351
3.0	1.407	3.232	3.604	14	0.646	1.762	2.368
4.0	1.129	2.709	3.425	15	0.627	1.702	2.335
5.0	0.934	3.116	3.245	16	0.621	1.664	2.295
6.0	0.803	2.188	2.997	17	0.619	1.681	2.247
7.0	0.720	2.002	2.817	18	0.627	1.654	2.265
8.0	0.675	1.992	2.695	19	0.652	1.710	2.217
9.0	0.640	1.901	2.514	20	0.619	1.608	2.161
10.0	0.626	1.747	2.446				

$$dr_{\max} = \max(dr_1, dr_2, dr_3) \tag{39}$$

3 Equivalent Static Load Analysis by the Presented Method

The shear wall-frame system under the triangular distributed load which represent the equivalent static loads is given in the Fig. 3.

In this case, the horizontal equilibrium equation is written as follows.

Table 6 Location of the maximum IDR for the first mode

k	ε	k	ε
0.0	1.0	11	0.272
1.0	0.832	12	0.261
2.0	0.617	13	0.249
3.0	0.496	14	0.240
4.0	0.430	15	0.231
5.0	0.389	16	0.223
6.0	0.359	17	0.215
7.0	0.336	18	0.208
8.0	0.316	19	0.202
9.0	0.300	20	0.196
10.0	0.285	30	0.152

Table 7 Location of the maximum IDR for the second mode

k	ε	k	ε
0.0	1.0	11	0.849
1.0	1.0	12	0.845
2.0	0.985	13	0.838
3.0	0.971	14	0.828
4.0	0.953	15	0.830
5.0	0.928	16	0.830
6.0	0.913	17	0.828
7.0	0.898	18	0.822
8.0	0.881	19	0.814
9.0	0.873	20	0.824
10.0	0.863	30	0.809

Table 8 Location of the maximum IDR for the third mode

k	ε	k	ε
0.0	1	11	0.707
1.0	0.981	12	0.702
2.0	0.937	13	0.699
3.0	0.885	14	0.699
4.0	0.833	15	0.695
5.0	0.797	16	0.688
6.0	0.771	17	0.693
7.0	0.748	18	0.688
8.0	0.737	19	0.688
9.0	0.726	20	0.687
10.0	0.714	30	0.671

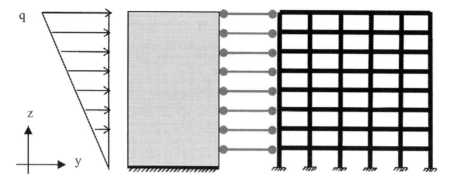

Fig. 3 The shear wall-frame system under the triangular distributed load

$$(EI)\frac{d^4y}{dz^4} - K_s\frac{d^2y}{dz^2} = q\frac{z}{H} \tag{40}$$

The boundary conditions of Eq. (40) are Eqs. (10), (11), (12) and (13).
To make the Eq. (40) dimensionless, the transform in Eq. (14) is used.

$$\frac{d^4y}{d\varepsilon^4} - k^2\frac{d^2y}{d\varepsilon^2} = q\frac{H^4}{EI}\varepsilon \tag{41}$$

The dimensionless parameter k is defined by the Eq. (16).
Equation (43) is obtained if the following definition is made in Eq. (41).

$$A = q\frac{H^4}{EI} \tag{42}$$

$$\frac{d^4y}{d\varepsilon^4} - k^2\frac{d^2y}{d\varepsilon^2} = A\varepsilon \tag{43}$$

The solution of the 4th order constant-coefficient nonhomogeneous differential equation in (43) is obtained as follows.

$$y(\varepsilon) = c_1 + c_2\varepsilon + c_3\cosh(k\varepsilon) + c_4\sinh(k\varepsilon) - A\frac{\varepsilon^3}{6k^2} \tag{44}$$

If the dimensionless boundary conditions in Eqs. (21), (22), (23) and (24) are applied in the Eq. (44), the displacement function is obtained as follows.

$$y(\varepsilon) = S_1[1 - \cosh(k\varepsilon)] + S_2\left[\varepsilon - \frac{1}{k}\sinh(k\varepsilon)\right] - A\frac{\varepsilon^3}{6k^2} \tag{45}$$

Here S_1 and S_2 are defined as below.

$$S_1 = A\left[\frac{\frac{1}{k^2} - \left[\frac{1}{k^3} - \frac{1}{2k}\right]\sin(k)}{k^2\cos(k)}\right] \tag{46}$$

$$S_2 = A\left[\frac{1}{k^4} - \frac{1}{2k^2}\right] \tag{47}$$

In the Equivalent Static Loads Method, the following equation can be written for the base shear force.

$$V_b = \frac{qH}{2} \tag{48}$$

Equation (49) is obtained by using the Eqs. (42) and (48).

$$A = \frac{2V_b H^3}{EI} \tag{49}$$

With the help of Eq. (27), EI is obtained as follows.

$$EI = Z_1^2 H^4 \frac{m}{T_1^2 h} \tag{50}$$

Equation (51) is written by using Eqs. (49) and (50).

$$A = \frac{2V_b T_1^2 h}{Z_1^2 H m} \tag{51}$$

The total mass of the building can be written as:

$$M = \frac{Hm}{h} \tag{52}$$

Equation (53) can be written by the help of Eq. (52)

$$A = \frac{2V_b T_1^2}{Z_1^2 M} \tag{53}$$

According to the Equivalent Static Loads Method, base shear force can be written as follows.

$$V_b = S_{a1} M \tag{54}$$

Here, S_{a1} represents the design spectral acceleration value for the first mode.
Equation (55) is obtained if the Eq. (54) is written in Eq. (53).

$$A = \frac{2S_{a1}T_1^2}{Z_1^2} \tag{55}$$

If the Eq. (56) is written instead of the square of the period and the Eq. (57) as known from the structure dynamics is written instead of the design spectral acceleration value, Eq. (58) is obtained.

$$T_1^2 = \frac{4\pi^2}{\omega^2} \tag{56}$$

$$S_{a1} = S_{d1}\omega^2 \tag{57}$$

$$A = \frac{8\pi^2 S_{d1}}{Z_1^2} \tag{58}$$

Maximum displacement is calculated by the equation below with the help of Eqs. (45), (46), (47) and (58).

$$y(1) = \frac{8\pi^2 S_{d1}}{Z_1^2}\left[\frac{\frac{1}{k^2} - \left[\frac{1}{k^3} - \frac{1}{2k}\right]\sin(k)}{k^2\cos(k)}\right][1 - \cosh(k)]$$
$$+ \frac{8\pi^2 S_{d1}}{Z_1^2}\left[\frac{1}{k^4} - \frac{1}{2k^2}\right]\left[1 - \frac{1}{k}\sinh(k)\right] - \frac{8\pi^2 S_{d1}}{Z_1^2}\frac{1}{6k^2} \tag{59}$$

Equation (59) can be written as below.

$$y(1) = v(k^2)S_{d1} \tag{60}$$

In Eq. (60), the change in v values according to k^2 is given in Table 9. Using the displacement function in Eq. (45), IDR is found as below.

$$\frac{1}{H}\frac{dy(\varepsilon)}{d\varepsilon} = -\frac{1}{H}S_1[k\sinh(k\varepsilon)] + S_2\frac{1}{H}[1 - \cosh(k\varepsilon)] - A\frac{\varepsilon^2}{2Hk^2} \tag{61}$$

If the derivation of the function is taken to find the location where the IDRs are maximum in Eq. (61), Eq. (62) is obtained.

$$\frac{1}{H^2}\frac{d^2y(\varepsilon)}{d\varepsilon^2} = -\frac{1}{H^2}S_1[k^2\cosh(k\varepsilon)] + S_2\frac{1}{H^2}[-k\sinh(k\varepsilon)] - A\frac{\varepsilon}{H^2k^2} \tag{62}$$

ε values that make the Eq. (62) zero, show where the IDR is maximum. These ε values are calculated by Eq. (62) for different k values and given in Table 10.
By using Eqs. (46), (47) and (58) in Eqs. (61), (63) is obtained.

Table 9 Displacement coefficient for the equivalent static loads method

k	ν	k	ν
0.5	2.255	11	1.751
1.0	2.232	12	1.734
2.0	2.147	13	1.735
3.0	2.057	14	1.722
4.0	1.977	15	1.713
5.0	1.910	16	1.703
6.0	1.866	17	1.701
7.0	1.829	18	1.706
8.0	1.803	19	1.697
9.0	1.780	20	1.687
10.0	1.762	30	1.670

Table 10 Location of the maximum IDR for equivalent static loads method

k	ε	k	ε
0.5	0.943	11	0.272
1.0	0.817	12	0.260
2.0	0.599	13	0.250
3.0	0.484	14	0.240
4.0	0.422	15	0.231
5.0	0.384	16	0.223
6.0	0.356	17	0.216
7.0	0.334	18	0.209
8.0	0.315	19	0.202
9.0	0.299	20	0.196
10.0	0.285	30	0.153

$$\frac{1}{H}\frac{dy(\varepsilon)}{d\varepsilon} = -\frac{1}{H}\frac{8\pi^2 S_{d1}}{Z_1^2}\left[\frac{\frac{1}{k^2} - \left[\frac{1}{k^3} - \frac{1}{2k}\right]\sin(k)}{k^2\cos(k)}\right][k\sinh(k\varepsilon)]$$

$$+ \frac{1}{H}\frac{8\pi^2 S_{d1}}{Z_1^2}\left[\frac{1}{k^4} - \frac{1}{2k^2}\right][1 - \cosh(k\varepsilon)] - \frac{8\pi^2 S_{d1}}{Z_1^2}\frac{\varepsilon^2}{2Hk^2} \qquad (63)$$

If $\varepsilon = \varepsilon$max is written in Eq. (63), the maximum IDR can be obtained as below.

$$\frac{1}{H}\frac{dy(\varepsilon)}{d\varepsilon} = \frac{\eta(k^2)S_{d1}}{H} \qquad (64)$$

The η values given in Eq. (64) are calculated and presented in Table 11 depending on k^2

Table 11 Coefficient of the maximum IDR for the equivalent static loads method

k	η	k	η
0.5	3.036	11	2.609
1.0	2.907	12	2.596
2.0	2.666	13	2.607
3.0	2.593	14	2.595
4.0	2.592	15	2.587
5.0	2.599	16	2.574
6.0	2.613	17	2.576
7.0	2.617	18	2.586
8.0	2.620	19	2.574
9.0	2.615	20	2.561
10.0	2.609	30	2.542

4   Contribution of the Axial Deformation

In the presented method, the effect of axial deformations can be approximately taken into account with the approach known from the literature [12]. To this end, K_{Sa} is written instead of K_S and it is calculated by the following equation [12].

$$K_{sa} = r K_s \tag{65}$$

Here, r is correction coefficient and it is found with the help of the equation below.

$$r = \frac{T_{s1}^2}{T_{s1}^2 + T_{a1}^2} \tag{66}$$

$$T_{s1}^2 = 16H^2 \frac{m}{h K_s} \tag{67}$$

$$T_{a1}^2 = 3.195H^4 \frac{m}{h D} \tag{68}$$

D is the global axial stiffness and it is calculated using the following equation.

$$D = E \sum_{i=1}^{p} A_{ci} t_i^2 \tag{69}$$

A_{ci} is the plan area of the i-th column, t_i shows the distance of the column to the center of gravity in the plan.

5   Numerical Examples

In order to investigate the convergence of the presented method to the Finite Element Method, two examples were solved by the presented method and the results were compared with ETABS.

One of the examples was a 7-storey reinforced concrete structure consist of frames, whereas the second example was a reinforced concrete structure having 15-storey shear wall-frame system.

5.1   Example 1

7-storey reinforced concrete structure of which the bearing system consists of frames, given in Fig. 4 was analyzed by using the presented method in this study. For this purpose, both the Equivalent Static Loads Method and Response Spectrum Analysis were applied to the example. The results obtained by the presented method were compared with the ones obtained by ETABS.

The height of each storey was 3 m. All columns were 45 cm/45 cm, beams were 25 cm/50 cm and the modulus of the elasticity is 32,000 Mpa. The mass of the storeys were taken as 748,800 kg. The structure was considered in the second seismic zone, local site class Z3 and the seismic load reduction factor was eight for the analysis according to Turkish Earthquake Code.

The comparisons of the presented method and ETABS for the values of maximum displacement and the maximum IDR obtained for the Equivalent Static Loads Method were given in Table 12.

As shown in Table 12, the ratio of the maximum difference between the proposed method and ETABS is 9.52%

Fig. 4 Plan of the frame system

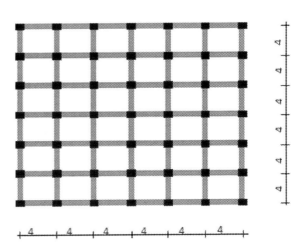

Table 12 The values of maximum displacement and the maximum IDR obtained for the equivalent static loads method (Example 1)

Equivalent static loads method		
	ETABS	Presented method
Max displacement (m)	0.0208	0.0205
Max IDR	0.001344	0.001472

Table 13 The values of maximum displacement and the maximum IDR obtained for the Response Spectrum Analysis (Example 1)

Response spectrum analysis		
	ETABS	Presented method
Max displacement (m)	0.0165	0.0156
Max IDR	0.00112	0.001139

The comparisons of the presented method and ETABS for the values of maximum displacement and the maximum IDR obtained for the Response Spectrum Analysis were given in Table 13.

As it is seen in Table 13, the maximum error according to the solutions by the Response Spectrum Analysis is 5.45%.

5.2   Example 2

15-storey reinforced concrete shear wall-frame structure seen in Fig. 5 was analyzed for the y direction by performing both the Equivalent Static Loads Method and the Response Spectrum Method. The results of the presented method and ETABS were compared.

Fig. 5 Plan of the shear wall frame system

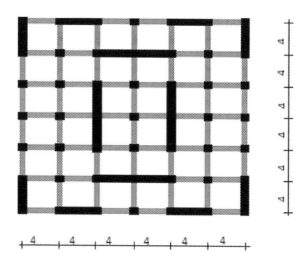

Table 14 Comparison of natural vibration periods for the first three modes (Example 2)

Periods (In Y Direction)		
Mode	ETABS	Presented method
1	0.749	0.772
2	0.196	0.183
3	0.089	0.073

Table 15 The values of maximum displacement and the maximum IDR obtained for the equivalent static loads method (Example 2)

Response spectrum analysis		
	ETABS	Presented method
Max displacement (m)	0.0299	0.03175
Max IDR	0.000812	0.000883

The height of each storey was 3 m, columns were 65 cm/65 cm and beams were 30 cm/60 cm, thickness of the shear wall, which was modeled by shell elements in ETABS was 50 cm. The modulus of the elasticity was 33,000 Mpa and the mass of the storeys were taken as 760,000 kg. The structure was considered in the second seismic zone, local site class Z3 and the seismic load reduction factor was six for the analysis according to Turkish Earthquake Code.

The comparison of the first three periods calculated for the structure was given in Table 14.

The comparison of the maximum displacement and the maximum IDR for Equivalent Static Loads Method was given in the Table 15.

It was seen that, in the analyzes made, both by the presented method and by ETABS the maximum IDR was occurred on the 8th floor level.

The maximum error in the results obtained by the Equivalent Static Loads Method was %8.74.

The comparisons of the maximum displacement and the maximum IDR results of two method obtained by the Response Spectrum Analysis were given in Table 16.

The maximum error in the results obtained by the Response Equivalent Static Loads Method was %9.16. Maximum IDR was occurred on the 8th floor level by the analysis made by the presented method and by ETABS.

Table 16 The values of maximum displacement and the maximum IDR obtained for Response Spectrum Analysis (Example 2)

Response spectrum analysis		
	ETABS	Presented method
Max displacement (m)	0.02086	0.0225
Max IDR	0.000575	0.000633

6 Conclusions

In this study, a method to determine the maximum displacement and the maximum interstory drift ratios by using tables has been proposed for the static and dynamic analysis of regular reinforced structures. By the method results of the analyses easily and quickly achieved. It has been observed that the Presented Method gave sufficient results with the Finite Element Method considering the examples that were solved at the end of the study. The approximation of the method is due to the approximation of the determination of the shear stiffness equation and the approximation of the axial deformations. Consequently, the proposed method can be used during the preliminary stage. Also, the method gives an idea about the behavior of the structure.

References

1. Chitty L (1947) On the cantilever composed of a number of parallel beams interconnected by cross bars. Philos Mag Londn Ser 7(38):685–699
2. Rosman R (1964) Approximate analysis of shear walls subject to lateral loads. Proc Am Concrete Inst 61(6):717–734
3. Rutenberg A, Heidebrecht AC (1975) Approximate analysis of asymmetric wall-frame structures. Build Sci 10(1):731–745
4. Murashev V, Sigalov E, Baikov VN (1976) Design of reinforced concrete structures. Mir Publishers, Moscow
5. Basu AK, Nagpal AK, Kaul S (1984) Charts for seismic design of frame-wall systems. J Struct Eng ASCE 110(1):31–46
6. Smith BS, Crowe E (1986) Estimating periods of vibration of tall building. J Struct Div ASCE 112(5):1005–1019
7. Smith BS, Yoon YS (1991) Estimating seismic base shears of tall wall-frame buildings. J Struct Eng ASCE 117(10):3026–3041
8. Mancini E, Savassi W (1999) Tall building structures unified plane panels behavior. Struct Des Tall Build 8:155–170
9. Ng SC, Kuang JS (2000) Triply coupled vibration of asymmetric wall–frame structures. J Struct Eng ASCE 126(8):982–987
10. Wang Y, Arnaouti C, Guo S (2000) A simple approximate formulation for the first two frequencies of asymmetric wall–frame multi-storey building structures. J Sound Vib 236(1):141–160
11. Swaddiwudhipong S, Lee SL, Zhou Q (2001) Effect of axial deformation on vibra-tion of tall buildings. Struct Des Tall Build 10:79–91
12. Zalka KA (2001) Simplified method for calculation of the natural frequencies of wall–frame buildings. Eng Struct 23(12):1544–1555
13. Hoenderkamp JCD (2002) Simplified analysis of asymmetric high-rise structures with cores. Struct Des Tall Build 11(2):93–107
14. Miranda E, Reyes CJ (2002) Approximate lateral drift demands in multistory buildings with nonuniform stiffness. J Struct Eng ASCE 128(7):840–849
15. Tarján G, László PK (2004) Approximate analysis of building structures with identical stories subjected to earthquakes. Int J Solids Struct 41(5–6):1411–1433
16. Miranda E, Akkar SD (2006) Generalized interstory drift spectrum. J Struct Eng ASCE 132(6):840–852
17. Clive LD, Harry EW (2007) Estimating fundamental frequencies of tall buildings. J Struct Eng ASCE 133(10):1479–1483

18. Georgoussis GK (2007) Approximate analysis of symmetrical structures consisting of different types of bents. Struct Des Tall Spec Build 16(3):231–247
19. Meftah SA, Tounsi A, El Abbas AB (2007) A simplified approach for seismic calculation of a tall building braced by shear walls and thin-walled open section structures. J Eng Struct 29(10):2576–2585
20. Laier JE (2008) An improved continuous medium technique for structural frame analysis. Struct Design Tall Spec Build 17(1):25–38
21. Rafezy R, Howson WP (2008) Vibration analysis of doubly asymmetric, three-dimensional structures comprising wall and frame assemblies with variable cross-section. J Sound Vib 318(1–2):247–266
22. Bozdogan KB (2009) An approximate method for static and dynamic analyses of symmetric wall-frame buildings. Struct Design Tall Spec Build 18(3):279–290
23. Takabatake H (2010) Two-dimensional rod theory for approximate analysis of building structures. Earthq Struct 1(1):1–19
24. Bozdogan KB (2011) A method for lateral static and dynamic analyses of wall-frame buildings using one dimensional finite element. Sci Res Essays 6(3):616–626
25. Wdowicki J, Wdowicka E (2012) Analysis of shear wall structures of variable cross section. Struct Des Tall Spec Build 21(1):1–15
26. Son HJ, Park J, Kim H, Lee HY, Kim DJ (2017) Generalized finite element analysis of high-rise wall-frame structural systems. Eng Comput 34(1):189–210
27. Stafford, Smith B, Kuster M, Hoenderkainp CD (1984) Generalized method for estimating drift in high-rise structures. J Struct Eng ASCE 110(7):1549–1562
28. Heidebrecht AC, Rutenberg A (2000) Applications of drift spectra in seismic design. In: Proceedings of 12WCEE, Auckland, NZ, New Zealand Society for Earthquake Engineering, Paper No. 209
29. Gülkan P, Akkar S (2002) A simple replacement for the drift spectrum. Eng Struct 24(11):1477–1484
30. Xie J, Wen Z (2008) A measure of drift demand for earthquake ground motions based on Timoshenko beam mode. In: The 14th world conference on earthquake engineering, Beijing, China
31. Khaloo AR, Khosravi H (2008) Multi-mode response of shear and flexural buildings to pulse-type ground motions in near-field earthquakes. J Earthq Eng 12(4):616–630
32. Yang D, Pan J, Li G (2010) Interstory drift ratio of building structures subjected to near-fault ground motions based on generalized drift spectral analysis. Soil Dyn Earthq Eng 30(11):1182–1197
33. Fardipour M, Lumantarna E, Lam N, Wilson J, Gad E (2011) Drift demand predictions in low to moderate seismicity regions. Aust J Struct Eng 15(3):195–206
34. Shodja AH, Rofooei FR (2014) Using a lumped mass, nonuniform stiffness beam model to obtain the interstory drift spectra. J Struct Eng ASCE 140(5)
35. Tekeli H, Atimtay E, Turkmen M (2015) An approximation method for design applications related to sway in RC framed buildings. Int J Civil Eng Trans A: Civil Eng 13(3):321–330
36. Chopra AK (2015) Dynamics of structures: theory and applications to earthquake engineering. Prentice Hall, Englewood Cliffs

Theoretical and Experimental Study on Multiparameter Criteria of Ultrasonic Method for Foundation Piles

Peng Huang, Zequan Yu, and Huiming Pan

Abstract As one of the important methods to survey the foundation pile integrity in most engineering industries in China, ultrasonic testing (UT) has such advantages as determining the position and scope of defects accurately and good operability. In this paper, on the basis of the in-depth study of China's industrial standards, main parameters for evaluating pile integrity in each standard were extracted, features and deficiencies of each parameter were analyzed in detail, and a multiparameter identification method of ultrasonically testing foundation pile integrity, which takes wave velocity, amplitude and basic frequency as the analytical parameters and outputs quantitative results, was proposed as an effective supplement to the ultrasonic testing of foundation pile integrity. This method has been applied to the result analysis of more than 800 engineering piles, and has obtained high consistency by comparing those results with core drilling results. Among them, $K_{(i)}$, the evaluation index of pile integrity with the pile complete and the buried acoustic pipe flat, is basically closed to the range $1 \leq K_{(i)} \leq 1.35$; when $K_{(i)} > 1.35$, all others, except for several points, are $K_{(i)}$ dispersion caused by head wave interpretation errors; when $0.85 \leq K_{(i)} < 1$, all acoustic lines within this range reflect slight or obvious abnormalities of sound velocity and amplitude and obvious distortion of waveforms, that is, this acoustic line has slight or obvious defects; when $K_{(i)} < 0.85$, all acoustic lines within this range reflect serious abnormalities of sound velocity and amplitude and obvious distortion of waveforms, that is, this acoustic line has serious defects. Conclusion: This method can effectively reduce the impact of changes in critical values of wave velocity and amplitude caused by the misinterpretation of the head wave position, thus providing a rapid and accurate evaluation for the ultrasonic testing of foundation pile integrity for later engineering practice reference.

P. Huang
Guangzhou Highway Co., Ltd., Guangzhou, China

Z. Yu (✉) · H. Pan
Guangdong Jianke Construction Engineering Quality Testing Center Co., Ltd., Guangzhou, China
e-mail: 244684443@qq.com

Guangdong Provincial Academy of Building Research Group Co., Ltd., Guangzhou, China

© The Author(s) 2023
S. Wang et al. (eds.), *Proceedings of the 2nd International Conference on Innovative Solutions in Hydropower Engineering and Civil Engineering*, Lecture Notes in Civil Engineering 235, https://doi.org/10.1007/978-981-99-1748-8_3

Keywords Foundation pile integrity · Acoustic parameters · Ultrasonic method · Multifactor probabilistic analysis

1 Introduction

The acoustic transmission method (ultrasonic method) is that the ultrasonic tester is used to measure the acoustic parameters of the pile concrete section point by point along the longitudinal axis of the pile, and determine the position, scope and extent of concrete defects by processing, analyzing and judging the test data, thereby inferring the concrete quality within the test range [1]. Ultrasonic testing (UT) of foundation pile integrity is mainly that pile concrete defects are qualitatively identified based on corresponding changes of wave velocity, amplitude and basic frequency in the received signals. The widely used probabilistic method can largely distinguish between accidental errors and negligent errors in concrete construction, but the UT method relies heavily on the on-site experience of test personnel and lacks automation and intelligence. After development in recent years, acoustic parameters develop from single factors to multiple factors, and judgment develops from being qualitative or empirical to quantitative. Concrete is a viscoelastic-plastic material that fluctuates to some extent in compactness, strength and other aspects, while the numerical statistics of acoustic parameters are basically in normal distribution; if there are pile defects (such as intercalated gouge, honeycomb, segregation) caused by the severe external environment or human errors, the concrete quality at the defect and the acoustic parameters of sound wave through the defective concrete will deviate from the normal distribution. However, concrete is not an ideal isotropic material, and fluctuations of strength and compactness in space caused by construction cannot be generalized as defects and deviations of different classes will also lead to fluctuations in the test values of acoustic parameters.

According to previous research results, there are great differences in the sensitivity of acoustic parameters such as wave velocity and amplitude to the quality of different forms of concrete. Wave velocity and amplitude reflect the elastic and plastic properties of concrete materials, reflectively. In terms of the multiparameter criteria of foundation piles tested by acoustic transmission method, Liu [2] conducted multi-angle and multi-level analysis and study on the pile foundation defects, and analyzed the characteristics of acoustic parameters corresponding to the common pile integrity defects tested by ultrasonic method. Zhang [3] introduced the advantages of the ultrasonic method in testing foundation piles and the theoretical basis of evaluating the quality of foundation piles using multiple synthetic criteria in fuzzy mathematics, and analyzed them based on engineering examples, proving that multivariate synthetic criteria can evaluate the quality of the foundation piles more accurately than univariate criteria. Zou [4] introduced a pile foundation test method that combines wavelet analysis with neural network, and conducted wavelet packet decomposition for the collected ultrasonic signals by wavelet analysis according to the ultrasonic propagation characteristics in the pile foundation;

the neutral work which is constructed after obtaining ultrasonic signal eigenvectors by wavelet analysis can effectively identify pile foundation defects and defect types.

In conclusion, it remains a difficult task to obtain the effective indexes for evaluating foundation pile integrity by analyzing single acoustic parameters with complex mathematical methods. In identifying the concrete defects of piles, it is more scientific and reasonable to comprehensively analyze the scope and severity of defects using multiple acoustic parameters.

2 Study on Relationship Between Concrete Strength and Ultrasonic Wave Velocity

In actual projects, specimens with concrete segregation, honeycomb, trench and other damages caused by external factors are usually characterized by low compressive strength. Moreover, concrete quality covers compressive strength, water-cement ratio, workability, durability, chlorine content, gas content and other indexes, while acoustic parameters obtained by UT can hardly directly reflect all the indexes of concrete. Thus, in this paper, a relationship was established between strength and wave velocity as one of the indexes to measure the quality of different concrete.

Assuming that concrete is an isotropic material composed of its composite materials and internal tiny cracks, it can be defined as follows using acoustic wave velocity [5]:

$$D_0 = 1 - \left(\frac{V_{p0}}{V_{pf}}\right)^2 \tag{1}$$

where, D_0—initial damage variable of concrete; V_{p0}—sound velocity of unloaded concrete; V_{pf}—sound velocity of concrete matrix (no damage).

The relationship between σ_c and D_0 is expressed by the power function curve [5]:

$$\sigma_c = A D_0^{-B} \tag{2}$$

where, A and B are parameters related to concrete E_c, v and ϕ, , and they are constants in the same concrete. The curve of the relationship between wave velocity and strength can be established by substituting (1) into (2):

$$\sigma_c = A(1 - C V_{p0}^2)^{-B} \tag{3}$$

In order to study the relationship between sound velocity and concrete failure strength, and establish the curve of the relationship between wave velocity and concrete strength, 450 * 450 * 150 mm concrete specimens were used in this test, which were cured under the temperature of 20 ± 5 °C and the relative humidity of

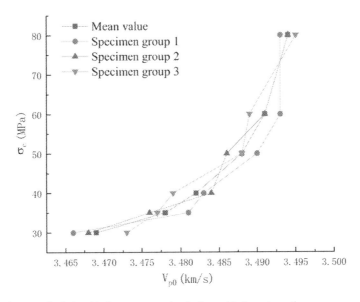

Fig. 1 Diagram of relationship between sound velocity and failure strength

90%. The concrete marks were C30, C35, C40, C50, C60 and C80, with three specimens in each group. The curve relationship was obtained by testing the compressive strength and wave velocity of the specimens (see Fig. 1).

Since the curve relationship obtained by (3) is not convenient to directly express the curve relationship of $\sigma_c - D_0$, the above formula is exponentially transformed and can be expressed as:

$$\sigma_c = A_1 e^{B_1 V_{p0}} \tag{4}$$

where, A_1 and B_1 are parameters related to concrete E_c, ν and ϕ, and are constants in the same concrete.

According to the above results, concrete failure strength is basically positively correlated with wave velocity. When the concrete strength level is greater than 40 MPa, the ultrasonic wave velocity in concrete significantly increases.

3 Theoretical Study on Multiparameter Synthetic Criteria and Pile Defect Identification

In Code for Testing of Building Foundation (DBJ/T 15-60-2019) [6], the depth profile integrity function of each pile is $I(j, i)$; in Technical Specifications for Foundation Piles Testing of Highway Engineering (JTG/T 3512-2020) [7], main discriminant features are from amplitude and wave velocity. In the two standards, whether the measured waveform is distorted or not is also taken as a supplementary basis for discriminant features. For the above problems, in the defect screening process in the first step, wave velocity, amplitude and basic frequency were selected to establish the main mathematical model of this algorithm.

Based on practical engineering experience, the sensitivity of single criteria to different types of defects is different, with different orders of magnitude. Common wave velocities are generally in 3800–4500 m/s; common amplitudes are generally in 90–120 dB; common basic frequencies are generally in 35–55 kHz.

Because acoustic parameters have different orders of magnitude, if weighting each parameter by analytic hierarchy process, a certain acoustic parameter may dominate the integrity evaluation index and changes of other parameters have little influence. Judged at the maximum value of wave velocity, amplitude and basic frequency by single parameters, the optimal concrete quality position can be determined. In this algorithm, each parameter was normalized, the acoustic line position was compared with the optimal concrete quality position in this profile to reflect the concrete quality level in this position, and the ratio of acoustic parameters of the acoustic line position to those of the optimal concrete quality position was limited to 0 to 1.

The multifactor probabilistic method adopts the three acoustic parameters of wave velocity V, amplitude A and frequency F. The amplitude and wave velocity are sensitive to defects, which is mainly reflected in the sharp decline of the two parameters, the basic frequency reflects waveform distortion, and obtaining one synthetic criterion critical value is considered to have defects.

Using the probability statistics in this algorithm and by the normal distribution principle, the measured values of acoustic parameters of defective concrete were mainly distributed in the outlier range and were less than the statistical critical value, then the ratio of normal concrete to statistical critical value of concrete was greater than or equal to 1, and those less than 1 were judged as outliers. The specific algorithm is established as follows:

$$K_{(i)} = \frac{V_i \cdot F_i \cdot A_i}{\frac{1}{n} \sum_{i=1}^{n} (V_i \cdot F_i \cdot A_i) - m\sigma_i} \tag{5}$$

where, V_i—the ratio of the actual wave velocity of the i-th acoustic line on the test profile to the maximum wave velocity on this profile; F_i—the ratio of the actual basic frequency of the i-th acoustic line on the test profile to the maximum basic frequency on this profile; A_i—the ratio of the actual amplitude of the i-th acoustic line on the test profile to the maximum amplitude on this profile; m—probability assurance coefficient; σ_i—standard deviation of $V_i \cdot F_i \cdot A_i$ for all calculated values of a single test profile; $K_{(i)}$—integrity evaluation index of this acoustic line.

The wave velocity, amplitude and basic frequency of the acoustic line were tested, and the integrity evaluation index of this acoustic line ($K_{(i)}$) was calculated. $K_{(i)} \geq 1$ indicates this acoustic line is complete; $K_{(i)} < 1$ indicates this acoustic line is defective. The lower the $K_{(i)}$, the higher the degree it deviates from the abnormal probability statistic, and the greater the degree of the defect.

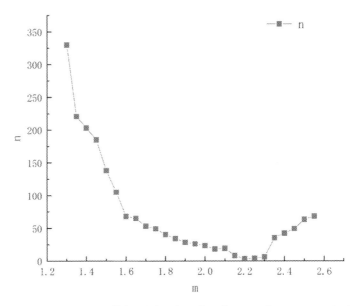

Fig. 2 Probability assurance coefficient (m) and number of non-conformance samples in acoustic line integrity results (n)

The acoustic line integrity function $I(j, i)$ was compared with the integrity evaluation index $K_{(i)}$ in this method. If $I(j, i) = 1$ and $K_{(i)} > 1$, it is denoted as a conformance sample. In this test, 836 acoustic lines containing complete concrete and defects were selected; the number of non-conformance samples in the integrity results of the two acoustic lines was reduced by adjusting the value of the probability assurance coefficient m, so as to determine the probability assurance coefficient m suitable for this method (see Fig. 2).

Based on the above calculation analysis, when $m = 2.2$, there were only nine non-conformance samples in the results of the two lines, with the accuracy rate reaching 98.9%.

In addition, the data tested and collected on the site are often affected by other sound sources or factors such as transducer vibration and aging, thereby resulting in the superposition of various sound source signals in ultrasonic data and the failure of the instrument to accurately and automatically identify the head wave position, and then an abnormal curve of acoustic parameters. There are many "spray waves" in the unprocessed waveform data graph. Interpretation based on the sound velocity-depth curve and amplitude-depth curve leads to a large number of outliers below the critical value, which makes it impossible to make interpretation accurately. After artificial interpretation of head wave, we can find that both sound velocity and amplitude are greater than the critical value, and the amplitude and wave velocity of the acoustic line on this profile are greater than the critical value (see Fig. 3).

As shown below, both waveforms belong to complete concrete without obvious distortion. It is hard to reflect the actual category of foundation pile integrity if

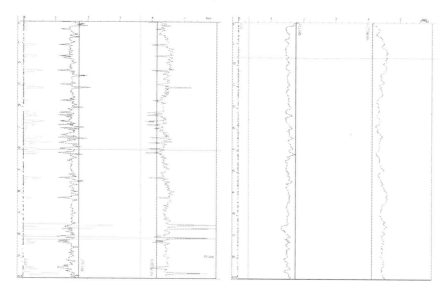

Fig. 3 Comparison between the waveform of head wave automatically interpreted by instrument and the waveform after manual adjustment

the unadjusted waveform is automatically interpreted by the traditional method. However, the concrete quality can be calculated by the integrity evaluation index $K_{(i)}$ in this paper (see Fig. 4).

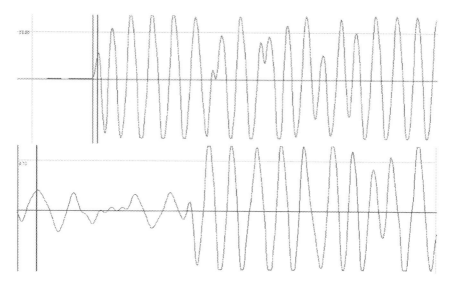

Fig. 4 Diagram of correct and abnormal interpretation of head wave of single wave

Different from traditional acoustic data, the impact of noise wave generated by other sound sources or factors such as transducer vibration and aging on the single wave curve is required. However, the above factors will not reduce the wave velocity, amplitude and basic frequency at the same time, namely $K_{(i)} < 1$. Therefore, the multi-parameter method is effective in identifying foundation pile integrity under the impact of the noise wave.

The above scatter plot was obtained according to the results of the integrity evaluation indexes of the measured 836 acoustic lines. The peak value of the samples in this test was 1.08; the integrity evaluation index $K_{(i)}$ was selected in the samples, and it was close to 1.0–1.35. When the integrity evaluation index $K_{(i)} > 1.35$, all others, except for several points, were $K_{(i)}$ discretions caused by head wave interpretation errors. When $0.85 \leq K_{(i)} < 1$, all acoustic lines within this range reflect slight or obvious abnormalities of sound velocity and amplitude and obvious distortion of waveforms, that is, this acoustic line has slight or obvious defects. When $K_{(i)} < 0.85$, all acoustic lines within this range reflect serious abnormalities of sound velocity and amplitude and obvious distortion of waveforms, that is, this acoustic line has serious defects (see Fig. 5).

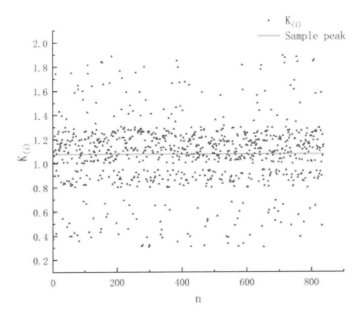

Fig. 5 $K_{(i)}$ scatter plot of acoustic line of the sample

4 Application of Multiparameter Synthetic Criteria of Ultrasonic Method

In an engineering, there was a 1# pile with the diameter of 1 400 mm and the length of 27.8 m, and three acoustic pipes were arranged at average in the reinforcement cage. The test results showed that the profile 1–2 was complete, the parameters of acoustic lines at the depth of 9.4–10.0 m on the profile 1–3 and 9.5–9.9 m on the profile 2–3 were seriously abnormal and the waveforms were distorted seriously, and the wave velocity-depth curve and amplitude-depth curve within this depth range were less than the critical value. Taking the depth of 9.6 m on the profile 1–3 as an example, the integrity evaluation index of this acoustic line $K_{(i)}$ was calculated. Tables 1 and 2 give a calculation of profile 1–3 and profile 2–3.

At 9.6 m of profile 1–3, the wave velocity of this acoustic line was 3 086 km/s, the amplitude was 78.1 dB, and the basic frequency was 25.5 kHz. Substituting them into (Formula 7), we obtained:

$$K_{(i)} = \frac{V_i \cdot F_i \cdot A_i}{\frac{i}{n} \sum_{i=1}^{n} (V_i \cdot F_i \cdot A_i) - 2.2\sigma_i} = 0.83 < 1$$

Based on the above calculation procedures, this was determined to be a serious defect (see Fig. 6).

After validation and test by drilled core method, concrete core samples were taken from the position 10 cm away from the center of the 2# acoustic pipe, and the concrete

Table 1 Integrity evaluation index of profile 1–3 within the depth range of 9.4–10.0 m	Depth (m)	Integrity evaluation index $K_{(i)}$
	9.4	0.91
	9.5	0.87
	9.6	0.83
	9.7	0.81
	9.8	0.85
	9.9	0.86
	10.0	0.93

Table 2 Integrity evaluation index of profile 2–3 within the depth range of 9.5–9.9 m	Depth (m)	Integrity evaluation index $K_{(i)}$
	9.5	0.91
	9.6	0.86
	9.7	0.84
	9.8	0.84
	9.9	0.87

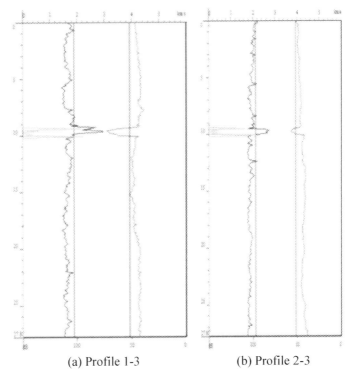

(a) Profile 1-3   (b) Profile 2-3

Fig. 6 Diagram of wave velocity and amplitude-depth curve of profiles 7 1–3 and 2–3

core samples were broken at about 9.5 m at the 8th time, with a length of 16 cm (see Fig. 7).

Fig. 7 Diagram of core sample drilled 10 cm away from the 2# acoustic pipe

5 Conclusion

Based on the study of China's industry standards, this paper extracted the main parameters for evaluating the pile integrity in each standard, analyzed the characteristics of each parameter, and completed the test on the basic theories and related parameters of UT of foundation pile integrity. On this basis, the method of comprehensively identifying the pile defect positions and severity according to sound velocity, amplitude and basic frequency was obtained by using the probability statistics method. A multiparameter synthetic identification method was proposed for the UT of foundation pile integrity, and the quantitative results calculated by this method could be used as an effective supplement to the UT of foundation pile integrity.

More than 800 acoustic lines were used for the statistics of integrity evaluation indexes, of which the integrity evaluation index $K_{(i)}$ with the pile complete and the buried acoustic pipe flat was basically close to $1 \leq K_{(i)} \leq 1.35$; when the integrity evaluation index $K_{(i)} > 1.35$, all others, except for several points, were $K_{(i)}$ dispersion caused by head wave interpretation errors; when $0.85 \leq K_{(i)} < 1$, all acoustic lines within this range reflected slight or obvious abnormalities of sound velocity and amplitude and obvious distortion of waveforms, that is, this acoustic line had slight or obvious defects; when $K_{(i)} < 0.85$, all acoustic lines within this range reflected serious abnormalities of sound velocity and amplitude and obvious distortion of waveforms, that is, this acoustic line had serious defects.

This method can effectively reduce the impact on the critical value changes caused by the misinterpretation of the head wave position, providing a method for rapidly and accurately interpreting the ultrasonic testing of foundation piles for later engineering practice reference.

References

1. Wu QZ (2009) Acoustic and dynamic testing technology of foundation pile. China Electric Power Press, Beijing
2. Liu YL, Zhou MR (2016) Corresponding parameter analysis on ultra-sonic transmission defect detection. China Build Mater Sci Technol 25(3):13–14
3. Zhang H, Hang GL (2010) Application of multivariate synthetic criteria in quality evaluation of foundation pile. J China Foreign Highway 30(5):204–207
4. Zou LL, Ye ZQ (2022) Pile foundation defect detection based on wavelet analysis and neural network. Nondestructive Testing 44(7):50–54
5. Zhao MJ, Wang K, Peng AH (2016) Identification technology with ultrasound imaging for hidden defects of concrete structure and its application. China Communications Press Co. Ltd, Beijing
6. DBJ/T 15-60-2019 code for testing of building foundation. Department of Housing and Municipal and Rural Construction, Guangzhou
7. JTG/T 3512-2020 Technical Specifications for Foundation Piles Testing of Highway Engineering. Ministry of Transport of the People's Republic of China, Beijing (2020)

Mechanical Properties and Micromechanism of Geopolymers to Replace Cement Stabilized Crushed Stone

Yongxiang Li, Lijuan Yang, Xiao Li, Yongfei Li, Qiang Zhang, and Shoude Pang

Abstract In order to realize the resource utilization of solid waste, the principle of alkali excitation is used to prepare geopolymers with fly ash, mineral powder and wet carbide slag as the main materials to replace part of the cement as the cementing material for the pavement base. Geopolymer-stabilized crushed stone was prepared by compounding cement and aggregate with geopolymer, and the unconfined compression strength, indirect tensile strength, compression rebound modulus, scour resistance and microscopic X-ray diffraction (XRD) and scanning electron microscopy (SEM) tests were carried out to study the effect of the change of geopolymer content on the mechanical properties of geopolymer-stabilized crushed stone and its mechanism. The test results show that when adding 30% geopolymer, the mechanical properties similar to those of cement can be obtained to a certain extent. XRD and SEM analysis showed that the geopolymer provided appropriate amount of silico-alumina and calcareous components to form calcium silicate hydrate (C–S–H) and calcium silicate (aluminum) hydrate (C–(A)–S–H) condensation. The glue can form a dense structure and increase the strength of the mixture.

Keywords Geopolymer · Stabilized crushed stone · Mechanical properties · Micromechanis

1 Introduction

Fly ash, mineral powder and wet carbide slag are industrial solid wastes, and improper treatment will cause environmental pollution. At present, the harmless treatment of industrial solid waste has become one of the problems that need to be solved when China implements "double carbon" measures. If it can be prepared into pavement base

Y. Li (✉) · L. Yang · X. Li
Inner Mongolia Agricultural University, No. 306 Zhaowuda Road, Saihan District, Hohhot, China
e-mail: lyxiang@imau.edu.cn

Y. Li · Q. Zhang · S. Pang
Inner Mongolia Jiaoke Road and Bridge Construction Co., Ltd, Huhhot, China

© The Author(s) 2023
S. Wang et al. (eds.), *Proceedings of the 2nd International Conference on Innovative Solutions in Hydropower Engineering and Civil Engineering*, Lecture Notes in Civil Engineering 235, https://doi.org/10.1007/978-981-99-1748-8_4

material, it can not only effectively solve the environmental pollution and ecological damage caused by a large number of solid wastes, but also provide a large number of raw materials for highway construction, and make solid wastes play a greater utilization value.

The comprehensive utilization of solid waste materials in road engineering is a research hotspot at present [1]. In 1980s, fly ash-coal gangue mixture was used in road base and subbase in the United States, and its application effect was confirmed by testing. Since the twenty-first century, the United States has mixed iron tailings crushed stone into building materials as the subbase and subgrade of asphalt concrete pavement, and achieved good application results [2]. In 2014, an airstrip was built with geopolymer concrete at West Wellcamp Airport in Brisbane, Australia [3]. In recent years, Shen [4] obtained the best mix ratio of steel slag, fly ash and phosphogypsum as road base materials, and found that its early strength was higher than that of lime-fly ash and lime-soil pavement base materials, and its long-term strength was much higher than that of cement stabilized aggregate, which could meet the relevant requirements of pavement base materials. Arulrajah [5] found that fly ash, slag and calcium carbide residue can be used to stabilize recycled building materials as road base or subbase through unconfined compression strength and compression rebound modulus tests. Hu [6] discussed the possibility of using fly ash and red mud geopolymer as the cementing material of crushed stone aggregate through unconfined compression strength, failure strain and drying shrinkage tests. The results showed that geopolymer stabilized aggregate had a stable application prospect in the application of flexible pavement. In China, the research on resource utilization of industrial solid waste started late. In 1990s, Maanshan Mine Research Institute applied iron tailings to pavement materials [7]. Chang'an University mixes lime and coal gangue with soil and uses it for road base [8]. In 1997, China built the first steel slag asphalt pavement test section [9]. In recent years, Beijing Research Institute of Mining and Metallurgy and Guangxi Pingguo Aluminum Company jointly developed a new type of red mud road base with good road performance using lime, fly ash, red mud and some additives as raw materials, which filled a domestic gap [10–12]. Qin and Liang [13] have also carried out the research on replacing part of cement with red mud and applied it in cement stabilized macadam base, which has certain reference value for the development of new materials for road base. Wang [14] through the tests of mechanical properties and crack resistance of cement stabilized macadam with three types of aggregate gradation, three cement dosages and fly ash, the results show that after adding fly ash, the early unconfined compression strength, flexural strength and splitting strength of cement stabilized macadam decrease, which helps to reduce early cracks. Xu [15] found that adding fly ash and slag to cement stabilized macadam can improve the mechanical properties and shrinkage characteristics of cement stabilized macadam through unconfined compression strength, indirect tensile strength. Chu [16] comprehensively analyzed the road performance of cement stabilized macadam mixture with iron tailings, and found that the addition of iron tailings can improve the unconfined compression strength, indirect tensile strength, water stability and frost resistance of cement stabilized macadam mixture. Liu [17] found that the mechanical properties of slag-based polymer and fly ash stabilized macadam are better than those of cement stabilized macadam in unconfined compression strength, splitting strength, compression rebound modulus.

Although some scholars and engineering units have carried out research and engineering demonstration on the preparation of pavement base materials from solid wastes, there are few technical researches on the preparation of pavement base materials from solid wastes at home and abroad based on the principle of geopolymer. In this paper, geopolymer cementitious materials prepared from fly ash, mineral powder and wet carbide slag by alkali excitation are used to replace part of cement stabilized macadam. The mechanical properties are evaluated by unconfined compression strength, indirect tensile strength, compression rebound modulus and scour resistance. The mineral composition of raw materials and microstructure changes of samples at different ages are analyzed by XRD analysis and SEM test, and the formation mechanism of geopolymer strength is analyzed, which provides a new method for comprehensive utilization of solid waste materials.

2 Experiment

2.1 Experiment Material

The raw materials used in the experiment are geopolymer cementing material, cement, aggregate, alkali activator and water.

Geopolymer cementitious materials include fly ash, mineral powder and wet carbide slag, and the main chemical compositions of effective minerals of the materials are shown in Table 1. The fly ash is low-calcium F-type fly ash with a median particle size of 12.72 μm; The mineral powder is S95 grade mineral powder mixed with 7% limestone, 15% fly ash and 78% slag. The moisture content of wet carbide slag is 33%.

Cement is taken from Jidong Cement Company, Hohhot, Inner Mongolia, and its strength grade is P.O42.5. The physical and mechanical properties of cement are tested according to the specification "Test Methods of Cement and Concrete for Highway

Table 1 Chemical compositions of materials

Composition	Fly ash	Mineral powder	WET carbide slag
SiO_2	44.9	29.9	3.20
Al_2O_3	42.7	18.6	1.34
CaO	4.74	34.6	93.4
Fe_2O_3	3.16	1.36	0.388
TiO_2	1.80	3.07	–
SO_3	–	2.73	0.498
MgO	–	6.92	–

Engineering" (JTG E30-2005), and all meet the requirements of the specification. See Table 2 for the test results.

Aggregates are divided into four grades, and the particle sizes of each grade are: 19–26.5 mm, 9.5–19 mm, 4.75–9.5 mm, 0–4.75 mm.

The alkali excited material is sodium hydroxide produced by Tianjin Zhiyuan Chemical Co., Ltd., which is a white translucent granular crystal.

2.2 Mix Proportion Design of Geopolymer Cementitious Materials

This study uses sodium hydroxide to stimulate fly ash, mineral powder and wet carbide slag to mix with water to prepare geopolymer cementitious materials. The total amount of materials is 2.58% with alkali, and the water-cement ratio is 0.5. The mix design is shown in Table 3. The setting time, fluidity and soundness indexes were measured, and the optimum formula was obtained by combining the compressive strength and drying shrinkage of geopolymer in 3d, 7d and 28d.

2.3 Performance Test of Geopolymer Cement Stabilized Crushed Stone

Mix proportion of Geopolymer Cement. It is necessary to control the cement dosage within a reasonable range to ensure certain mechanical properties and frost resistance and improve its crack resistance [18]. According to the recommended cement dosage value of cement stabilized material mixture ratio test in "Construction Guidelines for Highway Base and Subbase" (JTGT-F20-2015), 5% of total cement-geopolymer material dosage is proposed. Finally, the proportion of total cementitious materials is set at 5% and four groups of mixing ratios (Table 4) are determined for performance test.

Manufacture and testing of test pieces. Add a certain amount of water to the aggregate, mix well, and put it in a closed plastic bag for soaking. After soaking for four hours and within one hour before the specimen is formed, add geopolymer and cement materials with different mixing ratios, and stir well, and make the specimen within one hour; The model is made by Tiantong Tongda hydraulic part demoulding machine, and the mixture is poured into the test mold for three times and tamped evenly, and demoulded within 2–6 h. Immediately after weighing, put it in an airless plastic bag, seal it, tie the bag mouth tightly, and move it to a standard curing room with a temperature of 20 ± 2 °C and a relative humidity of over 95%; The specimen is placed on the iron frame, and the spacing is more than 10 cm. On the last day of the curing period, take out the specimen, observe whether the corners of the specimen

Table 2 Physical and mechanical properties of cement

	Fineness/%	Initial setting time/min	Final setting time/min	Soundness/mm	Rupture strength/MPa		Compressive strength/MPa	
					3 d	28 d	3 d	28 d
Test value	2.5	180	255	2.8	5.4	7.5	21.8	45.8
Specification value	≤ 10	≥ 45	≤ 600	≤ 5	≥ 3.5	≥ 6.5	≥ 17	≥ 42.5

Table 3 Geopolymer mix design

Sample number	Fly ash/%	Mineral powder/%	Wet carbide slag/%
A1	10	85	5
A2	0	88	12
A3	0	95	5
A4	20	75	5

Table 4 Design results of cement-geopolymer stabilized crushed stone mix proportions

Sample number	Proportion of geopolymer to cementitious materials (%)	Proportion of cement to cementitious materials (%)	Grading of aggregate
B1	0	100	0–4.75 mm
B2	30	70	4.75–9.5 mm
B3	60	40	9.5–19 mm 19–31.5 mm
B4	90	10	= 41%:14%:24%:21%

are worn or missing, and measure the quality. Then, soak the specimen in water at about 20 °C so that the water surface is about 2.5 cm above the top of the specimen.

In this paper, the unconfined compression strength, indirect tensile strength, compression rebound modulus and scour resistance tests are carried out according to "Test Methods of Materials Stabilized with Inorganic Binders for Highway Engineering" (JTG E51-2009). Cylindrical specimens with a height of 150 mm and a diameter of 150 mm are prepared, and the number of matched reference specimens in each group is 9. In the unconfined compressive strength test, the specimen should be cured for 7 days, 14 days and 28 days. During the test, the specimen should be immersed in water for one day in advance, and the universal testing machine should be used for the test. Indirect tensile strength and compressive resilience modulus test shall be conducted according to "Construction Guidelines for Highway Base and Subbase" (JTGT-F20-2015) after the specimen is cured to 28 days, 60 days and 90 days. The curing age of the anti-erosion test is 28 days, after which it is soaked for 24 h. Using the anti-erosion tester, the peak impact force is 0.5 MPa, the scouring frequency is 10 Hz, and the scouring time is 30 min.

2.4 Microscopic Analysis

X-ray diffraction (XRD) was used to analyze the phase composition of the four raw materials. The parameters were: 40 kV, 100 mA, Cu target, scanning speed 5°/min. The scanning range is 10–90; The instruments used for SEM are Hitachi S-4800 Field Emission Scanning Electron Microscope in Japan and BRUKER Energy

Spectrometer in Germany. The microscopic area of the sample is scanned by line and plane, and the distribution of elements, the types of elements, their atomic ratio and weight ratio are analyzed.

3 Results and Discussion

3.1 Mix Proportion Design of Geopolymer Cementitious Materials

The performance indexes of geopolymer cementitious materials with different mixing ratios are shown in Fig. 1. For cementitious materials, the setting time has an important influence on the setting degree and the strength of the mixture. The initial setting time should not be too early and the final setting time should not be too late. The fluidity affects the plasticity of cementitious materials, and the drying shrinkage and compressive strength are important indexes to examine the crack resistance and mechanical properties. As can be seen from Fig. 1, sample A1 has moderate initial setting time and final setting time, good fluidity and stability, low drying shrinkage coefficient and high compressive strength at 3 d, 7 d and 14 d; The initial setting time of sample A2 is relatively long, the final setting time is short, and the stability is poor. The initial setting time of sample A3 is short and its stability is poor. A4 sample has a long final setting time, poor fluidity and shrinkage. Considering all performance factors, sample A1 is selected as the mixture ratio of geopolymer cementing material.

3.2 Performance Test of Geopolymer Cement Stabilized Crushed Stone

Unconfined Compression Strength. The influence of geopolymer content change on the compressive strength of materials is shown in Fig. 2; It can be seen from Fig. 2 that, with the increase of geopolymer content, the change law of the compressive strength of the mixture at different ages is basically the same, with a slight increase at first and then a decrease. When the geopolymer content is 30%, the unconfined compression strength reaches the maximum value, and the unconfined compression strength at 7 d, 14 d, and 28 d is increased by 3.8%, 2.51%, and 1.46% compared with that at zero. When the content of geopolymer increased to 60%, the unconfined compression strength decreased sharply, and reached the minimum when the content of geopolymer was 90%, and the unconfined compression strength was only 2.40 MPa at 7 days. According to the provisions of the 7-day-old unconfined compressive strength standard of cement stabilized materials in "Technical Guidelines for Construction of Highway Roadbases" (JTG/T F20—2015), when the dosage is 30%

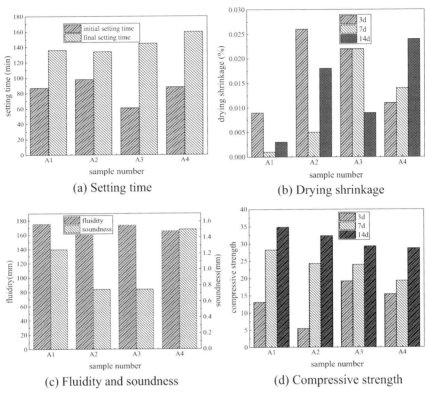

Fig. 1 Performance index of geopolymer cementitious materials with different mixing proportions

and 60%, it can meet the technical requirements of the heavy traffic volume of the second grade and below.

It can be seen from Table 1 that geopolymer materials mainly provide silicon, aluminum and calcium. Si can promote the strength of geopolymer, and Al can promote the polymerization rate. Proper Si/Al ratio can increase the strength of geopolymer, and a proper amount of calcium component can play a better role in strengthening, thus generating C–S–H gel in the system, providing more nucleation sites for the formation of geopolymer, and making the silicon and aluminum components in raw materials better dissolve out, thus accelerating the gelation process [19]. It shows that 30% geopolymer can provide high-quality Si/Al ratio, and ensure that the mineral components in cement can be fully hydrated, coagulated and hardened. With appropriate dosage of alkali activator, it can fully ensure that geopolymer can give full play to the potential activity of calcium component and silicon-aluminum component in the strength process. However, the Si/Al ratio of geopolymer is out of balance with the increase of content, and the calcium component and silicon-aluminum component can't give full play to their potential activity, which leads to a significant decrease in the strength of the mixture.

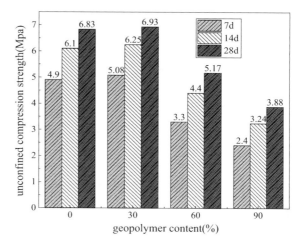

Fig. 2 Effect of geopolymer content change on the unconfined compression strength of materials

Indirect Tensile Strength. The influence of geopolymer content on the indirect tensile strength of materials is shown in Fig. 3; It can be seen from Fig. 3 that, with the increase of geopolymer ratio and the decrease of cement ratio, the change law of tensile strength at different ages is the same, showing a decreasing trend. When the content of geopolymer is 30%, the indirect tensile strength of the mixture of 28 d, 60 d and 90 d is only 0.4%, 0.8% and 1.8% lower than that of 0%. When the content of geopolymer is increased to 60% and 90%, the indirect tensile strength is obviously reduced. According to the requirements of "Specifications for Design of Highway Asphalt Pavement" (JTG D50-2017) for traffic grade, cumulative equivalent axle number Ne, and tensile strength structural coefficient, when the local polymer content is 30%, it meets the requirements of heavy traffic of grade II and below highways for pavement base.

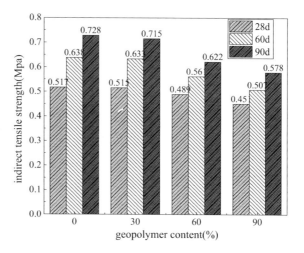

Fig. 3 Effect of geopolymer content change on indirect tensile strength of materials

The indirect tensile strength of the mixture with 30% content is approximately equal to that of the mixture with 0% content of geopolymer, because calcium oxide (CaO), silica (SiO_2) and other components in geopolymer react to form hydrated calcium aluminosilicate (C–(A)–S–H) gel [20], which can replace dicalcium silicate (C_2S) and tricalcium silicate in some cement. When the content of geopolymer increases, the calcium-based compounds in cement, mainly dicalcium silicate (C_2S), which has significant influence on the strength formation, decrease obviously, and their strength under the maximum pressure can be resisted.

Compression Rebound Modulus. The influence of geopolymer content on the compression rebound modulus of materials is shown in Fig. 4. It can be seen from Fig. 4 that at different ages, the compression rebound modulus of the mixture shows the same trend, and decreases with the increase of geopolymer content. When the content of geopolymer increases from 0 to 30%, the compression rebound modulus of the mixture decreases by 7.6%, 7.9% and 5.5% at 28 d, 60 d and 90 d, respectively, but it can be seen from Fig. 2 that the unconfined compression strength of the mixture increases by 2.8% at 28 d. This shows that adding a certain proportion of geopolymer can improve the compressive strength of the mixture, at the same time, the ability to resist vertical deformation and the crack resistance of cement stabilized macadam. However, on the premise of ensuring the strength and cementation, the elastic modulus of hydration products produced by polymer and cement is lower than that of cement hydration products. When geopolymer with low elastic modulus is added, the strength and flexibility of geopolymer mixture are improved, the rigidity is decreased, and the shrinkage deformation ability is improved. In this paper, referring to the experimental data of a large number of scholars and the calculation and analysis of Qu [21] optimum value range of compression rebound modulus of semi-rigid materials, when adding 30% geopolymer, the results are in line with the optimum value range of compression rebound modulus of semi-rigid base materials in laboratory test, and it is suitable for the performance requirements of heavy traffic volume of second grade and below.

Fig. 4 Effect of geopolymer content change on compressive resilience modulus of materials

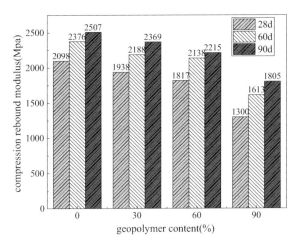

Scour Resistance. The influence of geopolymer content on scour resistance of materials is shown in Fig. 5. From Fig. 5, it can be seen that the cumulative erosion amount and erosion loss of the specimen increase gradually with the increase of geopolymer content, and the scour resistance of the specimen decreases gradually. Compared with 0% content, the erosion loss of 30% content only increases by 0.2%, and when the content increases to 90%, the erosion quality loss is 2.3 times that of 0% content. This shows that when the content of cement and geopolymer cementitious materials is increased to 30%, the difference between the cohesiveness and mechanical strength of cement and geopolymer cementitious materials is very small. When the scouring is carried out to a certain extent, the fine particles on the surface are gradually lost, and the skeleton composed of coarse particles plays a major role in bearing capacity. However, with the increase of the content of geopolymer to 90%, the mechanical properties of the mixture are greatly reduced, and the cohesiveness of geopolymer cementitious materials is insufficient, which leads to the double increase of erosion loss. According to Xiong and Gao [22] and others' research on the control index of anti-erosion performance of cement stabilized macadam base, it is found that when the content of geopolymer is 30%, it meets the requirements of heavy traffic of grade II and below highways for pavement base.

Through the unconfined compression strength test, it can be seen that the strength is the highest when the geopolymer content is 30%, and the tensile strength, compression rebound modulus and scour resistance test are similar to the mechanical properties of pure cement, and all the properties meet the requirements of heavy traffic of grade II and below highways for pavement base. Therefore, it is feasible to use geopolymer to replace part of cement, and comprehensive utilization of resources can be achieved.

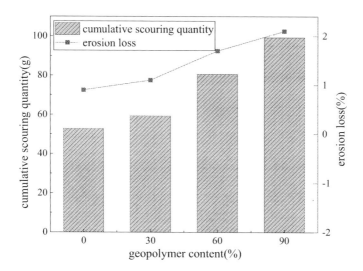

Fig. 5 Effect of geopolymer dosage change on the erosion resistance of the material

3.3 Formation Mechanism of Strength

Material Composition Analysis. The X-ray diffraction patterns of mineral powder, fly ash, wet carbide slag and cement are shown in Fig. 6.

It can be seen from Fig. 6a that the diffraction peaks at $2^{\theta} = 16.16°$, $26.38°$, $29.18°$, $31.08°$, $39.18°$, $43.04°$, $57.26°$, $60.38°$ are mainly quartz (SiO_2) crystals, and their diffraction peaks are high in intensity.

It can be seen from Fig. 6b that $2^{\theta} = 16.42°$, $25.93°$, $33.19°$, $35.24°$, 36.52, $50.14°$, $60.36°$ are characteristic peaks of mullite, and it can be known that the main component dried at 115 °C is mullite. Mullite mainly comes from the decomposition products of kaolin, illite and other clay minerals in coal, especially illite, which is a typical clay mineral rich in iron, potassium, sodium and magnesium. When the temperature is slightly higher, it begins to decompose into aluminosilicate. There are obvious Si diffraction peaks at $2^{\theta} = 27.97°$, $20.87°$, $50.15°$, and a wide diffraction characteristic peak at 20–35°, which indicates that there is vitreous material in fly ash,

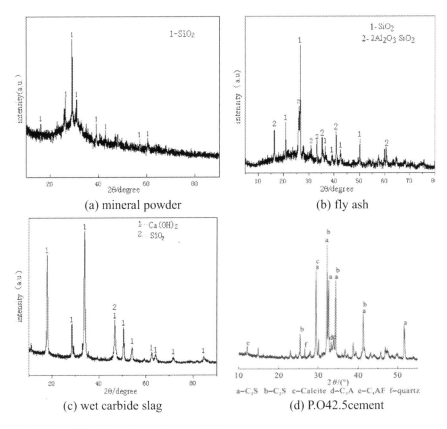

(a) mineral powder   (b) fly ash

(c) wet carbide slag   (d) P.O42.5cement

Fig. 6 X-ray diffraction patterns of raw materials

which can form a three-dimensional network structure with stable structure when it reacts with alkali activator.

It can be seen from Fig. 6c that the phase represented by the diffraction peaks at $2^\theta = 17.82°, 28.48°, 33.9°, 46.9°, 50.56°, 54.16°, 62.38°$ is mainly $Ca(OH)_2$, and a small amount of quartz is contained at $2^\theta = 46.9°$. (SiO_2) phase crystal.

It can be seen from Fig. 6d that the main mineral components of P.O42.5 cement are tricalcium aluminate (C_3A), tetracalcium ferrate (C_4AF), dicalcium silicate (C_2S) and tricalcium silicate (C_3S).

Microcosmic Appearance. SEM *photos are* shown in Fig. 7. Figure 7a, b are SEM photos of samples with 0% and 30% content after curing for 7 days, respectively, and the magnification is 1000 times. It can be seen that there are large and small pores on the surface of the mixture, which can be better combined with the binder, and the embedding capacity among the particles will also be improved, which is conducive to increasing the strength of the mixture. Figure 7c, d are SEM photos of 30% doped samples at 7 d and 28 d, respectively, with a magnification of 5000 times. It can be seen that there are strip-shaped particles on the surface and the surface is smooth. This is because at the initial stage of hydration reaction, after cement is mixed with water, C_3A reacts quickly, and a large number of flaky calcium aluminate hydrate are quickly generated, which are connected with each other to form a structure with early strength. With the increase of age, the slow dicalcium silicate (C_2S) in the hydration reaction of cement also began to participate in the reaction. At the same time, the hydrated calcium silicate (C–S–H) and hydrated calcium aluminosilicate (C–(A)–S–H) gels produced by the geopolymer reaction wrapped the aggregate particles completely, and no exposed aggregate particles were seen.

4 Conclusion

(1) Through comprehensive analysis of fluidity, soundness, setting time and drying shrinkage of geopolymer cementitious materials with different proportions, sample A1 (85% mineral powder, 10% fly ash and 5% wet carbide slag) is the best proportion of geopolymer cementitious materials.

(2) When 30% geopolymer is used instead of cement, the unconfined compression strength of the mixture can be appropriately improved, and its compression rebound modulus can be reduced to a certain extent; And can obtain a mixture similar to the indirect tensile strength and scour resistance of cement stabilized macadam.

(3) A proper amount of silica-alumina components and calcium components in the geopolymer can generate hydrated calcium silicate (C–S–H) and hydrated calcium silicate (C–(A)–S–H) gel under the action of alkali excitation, thus creating conditions for accelerating coagulation and improving the strength of the mixture.

(4) When proper amount of geopolymer is added, the mechanical properties of the mixture can meet the anti-scour ability, meet the requirements of heavy traffic

(a) B1 7 d×1000   (b) B2 7 d×1000

(c) B2 7 d×5000   (d) B2 28 d×5000

Fig. 7 SEM pictures of sample B1 and sample B2

of grade II and below highways for pavement base and have good anti-cracking, which can provide some technical support for engineering practice.

Acknowledgements This study was financially supported by Inner Mongolia Autonomous Region Transportation Science and Technology Project NJ-2021-15; Inner Mongolia Autonomous Region Science and Technology Plan Project 2020GG0257; National Natural Science Foundation of China (52169023). We are very grateful to the reviewers for taking valuable time to make constructive comments.

References

1. Zhu KJ (2021) Research and application of key technology of industrial solid waste in roadbed engineering. Hebei University of Engineering, Handan (in Chinese)
2. Chavva P (2002) Evaluation of strength, swell and shrinkage characteristics of chemically treated soils from north Texas. The University of Texas at Arlington
3. Glasby T, Day J, Genrich R (2015) EFC geopolymer concrete aircraft pavements at Brisbane West Wellcamp Airport. Concrete Institute of Australia Conference 2015, pp 1–9. Melbourne, Victoria, Australia

4. Shen W (2009) Investigation on the application of steel slag–fly ash–phosphogypsum solidified material as road base material. J Hazard Mater 164(1):99–104
5. Arulrajah A (2016) Stabilization of recycled demolition aggregates by geopolymers comprising calcium carbide residue fly ash and slag precursors. Constr Build Mater 11:864–873
6. Hu W (2018) Mechanical property and microstructure characteristics of geopolymer stabilized aggregate base. Constr Build Mater 191:1120–1127
7. Li H (2009) Study on road performance of cement stabilized tailing sand base. Hebei University of Technology, Tianjin (in Chinese)
8. Xiao XJ (2008) Application and research status of coal gangue in road engineering in China. Sci Technol Innov Herald 4(32):12–13
9. Yu HY (2020) Review on China's pavement engineering research 2020. China J Highway Transp 33(10):1–66
10. Xie Y (2002) Study on preparation and forming technology of road base material of red mud. Mining Metall 11(1):4–7
11. Qi JZ (2005) Experiment research on road base material of red mud. J Highway Transp Res 6:30–33
12. Yang JK (2006) Engineering application of red mud high grade pavement base material. China Municipal Eng 5:7–9
13. Qin M (2008) Research on properties of red mud cement stabilized macadam base. In: Proceedings of the 2nd national symposium on environmental geotechnical engineering and geosynthetics technology. Changsha: Journal of Hunan University, pp 15–18
14. Wang Y (2011) Study on the function of the bray stone stabilization with cement adding fly ash in the road. Chang'an University, Xi'an (in Chinese)
15. Xu OM (2019) Influence of fly ash and granulated blast furnace slag on strength and shrinkage characteristics of cement stabilized crushed stone. J Guangxi Univ (Natural Science Edition) 44(02):509–515
16. Chu F (2022) Study on the influence of iron tailings sang cement stabilized macadam mixture. J Wuhan Univ Technol (Transportation Science & Engineering) 1–7
17. Liu JL (2021) Research on road performance of slag geopolymer stabilized soil and stabilized macadam. Zhengzhou University, Zhengzhou (in Chinese)
18. Bai Y (2014) Glass fiber cement stabilized aggregate performance study. Chang'an University, Xi'an (in Chinese)
19. Zhang DW (2020) Review on property of geopolymer binder and its engineering application. J Archit Civil Eng 37(05):13–38
20. Yang SL (2021) Study on preparation and fiber reinforcement of geopolymer concrete based on fly ash and slag. Chang'an University, Xi'an (in Chinese)
21. Zhou Y (2017) Study on compressive strength and scour resistance of coal gangue mixture for road subbase construction. China Standardization 4:255–256
22. Li JY (2017) A brief review of research progression on shrinkage of cement-stabilized macadam. Transp Sci Technol 6:103–106

Rheological Properties of Composite Modified Asphalt with Direct Coal Liquefaction Residues

Yongxiang Li, Xiatong Kang, Qi Gao, and Yongjie Jia

Abstract In order to solve the problem of high-value utilization of coal-to-oil residual direct coal liquefaction residual asphalt, it is compounded with SBS and aromatic oil to modify the matrix asphalt, and 9 compounding schemes are designed using orthogonal experimental methods. Dynamic frequency sweep tests using DSR and a simplified Carreau equation model fitted to the complex viscosity to obtain its zero shear viscosity; the creep recovery rate, irrecoverable creep flexibility and irrecoverable creep flexibility difference of each modified asphalt were determined by MSCR at different temperatures and stress levels, and the high temperature rheological properties of 9 composite modified asphalts were evaluated by grey correlation analysis of zero shear viscosity and high temperature rheological parameters. Bending beam rheological experiments were carried out on the aged composite modified asphalt to analyse its low temperature rheological properties based on the viscoelastic parameters and linear fitting of the Burgers model. The results show that: The high temperature deformation resistance of DCLR composite modified asphalt are better than the matrix asphalt, the most influential modifier is SBS, and the higher the dose, the stronger the high temperature deformation resistance. The unrecoverable creep flexibility $J_{nr3.2}$ at 70 °C can better respond to the high temperature performance of asphalt, the ratio of 9% DCLR + 4% SBS + 2% aromatic oil DCLR composite modified asphalt with the best high temperature performance. Burgers model can better reflect the creep process of asphalt, DCLR composite modified asphalt has some defects in low temperature performance, the higher the dose of DCLR, the poorer the low temperature performance of the composite modified asphalt. The low temperature sensitivity of DCLR composite modified asphalt has been reduced, low temperature crack resistance has been slightly enhanced.

Keywords Direct coal liquefaction residues · High and low temperature rheological properties · Grey theory · Burgers model · Creep properties

Y. Li (✉)
Inner Mongolia Agricultural University, No. 306 Zhaowuda Road, Saihan District, Hohhot, China
e-mail: lyxiang@imau.edu.cn

X. Kang · Q. Gao · Y. Jia
College of Energy and Engineering, Inner Mongolia Agricultural University, Hohhot, China

© The Author(s) 2023
S. Wang et al. (eds.), *Proceedings of the 2nd International Conference on Innovative Solutions in Hydropower Engineering and Civil Engineering*, Lecture Notes in Civil Engineering 235, https://doi.org/10.1007/978-981-99-1748-8_5

1 Introduction

The direct coal liquidation is a technology for the clean conversion of coal into light fuel oil and high value-added chemical products. It has become an important initiative for the clean and efficient utilization of coal, which is conducive to the strategic restructuring of China's coal industry [1]. The direct coal liquefaction residue (DCLR) produced in the process is nearly 1/3 of the original coal, which not only causes waste of resources, but also imposes a burden on the natural environment due to its difficult degradability. The composition and properties of DCLR are similar to those of lake asphalt, with a high asphaltene and gum content and a low saturated and aromatic component. It has the advantages of economy and environmental friendliness as a modifier. And it can change the ratio of asphalt viscoelastic components and improve the high temperature rheological properties of asphalt. However, it also has a negative impact on the low temperature and fatigue properties of asphalt [2].

To enhance the low-temperature performance of DCLR modified asphalt, it was found [3, 4] that the incorporation of SBS can make up for the shortcomings of the low-temperature performance of DCLR modified asphalt to a certain extent. Aromatic oil can promote the reasonable combination of asphaltene, resins, aromatics and saturates in asphalt to form a stable colloidal structure. And it can promote the compatibility of SBS and asphalt to improve the low-temperature performance of asphalt. In this paper, based on the literature studies [5–10]. DCLR, SBS and aromatic oils were compounded to modify the matrix asphalt. Nine compounding schemes were designed using orthogonal test method and dynamic frequency scan test was performed using dynamic shear rheometer (DSR). The zero-shear viscosity (ZSV) was then obtained by fitting a simplified Carreau equation model to the complex viscosity. The creep recovery rate R and the irrecoverable creep flexibility J_{nr} of each modified asphalt were determined by multiple stress creep and recovery (MSCR) test. And the high-temperature rheological properties of the nine composite modified asphalts were evaluated by gray correlation analysis between ZSV and high-temperature rheological parameters. Bending beam rheometer (BBR) tests were conducted on the aged composite modified asphalt, based on the viscoelastic parameters of Burgers model and linear fitting, to analyze its low temperature rheological properties.

2 Experimental Materials and Specimen Preparation

2.1 Test Materials

The selected Tokai brand 90# matrix asphalt, the conventional index test results are shown in Table 1; the selected DCLR performance is shown in Table 2; the selected modifier is a linear YH-791 SBS modifier with a block ratio of 3/7, the

Table 1 Technical index of 90# matrix asphalt

Test project	Technical requirements	Results	Testing standards
Penetration (25 °C, 5 s, 100 g) (0.1 mm)	80–100	82.6	T0604-2011
Softening point (°C)	≥ 43	52.5	T0606-2011
Viscosity 60 °C (Pa s)	≥ 140	157	T0625-2011
10 °C Ductility (cm)	≥ 20	34.0	T0605-2011
Relative Density 15 °C (g/cm^3)	Real measurement	1.031	T0603-2011
TFOT (or RTFOT) post-residuals			
Quality changes (%)	≤ ± 0.8	0.25	T0610-2011
Residual penetration ratio (25 °C, 5 s, 100 g) (%)	≥ 57	65.4	T0604-2011
Residual ductility 10 °C (cm)	≥ 8	8.2	T0605-2011

Table 2 Performance of DCLR

Test project	Density (g/cm^3)	25 °C Penetration (0.1 mm)	10 °C Ductility (cm)	Softening point (°C)
Results	1.24	4.8	1.8	175

basic performance is shown in Table 3; the selected 2#-38 aromatic oil is used as a compatibilizer for the composite modified asphalt.

2.2 Method

(1) Orthogonal tests were designed to prepare nine DCLR composite modified asphalts, and the results are shown in Table 4.

(2) High temperature rheological test

Dynamic frequency scan test using DSR. Dynamic shear load loads at low strain levels were applied to asphalt samples at loading frequencies of 0.1–100 rad/s and at 46–82 °C (temperature step of 6 °C). And a simplified Carreau equation model is fitted to the complex viscosity to analyze its high-temperature stability.

MSCR tests were performed on each asphalt after short-term aging using DSR apparatus at 64 and 70 °C with a holding time of 15 min.

(3) Low temperature rheological test

Fatigue testing was performed by strain scanning, using strain control mode. The test temperature was 15 °C, the strain was 1%, the number of loading was 50,000, and the frequency was 10 Hz. A low temperature bending beam rheometer was used to test the creep modulus of stiffness S and creep rate m at a load loading time of 60 s.

Table 3 Basic properties of SBS

Brand	Structure	Block ratio S/B	Volatiles (%)	Oil-filling rate (%)	Ash (%)	Tensile strength (MPa)	Elongation (%)	Permanent distortion (%)	Shore hardness (HA)	Melt mass-flow rate (g/10 min)
YH-791	Linear	30/70	≤ 0.7	0	≤ 0.2	≥ 15	≥ 700	≤ 40	≥ 68	0.5–2.5

Table 4 Composite modified asphalt orthogonal test table

Test serial number	A (DCLR doping/%)	B (SBS doping/%)	C (Aromatic oil doping/%)
1	5	2	2
2	5	3	4
3	5	4	6
4	7	2	4
5	7	3	6
6	7	4	2
7	9	2	6
8	9	3	2
9	9	4	4

The specimens were asphalt aged by RTFOT short-term and then by PAV long-term at −12, −18 and −24 °C.

3 High Temperature Rheological Test Results and Analysis

3.1 Asphalt ZSV Fitting Results

The ZSV of asphalt can better characterize the storage capacity of asphalt at high temperatures, and the higher its value, the better its high temperature stability. However, the measurement conditions of ZSV are demanding, requiring very low frequency or shear rate, which is difficult to be satisfied by existing test instruments. Therefore, the widely used Carreau model was adopted to fit the measured frequency band to obtain ZSV, whose equation is shown in Eq. (1), and the fitting results are shown in Fig. 1.

$$\eta = \frac{\eta_0}{[1 + (k\omega)^2]^{\frac{m}{2}}} \tag{1}$$

where η is the complex viscosity; η_0 is zero shear viscosity; ω is the frequency; k is a constant, which is a material parameter with a time scale; and m is a constant, which is a dimensionless material parameter.

From Fig. 1, it can be seen that the ZSV values of the composite modified asphalt are higher than those of the matrix asphalt at all temperatures. This indicates that the composite modifier can improve the high-temperature elastic recovery of asphalt; the ZSV value of No.6 composite modified asphalt is the highest at all temperatures, indicating that its high-temperature performance is more excellent.

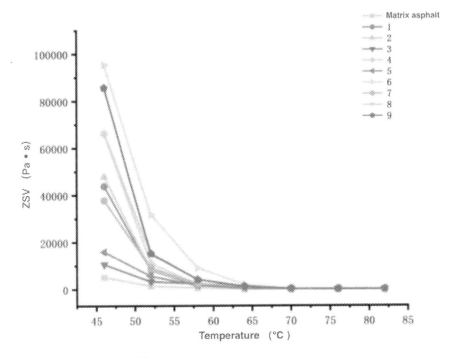

Fig. 1 ZSV of asphalt at different temperatures

3.2 MSCR Test Results

The main evaluation indicators R and J_{nr} were obtained from the MSCR test and the results are presented in Table 5. The polar differences were calculated using the analysis of range (ANOR) to determine the order of influence of the factors. Where k_i denotes the arithmetic mean of the test results obtained at factor level i in any column, and the polar difference R = max{k_1, k_2, k_3}-min{k_1, k_2, k_3}. The results of ANOR of R are presented in Table 6, and the results of ANOR of J_{nr} are presented in Table 7.

As can be seen from Tables 5 and 6, the R of both the composite modified asphalt 1–9 and the matrix asphalt decreased significantly when the temperature was increased to 70 °C. This indicates that the increase in temperature decreases the elastic properties of the asphalt; $R_{0.1}$ is significantly higher than $R_{3.2}$, which shows that it is more difficult for the asphalt to recover from deformation at high stress levels; the greatest effect on the R value is for SBS, and the higher its admixture, the greater the R value.

As can be seen from Table 7, J_{nr} increased for all specimens when the temperature was increased to 70 °C. This indicates that the higher the temperature, the more viscous the asphalt and the weaker the elastic recovery; $J_{nr3.2}$ was higher than $J_{nr0.1}$ for all specimens, indicating that the irrecoverable creep deformation was greater at

Table 5 MSCR test results

Test serial number	Matrix asphalt	1	2	3	4	5	6	7	8	9
64 °C $R_{0.1}$	2.44	7.44	9.00	23.32	7.48	16.50	15.04	9.07	28.14	25.06
70 °C $R_{0.1}$	1.18	2.65	4.13	18.94	4.90	8.32	6.80	6.42	10.94	15.40
64 °C $R_{3.2}$	0.91	2.33	4.25	11.66	3.07	7.41	6.54	3.61	7.25	15.33
70 °C $R_{3.2}$	0.28	0.66	0.81	5.85	1.56	2.67	2.07	0.86	2.03	5.61
64 °C $J_{nr0.1}$	2.71	3.49	1.76	1.75	1.97	1.88	1.08	2.11	0.96	0.90
70 °C $J_{nr0.1}$	6.12	4.31	4.40	3.85	4.31	4.68	2.89	4.34	2.74	2.28
64 °C $J_{nr3.2}$	2.94	3.77	2.04	2.27	2.25	2.32	1.33	2.47	1.41	1.12
70 °C $J_{nr3.2}$	6.94	4.92	5.15	5.24	4.99	5.67	3.44	5.48	3.56	2.94

Table 6 Results of ANOR of R

	Modifiers	A	B	C
64 °C $R_{0.1}$	k_1	13.25	8.00	16.87
	k_2	13.01	17.88	13.85
	k_3	20.76	21.14	16.30
	R	7.75	13.14	3.03
70 °C $R_{0.1}$	k_1	8.57	4.66	6.80
	k_2	6.67	7.80	8.14
	k_3	10.92	13.71	11.23
	R	4.25	9.06	4.43
64 °C $R_{3.2}$	k_1	6.08	3.00	5.37
	k_2	5.67	6.30	7.55
	k_3	8.73	11.18	7.56
	R	3.06	8.17	2.19
70 °C $R_{3.2}$	k_1	2.44	1.03	1.59
	k_2	2.10	1.84	2.66
	k_3	2.83	4.51	3.13
	R	0.73	3.48	1.54
Factor priorities	64 °C $R_{0.1}$:BAC	70 °C $R_{0.1}$:BCA	64 °C $R_{3.2}$:BAC	70 °C $R_{3.2}$:BCA
Optimum ratio	70 °C $R_{0.1}$:$A_3B_3C_1$	70 °C $R_{0.1}$:$A_3B_3C_3$	64 °C $R_{3.2}$:$A_3B_3C_3$	70 °C $R_{3.2}$:$A_3B_3C_3$

high stress levels; the higher the doping of DCLR with SBS, the smaller the value of J_{nr}.

In summary, the high temperature resistance to deformation of the composite modified asphalt is better than that of the matrix asphalt, and the modifier with the greatest influence on the modified properties is SBS, and the higher the SBS admixture, the stronger its high temperature resistance to deformation. This is probably

Table 7 Results of ANOR of J_{nr}

	Modifiers	A	B	C
64 °C $J_{nr\,0.1}$	k_1	2.33	2.52	1.84
	k_2	1.64	1.53	1.54
	k_3	1.32	1.24	1.91
	R	1.01	1.28	0.37
70 °C $J_{nr0.1}$	k_1	4.19	4.32	3.31
	k_2	3.96	3.94	3.66
	k_3	3.12	3.01	4.29
	R	1.07	1.31	0.98
64 °C $J_{nr3.2}$	k_1	2.69	2.83	2.17
	k_2	1.97	1.92	1.80
	k_3	1.67	1.57	2.35
	R	1.03	1.26	0.55
70 °C $J_{nr3.2}$	k_1	5.10	5.13	3.97
	k_2	4.70	4.79	4.36
	k_3	3.99	3.87	5.46
	R	1.11	1.26	1.49
Factor priorities	64 °C $J_{nr0.1}$: BAC	70 °C $J_{nr0.1}$: BAC	64 °C $J_{nr3.2}$: BAC	70 °C $J_{nr3.2}$: CBA
Optimum ratio	64 °C $J_{nr0.1}$: $A_3B_3C_2$	70 °C $J_{nr\,0.1}$: $A_3B_3C_1$	64 °C $J_{nr3.2}$: $A_3B_3C_2$	70 °C $J_{nr3.2}$: $A_3B_3C_1$

because when SBS reaches a certain concentration, its thermoplastic nature enables it to form a stable three-dimensional mesh structure in the asphalt [11, 12]. The addition of aromatic oils enhances this mesh structure [13], thus improving the high temperature deformation resistance of the composite modified asphalt. From the optimum ratio of each index, it can be seen that the high temperature performance of the composite modified asphalt is best when both DCLR and SBS are at the highest dose, but the optimum dose of aromatic oil cannot be determined, and further research and analysis are still needed.

3.3   Grey Correlation Analysis

Due to the number and complexity of the indicators used to evaluate the high temperature performance of asphalt and the large dispersion of the data for each indicator. It is difficult to carry out mathematical and statistical processing to find out the main indicators for evaluating the high temperature performance of asphalt. Grey theory introduces the correlation degree into the system analysis, which can solve the shortcomings of mathematical and statistical methods in the analysis of the large amount of calculation and many samples. The correlations between R and J_{nr} at different temperatures and stress levels obtained by coupling MSCR tests, combined with the variation pattern of ZSV indicators, were quantitatively compared and analysed, leading to an in-depth analysis of the primary and secondary relationships between them. The results of the ZSV (64 °C) fitted by the Carreau model were chosen as the reference series and the MSCR test results were used as the comparison series to obtain the grey correlations of the factors, as follows.

(1) Grey Relational Coefficient

Let the set of grey correlation factors be $Z_{m \times n}(z_1, z_i)$, $z_1(m)$ is the reference column and is the comparison column. The number of grey-off links is

$$\xi = \left| \frac{\min\limits_{i=2,n} \cdot \min\limits_{t=1,n} \Delta_i(t) + \rho \max\limits_{i=2,n} \cdot \max\limits_{t=1,n} \Delta_i(t)}{\Delta_i(t) + \rho \max\limits_{i=2,n} \cdot \max\limits_{t=1,n} \Delta_i(t)} \right| \tag{2}$$

where $t = 1, 2, \ldots, n$; $\Delta_i(t) = |z_1 - z_i|$; ρ is a resolution factor generally taken as 0.5.

(2) Grey correlation entropy

The distribution density function P_h is noted as:

$$P_h \overset{\Delta}{=} \xi \bigg/ \sum_{t=1}^{n} \xi \tag{3}$$

The grey correlation entropy of z_i is:

$$H(x) \overset{\Delta}{=} - \sum_{t=1}^{n} P_h \ln P_h \tag{4}$$

(3) Grey entropy correlation

The grey entropy correlation of z_i is

$$E(z_i) \overset{\Delta}{=} H(x) \big/ H_{max} \tag{5}$$

where $H_{max} = \ln m$ is the maximum value of the difference information column consisting of m elements. The larger $E(z_i)$ is, the greater the correlation between the comparison sequence z_i and the reference sequence z_1.

The parameters of the grey correlation analysis are shown in Table 8 and the dimensionless results and grey correlations are shown in Tables 9 and 10.

As can be seen from Table 10, the ash correlation between each index and ZSV is high, all above 0.97, with Jnr3.2 of asphalt at 70 °C having the highest ash correlation with ZSV, followed in order by Jnr0.1 at 64 °C, Jnr0.1 at 70 °C, R0.1 at 64 °C, Jnr0.1

Table 8 Corresponding parameters of grey correlation analysis

Parameters	Serials	Matrix asphalt	1	2	3	4	5	6	7	8	9
ZSV	Z_1 (64 °C)	118.57	273.55	978.17	563.27	532.02	563.14	1920.69	1418.34	1137.34	1210.08
$R_{0.1}$	Z_2 (64 °C)	2.44	7.44	9.00	23.32	7.48	16.5	15.04	9.07	28.14	25.06
	Z_3 (70 °C)	1.9	2.65	4.13	18.94	4.9	8.32	6.8	6.42	10.94	15.4
$R_{3.2}$	Z_4 (64 °C)	0.91	2.33	4.25	11.66	3.07	7.41	6.54	3.61	7.25	15.33
	Z_5 (70 °C)	0.25	0.66	0.81	5.85	1.56	2.67	2.07	0.86	2.03	5.61
$J_{nr0.1}$	Z_6 (64 °C)	2.71	2.49	1.76	1.75	1.97	1.88	1.08	2.11	0.96	0.90
	Z_7 (70 °C)	5.06	4.31	4.40	3.85	4.31	4.68	2.89	4.34	2.74	2.28
$J_{nr3.2}$	Z_8 (64 °C)	2.94	2.77	2.04	2.27	2.25	2.32	1.33	2.47	1.41	1.12
	Z_9 (70 °C)	5.69	4.92	5.15	5.24	4.99	5.67	3.44	5.48	3.56	2.94

Table 9 Nondimensionalize calculation results

Parameters	Serials	Matrix asphalt	1	2	3	4	5	6	7	8	9
ZSV	Z_1 (64 °C)	0.136	0.170	0.236	0.146	0.112	1.539	1.302	1.405	1.209	0.402
$R_{0.1}$	Z_2 (64 °C)	0.314	0.519	0.330	0.374	0.295	1.414	1.109	1.324	1.045	0.388
	Z_3 (70 °C)	1.122	0.627	0.514	0.682	0.362	0.999	1.132	0.975	1.094	0.754
$R_{3.2}$	Z_4 (64 °C)	0.646	1.625	2.356	1.870	2.615	0.994	0.991	1.085	1.113	1.409
	Z_5 (70 °C)	0.611	0.521	0.610	0.492	0.697	1.119	1.109	1.076	1.060	0.693
$J_{nr0.1}$	Z_6 (64 °C)	0.646	1.150	1.035	1.188	1.194	1.068	1.204	1.109	1.204	1.081
	Z_7 (70 °C)	2.204	1.048	0.846	1.049	0.925	0.613	0.744	0.636	0.731	1.067
$J_{nr3.2}$	Z_8 (64 °C)	1.627	0.632	0.799	0.579	0.384	1.198	1.117	1.181	1.164	0.825
	Z_9 (70 °C)	1.305	1.961	1.361	1.163	0.908	0.545	0.705	0.674	0.756	2.198

Table 10 Grey relational calculation results

Parameters	$R_{0.1}$	$R_{3.2}$	$J_{nr0.1}$	$J_{nr3.2}$
Grey correlation entropy	2.2724	2.2520	2.2572	2.2405
Grey correlation	0.9869	0.9781	0.9803	0.9730

at 64 °C, R3.2 at 64 °C, R0.1 at 70 °C and R3.2. This indicates that the 70 °C non-recoverable creep flexibility Jnr3.2 can better respond to the high temperature performance of the asphalt and the best ratio for high temperature performance is 9% DCLR + 4% SBS + 2% aromatic oil.

4 Low Temperature Rheological Test Results and Analysis

4.1 BBR Test Data and Burgers Model Parameters

As a single consideration of the modulus of stiffness S or creep rate m to evaluate the low temperature performance of asphalt is more one-sided, the BBR test data (−18 °C) was subjected to Burgers fitting. The results are shown in Table 11, and the results of ANOR of Burgers model parameters are shown in Table 12. The Burgers model equation is given in Eq. (6).

$$y = \frac{1}{E_1} + \frac{1 - e^{\frac{-tE_2}{\eta_2}}}{E_2} + \frac{t}{\eta_1} \qquad (6)$$

where E_2 is the instantaneous elastic parameter, E_2 is the delayed elastic parameter, η_1 is the viscous flow parameter and η_2 is the delayed viscous parameter. η_1 reflects the

Table 11 Fitting results of low temperature viscoelastic parameters for asphalt PAV state

Temperature (°C)	Parameters	E_1 (MPa)	E_2 (MPa)	η_1 (MPa/s)	η_2 (MPa/s)	λ	J_C	R^2
−18	Matrix sphalt	665.40	914.35	134,856.72	30,258.90	202.67	409.92	0.9998
	1	683.26	1071.37	224,592.99	38,467.40	328.71	451.77	0.9994
	2	714.67	1061.92	212,430.06	36,921.45	297.24	463.35	0.9997
	3	591.00	1082.39	185,422.18	44,449.60	313.74	400.39	0.9978
	4	718.03	1063.08	278,645.51	80,549.72	388.07	442.63	0.9998
	5	810.53	1256.15	193,600.17	30,278.60	238.86	537.27	0.9999
	6	870.45	1461.19	284,429.05	55,517.90	326.76	583.10	0.9997
	7	795.55	1279.67	250,539.73	35,058.82	314.93	537.53	0.9980
	8	809.00	1312.65	262,706.17	50,402.89	324.73	536.53	0.9995
	9	639.81	943.79	208,429.93	82,553.24	325.77	377.02	0.9966

Table 12 Results of ANOR of burgers model parameters

Parameters	Modifiers	A	B	C
η_1	k_1	190,626.59	201,569.32	192,998.87
	k_2	219,222.62	251,259.41	224,240.45
	k_3	265,891.65	222,912.13	258,501.54
	R	75,265.06	49,690.09	65,502.67
λ	k_1	276.21	281.06	252.15
	k_2	313.56	343.90	322.39
	k_3	322.14	286.94	337.36
	R	45.93	62.85	85.20
J_C	k_1	441.68	464.47	494.91
	k_2	460.10	477.31	462.90
	k_3	552.39	512.38	496.36
	R	110.71	47.91	33.46
Factor Priorities	η_1:ACB		Λ:CAB	J_C:ABC
Optimum ratio	$A_1B_1C_1$		$A_1B_1C_1$	$A_1B_1C_2$

deformation capacity of the asphalt, the smaller the η_1, the better the low temperature performance of the asphalt [14].

The relaxation time λ of asphalt represents the ability of stress dissipation, the shorter the λ, representing the more rapid dissipation of stress within the asphalt, the better the low temperature performance [15]. Calculated as shown in Eq. (7).

$$\lambda = \eta_1 / E_1 \tag{7}$$

The low temperature flexibility parameter J_C reflects the viscoelastic properties of the asphalt, the smaller the J_C, the higher the proportion of viscous components of the asphalt, the better the low temperature performance [16]. The calculation is given in Eq. (8).

$$J_C = \frac{1}{J_V}\left(1 - \frac{J_E + J_{DE}}{J_E + J_{DE} + J_V}\right) \tag{8}$$

where $J_V = \frac{t}{\eta_1}$, $J_E = \frac{1}{E_1}$, $J_{DE} = \frac{1}{E_2}\left(1 - e^{\frac{-tE_2}{\eta_2}}\right)$.

From Tables 11 and 12 can be seen, Burgers fitting accuracy R^2 are above 0.99, indicating that the model can better reflect the creep process of asphalt. The composite modified asphalt, η_1, λ and J_C are basically larger than the matrix asphalt, and the higher the DCLR dose, the worse the low temperature performance of the composite modified asphalt. This indicates that the performance of the composite modified asphalt in low temperature performance has some defects. This is mainly due to the high proportion of asphalt in DCLR, which cross-linked with the matrix asphalt and increased the flow resistance of the modified asphalt [17]. The ratio of 5% DCLR + 2% SBS + 2% aromatic oil is better.

4.2 Low Temperature Sensitivity Analysis

The creep stiffness and creep rate versus temperature were regressed for nine asphalt tests and the fitted results are shown in Table 13, and the results of ANOR of Burgers model parameters are shown in Table 14, with the linear regression equation in Eq. (9).

$$\lg S = SA_S \cdot T + C \tag{9}$$

where S represents the creep stiffness; SA_S represents the slope of the equation; C represents the temperature; and C is a constant.

As can be seen from Table 13, the linear fit correlation coefficients are all above 0.94, which is a good fit; the values of the composite modified bitumen are equal to or somewhat smaller than the $|SA_S|$ values of the matrix asphalt, which indicates that the low temperature sensitivity of the composite modified asphalt has been reduced and the low temperature crack resistance has been slightly enhanced; as can be seen from Table 14, the best low temperature sensitivity is 9% DCLR + 4%/6% SBS + 3% aromatic oil.

Table 13 Fitting results of low temperature sensitivity for asphalt

Test serial number	Intercept	Slope	$|SA_S|$	R^2
Matrix asphalt	1.02	−0.076	0.076	0.975
1	1.43	−0.068	0.068	0.952
2	1.46	−0.068	0.068	0.948
3	1.32	−0.073	0.073	0.943
4	1.42	−0.071	0.071	0.941
5	1.33	−0.077	0.076	0.951
6	1.53	−0.068	0.068	0.977
7	1.58	−0.063	0.063	0.944
8	1.68	−0.058	0.058	0.962
9	1.43	−0.072	0.072	0.945

Table 14 Results of ANOR of SA_S

Parameters	Modifiers	A	B	C
$|SA_S|$	k_1	0.0707	0.0723	0.0717
	k_2	0.0733	0.0673	0.0663
	k_3	0.063	0.0673	0.069
	R	0.0103	0.005	0.0053
Factor priorities	ACB			
Optimum ratio	$A_3B_2C_2/A_3B_3C_2$			

5 Conclusions

(1) At 46–82 °C, the ZSV values fitted by the Carreau model were higher than those of the matrix asphalt, and the highest ZSV value was obtained with 9% DCLR + 4% SBS + 2% aromatic oil. The higher the amount of SBS, the stronger the high temperature deformation resistance. Based on the grey correlation entropy analysis, the non-recoverable creep flexibility $J_{nr3.2}$ at 70 °C can better reflect the high temperature performance of the asphalt, and the DCLR composite modified asphalt with 9% DCLR + 4% SBS + 2% aromatic oil has the best high temperature performance.

(2) The accuracy of the Burgers model is above 0.99, which indicates that the model can better reflect the creep process of asphalt. The DCLR composite modified asphalt with a ratio of 5% DCLR + 2% SBS + 2% aromatic oil has the best temperature performance. The low temperature sensitivity of the DCLR composite modified asphalt is reduced, and the low temperature crack resistance is slightly enhanced.

Acknowledgements This study was financially supported by National Natural Science Foundation of China (52169023), Inner Mongolia Autonomous Region Science and Technology Program Project 2020GG0257, Inner Mongolia Autonomous Region Transportation Science and Technology Project NJ-2020-13. I would like to thank anonymous reviewers for their constructive comments, which helped improve the clarity and completeness of this paper.

References

1. Liu ZY, Shi SD, Li YW (2010) Coal liquefaction technologies—development in China and challenges in chemical reaction engineering. Chem Eng Sci 65(1):12–17
2. Ji J, Xu XQ (2021) Research on performance of direct coal liquefaction residue modified asphalt mortar. J Fuel Chem Technol 49(8)
3. Xiao QY, Zhao P, Lin YH, Zhang MQ (2019) Study on high temperature rheological performance waste engine oil modified asphalt under thermal aging. Bull Chin Ceramic Soc 38(11):3597–3604 (in Chinese)
4. Cheng PF, Li JH, Kou HY (2019) Study on the improvement of the performance of cold recycled asphalt mixes by the dosing of light oil recycler. J China Foreign Highway 39(01) (in Chinese)
5. Luo R, Xu Y, Liu HQ (2018) Rheological mechanical properties of DCLR-modified asphalt binders. Chin J Highway Transp 31(6):165–171 (in Chinese)
6. Shi YF, Yao H, Xu SF (2015) Influence of direct coal liquefaction residues on viscoelastic properties of asphalt mortar. J Traffic Transp Eng 15(4):1–8 (in Chinese)
7. Liu ST, Cao WD, Shang SJ (2010) Analysis and application of relationships between low-temperature rheological performance parameters of asphalt binders. Constr Build Mater 24(4):471–478
8. Liu Z, Xuan M, Zhao ZH (2003) A study of the compatibility between asphalt and SBS. J Pet Sci Technol 21:1317–1325
9. Xing HP (2018) Performance research on asphalt and mixture modified with SEBS. Chang'an University (in Chinese)
10. Zhai XG, Chen B, Ding LT (2019) Effect of aromatic oil on road performance of direct-to-plant SBS modifier. Highway Eng 44(06):57–61 (in Chinese)
11. Giovanni P, Sara F, Filippo M (2015) A review of the fundamentals of polymer-modified asphalts: asphalt/polymer interactions and principles of compatibility. Adv Coll Interface Sci 224:72–112
12. Schaur A, Unterberger S, Lackner R (2017) Impact of molecular structure of SBS on thermomechanical properties of polymer modified asphalt. Eur Polym J 96:256–265
13. Luo W, Zhang Y, Cong P (2017) Investigation on physical and high temperature rheology properties of asphalt binder adding waste oil and polymers. Constr Build Mater 144:13–24
14. Wang K, Hao PW (2016) Analysis of low temperature properties and viscoelasticity of asphalt for BBR test. J Liaoning Univ Eng Technol (Natural Science Edition) 35(10):1138–1143
15. Li B, Zhang XJ, Li JX, Yang KH (2021) Evaluation of the low temperature performance of hard asphalt based on Burgers model. J Constr Mater 24(05):1110–1116
16. Li XL (2013) Unified evaluation index for high-and-low temperature properties of asphalt based on rheological theory. Harbin: Harbin Institute of Technology (in Chinese)
17. Ji J, Xu XQ, Xu Y, Wang Z (2021) Study on the performance of modified bitumen slurry from coal direct liquefaction residue. J Fuel Chem 49(08):1095–1101

The Influence of Coating Material and Thickness on the Corrosion Degree of Q345 Steel

Li Wan, Xiang Pan, Lizhen Huang, Baotao Huang, Cai Yang, and Yiming Du

Abstract In order to investigate the anti-corrosion effect of coated steel of steel bridge, Q345 steel plate specimens with three types of coatings, including zinc coating, aluminum coating and zinc-aluminum coating, are produced by the arc spraying technology. In the present study, chlorine corrosion tests are performed to investigate the influence law of different coating material and its thickness on the corrosion degree. Then the calculation results of two corrosion indicators are compared and analyzed. It is shown that the two corrosion indicators reflect the same corrosion law of three kinds of coating steel. The corrosion of all coated specimens is obviously severe in the early stage and gradually gentle in the later stage. It is also found that during the whole corrosion cycle, the corrosion rate of aluminum coating is smaller and change slower than the other two kinds of coatings, whereas the coating thickness of 200 μm of aluminum coating changes significantly. Therefore, aluminum coating is recommended as a priority, and the recommended coating thickness range of which is 100–150 μm.

Keywords Chloride environment · Corrosion rate · Anti-corrosion coating · Steel bridge

L. Wan
Capital Construction Management Office, Hubei Engineering University, Xiaogan, China

X. Pan
School of Civil Engineering, Architecture and the Environment, Hubei University of Technology, Wuhan, China

L. Huang (✉) · C. Yang · Y. Du
School of Civil Engineering, Hubei Engineering University, Xiaogan, China
e-mail: hlizhen@hbeu.edu.cn

B. Huang (✉)
School of Naval Architecture and Maritime, Zhejiang Ocean University, Zhoushan, China
e-mail: 7247480@qq.com

© The Author(s) 2023
S. Wang et al. (eds.), *Proceedings of the 2nd International Conference on Innovative Solutions in Hydropower Engineering and Civil Engineering*, Lecture Notes in Civil Engineering 235, https://doi.org/10.1007/978-981-99-1748-8_6

1 Introduction

Owing to the advantages of high strength, light weight, good plasticity and tough-
ness, and the continuous growth of steel production in recent years, steel structures
have good prospects in the field of civil engineering. Now steel bridge plays a very
important role in transportation power in China, but its corrosion in chloride salt
environment has become one of the key problems affecting the durability of steel
bridges. The corrosion of steel is more serious in the environment with high humidity
and aggressive media, among which chloride ions is especially prominent. Therefore, protective measures such as rust removal, painting and coating must be taken to
ensure the safety of the steel structures. Meanwhile, steel structures must also be regularly maintained, which spends a certain amount of maintenance costs. Experiment
studies are carried out involved the corrosion behavior of steel in different marine
areas and laboratory simulated marine environments. These studies mainly focus on
the corrosion mechanisms of different types of steel as well as the components and
properties of corrosion products. And whereas, the research on the corrosion rate
of steel, especially coated steel, is yet relatively few. It is not only unhelpful for
the accurate evaluation of the durability of steel bridges, but also for the reasonable
selection of coating type and thickness. Therefore, it is very important to investigate
the corrosion rate of coated steel adopted in steel bridges.

Corrosion rate is an important index to study the corrosion law of steel. Huang [1]
and Zhu [2] point out that under the same corrosion cycles, the change rule of corrosion rate of different kinds of carbon steel is basically the same through analyzing
the corrosion data. In addition, Schumacher's empirical formula is applied to fit the
corrosion data, and it is proposed that the slope of the linear part of the formula
can be approximated as the long-term corrosion rate of carbon steel. However, this
experimental formula does not take into account the influence of rust layer and other
factors on corrosion rate. Yu [3], Pour-Ghaz [4, 5] and Pradhan [6] study on the corrosion rate of rebar in concrete in chloride environment by using self-made corrosion
rate test equipment, and then the corrosion rate model of steel bar in corresponding
environment is respectively established. The corrosion behavior of steel plated with
galvanized, aluminum and zinc-aluminum alloy materials is studied in Refs. [7–9]. It
is shown that these three plating layers can restrict the corrosion rate and have a good
corrosion protection effect for steel. The practice of Arc spraying is well established
because of its high strength, excellent coating performance and high efficient energy
use. Now, it is few report about the anti-corrosion behavior of arc thermal spraying
coatings. So it is urgently needed to study.

In this paper, the corrosion rate of Q345 steel coated by arc thermal spraying is
studied based on chloride corrosion test. Then the effect of different coating type and
thickness on the corrosion rate is also discussed. Finally, the existing representative
corrosion rate models are evaluated by comparison.

2 Experimental Test

By using a 3.5% sodium chloride solution, corrosion tests are carried out in the environment of constant temperature and humidity (temperature is 20 °C, humidity is 95%). The experimental cycle were 7, 15, 30, and 60 days. The shape of steel specimens is rectangular plate with a size of 50 × 25 × 2 (mm). Three coating types are designed in the experiments, which are zinc coating, aluminum coating and zinc-aluminum coating. Three coating thickness are made for each type, respectively, 100 μm, 150 μm, 200 μm.

The preparation of the coating is in accordance with GB/T9 793–1997 "thermal spraying zinc, aluminum and their alloys for metal and other inorganic coatings" and other relevant standards. After each corrosion cycle, the corresponding specimen is removed and the corrosion products on the surface of the specimen are cleaned. The floating rust on the surface of the specimen is scraped off with a blade. Then 1000 ml solution is made for pickling and rust removal of specimens, which mixed with 500 ml hydrochloric acid, 3.5 g hexamethylenetetramine and distilled water. Specimens are put into anhydrous alcohol to dehydrate and to blow dry. The specimens are taken out of the oven after 24 h, and the weight of the corroded specimens are measured.

3 Results and Discussion

3.1 Test Results

Two indicators of corrosion degree are calculated in this paper, which are corrosion rate and cumulative loss of thickness, respectively.

(1) Corrosion rate

The weightlessness method is the most intuitive and reliable method to evaluate the corrosion resistance of metals, which can reflect the macroscopic corrosion rate of metals [10]. It is assumed that the specimen is uniformly corroded. The corrosion rate is calculated by

$$W = (W_0 - W_t) \times \frac{10^6}{[2(a \times b + b \times c + a \times c)t]} \tag{1}$$

where W is the corrosion rate. W_0 is the mass of the specimen before corrosion. W_t is the mass of the specimen after corrosion. a, b and c are the length, width and height of the specimen, respectively.

(2) Cumulative loss of thickness

The other indicator is cumulative loss of thickness [11]. The formula is given by

Table 1 Results of corrosion rate of zinc coating

Corrosion time (day)	Coating thickness (μm)	Mass before test (g)	Mass after test (g)	Corrosion rate $g/(m^2\,h)$	Cumulative loss of thickness (μm)
7	100	19.452	17.369	4.428	95.375
	150	19.120	16.675	5.199	111.981
	200	24.318	21.888	5.166	111.264
15	100	18.312	17.097	1.205	151.007
	150	20.508	18.805	1.690	189.973
	200	25.255	23.929	1.315	171.978
30	100	18.257	16.852	0.697	215.324
	150	18.635	16.963	0.829	266.514
	200	23.983	22.325	0.822	247.894
60	100	18.029	16.192	0.456	299.420
	150	20.235	16.855	0.838	421.261
	200	21.397	18.668	0.677	372.833

$$D_i = \frac{Wt_i}{\rho} \tag{2}$$

$$D = \sum D_i \tag{3}$$

Where D is the cumulative loss of thickness of the specimen. D_i is the loss of thickness of the corrosion cycle i. W is the corrosion rate. t_i is the time of the corrosion cycle i. ρ is the density of steel, the value is 7.8 g/cm^3 in this paper.

The data recorded in the tests are introduced into these two equations, and the results of three coating types, which are zinc coating, aluminum coating and zinc-aluminum coating with different thicknesses, are obtained under each corrosion cycle. The calculation results are shown in Tables 1, 2 and 3.

3.2   Analysis and Discussion

Based on the obtained calculation results in Tables 1, 2 and 3, the curves of the corrosion degree indicators with time are drawn by Origin, as shown in Figs. 1 and 2, respectively.

As shown in Figs. 1 and 2, it can be seen that:

The corrosion law of three coating types is basically the same in Fig. 1. In the early stage, the corrosion of various coated specimens is severe, and then decreases rapidly. When the corrosion continues for 30 days, the corrosion rate tend to be gentle. During the whole cycle, the corrosion rate curves of zinc coating specimens

Table 2 Results of corrosion rate of aluminum coating

Corrosion time (day)	Coating thickness (μm)	Mass before test (g)	Mass after test (g)	Corrosion rate g/$(m^2 h)$	Cumulative loss of thickness (μm)
7	100	16.689	15.969	1.530	32.952
	150	18.079	17.402	1.439	30.998
	200	19.736	17.500	4.753	102.381
15	100	16.495	15.676	0.812	70.421
	150	18.103	16.878	1.215	87.057
	200	19.480	18.578	0.894	143.651
30	100	16.486	15.838	0.322	100.122
	150	18.256	17.641	0.305	115.217
	200	19.429	17.740	0.838	220.971
60	100	16.933	16.129	0.199	136.935
	150	17.844	16.781	0.264	163.889
	200	20.177	19.916	0.065	232.921

Table 3 Results of corrosion rate of zinc-aluminum coating

Corrosion time (day)	Coating thickness (μm)	Mass before test (g)	Mass after test (g)	Corrosion rate g/$(m^2 h)$	Cumulative loss of thickness (μm)
7	100	19.071	18.460	1.299	27.976
	150	20.238	19.674	1.199	25.824
	200	21.466	20.672	1.689	36.386
15	100	18.701	16.651	2.034	121.841
	150	19.481	16.460	2.996	164.118
	200	20.032	17.356	2.655	158.929
30	100	19.294	17.768	0.757	191.712
	150	20.355	19.475	0.437	204.411
	200	21.770	19.896	0.929	244.704
60	100	19.328	17.649	0.416	268.574
	150	19.856	18.226	0.404	279.014
	200	21.177	19.399	0.441	326.084

and zinc-aluminum coating specimens fluctuate greatly, while the corrosion rate of aluminum coating specimens is the smallest, continuing to decline.

It is not difficult to find that like the corrosion rate, the growth of corrosion loss of thickness is evident in the early stage and smooth in the late stage in Fig. 2. The cumulative loss of thickness of zinc coating is the largest, zinc-aluminum coating

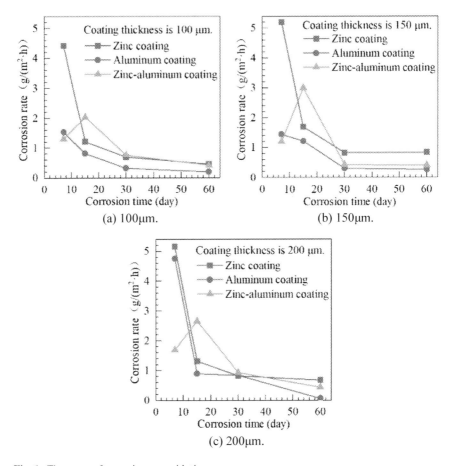

Fig. 1 The curve of corrosion rate with time

comes second, and aluminum coating is smallest. In other words, the corrosion resistance of aluminum coating is the best.

In addition, compared with other thickness, it is more severe corrosion in all aluminum coating specimens with 200 μm at the initial stage of corrosion, and the corrosion loss of thickness is bigger than 100 and 150 μm. Therefore, aluminum coating is recommended to be the optimal solution, and the coating thickness ranges from 100 to 150 μm.

4   Conclusions

This paper focuses on the influence of different coating categories and thickness on the corrosion resistance of Q345 steel. Corrosion tests are performed by using 3.5%

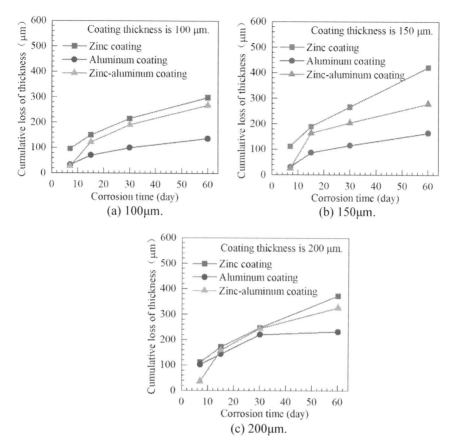

Fig. 2 The curve of cumulative loss of thickness with time

sodium chloride solution, the results are present. Then the corrosion degree is determined by two indicators for comparison. Based on experimental studies associated with analytical studies, the key conclusions are as follows:

(1) It can be shown that the corrosion evolution law of the three coating types is consistent according to the two indicators, the corrosion is severe in the early stage and gentle in the late stage.

(2) The corrosion resistance of aluminum coating is better than that of Zinc coating and Zinc-aluminum coating. The corrosion rate and the cumulative loss of thickness of aluminum coating are smaller. It is recommended that aluminum coating is preferred, and the reasonable thickness range of coating is 100–150 μm.

Acknowledgements This research was funded by [Natural Science Foundation of Hubei] grant number [2022CFB547], [Natural Science program of Xiaogan] grant number [XGKJ2022010097], [Innovation and Entrepreneurship Training Program for university students] grant number [202110528019, 202210340048].

References

1. Huang GQ (2001) Corrosion behavior of carbon steels immersed in sea areas of China. Corrosion Sci Protection Technol 13(2):81–84, 88
2. Zhu XR, Huang GQ, Lin LY et al (2005) Research progress on the long period corrosion law of metallic materials in seawater. J Chin Soc Corrosion Protection 25(3):142–148
3. Yu B, Ling GZ, Liu JB et al (2019) Probabilistic prediction model of steel corrosion rate in concrete. Bull Chin Ceramic Soc 38(11):3385–3391
4. Pour-Ghaz M, Isgor OB, Ghods P (2009) The effect of temperature on the corrosion of steel in concrete. Part 1: simulated polarization resistance tests and model development. Corrosion Sci 51(2):415–425
5. Pour-Ghaz M, Isgor OB, Ghods P (2009) The effect of temperature on the corrosion of steel in concrete. Part 2: model verification and parametric study. Corrosion Sci 51(2):426–433
6. Pradhan B, Bhattacharjee B (2009) Performance evaluation of rebar in chloride contaminated concrete by corrosion rate. Constr Build Mater 23(6):2346–2356
7. Wang LX, Wang JJ, Yu JH, Qi HQ (2022) Corrosion performance of arc-sprayed Zn coating in simulated acid rain environment. Welding Joining 06:49–53
8. Wang Y, Zhao XN, Wang XD et al (2018) Aluminizing steel surface for providing protection in Marine environment: a technological review. Mater Rep 32(21):3805–3813+3822
9. Su X (2015) Study on corrosion behavior of Zn-Al coating in typical simulated environment. Wuhan University of Technology, Wuhan
10. Ministry of Railways: TB/T 2375-1993 (1993) Test method for periodic immersion corrosion of weathering steels for railway, Beijing
11. Gong TTL (2015) A study on correlation between typical metal materials marine environmental atmospheric exposure test and practical accelerated corrosion test. Guangdong Ocean University, Zhanjiang

Analysis of Influence Range of Sudden Change of Rock Mass Grade on Surrounding Rock Stability in Shallow Tunnel Construction

Fubin Wang and Kui Yu

Abstract In this paper, the construction process of the diversion tunnel of Er-Jia-Gou Reservoir in Harbin is numerically simulated. Based on Biot consolidation theory and porous elastic medium theory, considering the influence of groundwater seepage during the construction process, a three-dimensional fluid solid coupling model of the tunnel is established to simulate the seepage field, stress field and displacement field changes during the construction process. According to the calculation results, the distribution of pore water pressure and stress around the tunnel during the construction process is analyzed, and the variation rules of pore water pressure, stress and the displacement of the vault and arch bottom of the tunnel are obtained when the shallow tunnel construction passes through the sudden change area of rock mass. Finally, the influence range of the sudden change area of rock mass grade in the tunnel construction process is determined: pore water pressure, stress and vertical displacement will be affected at 20 m from the sudden change area. The results of the paper can provide reference for the safe construction of shallow tunnels.

Keywords Shallow buried tunnel · Rock mass grade · Numerical simulation · Scope of influence

1 Introduction

In case of sudden change of rock mass grade during the construction of tunnel and other underground projects, instability collapse, water and mud inrush and other accidents are very easy to occur due to the influence of surface water and groundwater seepage, which pose a huge threat to construction safety and the surrounding environment, and adversely affect the development of tunnel construction [1]. Therefore, it is of great significance to analyze the stability during tunnel construction. Some experts and scholars have carried out relevant research. For example, Shiau, Junping et al. studied the minimum bearing pressure required for the stability of the double circular

F. Wang (✉) · K. Yu
School of Hydraulic and Electric Power, Heilongjiang University, Harbin 150080, China
e-mail: silvan1997@163.com

© The Author(s) 2023
S. Wang et al. (eds.), *Proceedings of the 2nd International Conference on Innovative Solutions in Hydropower Engineering and Civil Engineering*, Lecture Notes in Civil Engineering 235, https://doi.org/10.1007/978-981-99-1748-8_7

tunnel through finite element numerical calculation [2]; Xiongyu Hu et al. studied the influence of the relative depth of the tunnel and the density of granular soil on the stability and collapse mechanism of shallow tunnels by using the discrete element method (DEM) [3]. Jinjie Zhou et al. analyzed the influence of groundwater seepage on surface settlement in the shallow underground excavation section of the north entrance of Hangzhou Zizhi Tunnel by combining on-site monitoring data analysis with numerical simulation [4]; Liyuan Wei et al. conducted model tests on Qingdao Jiaozhou Bay undersea tunnel, and obtained the variation rules of seepage field and displacement field values [5]; Kezhong Wang et al. conducted finite element analysis on the Rizhao Shushui East Diversion Tunnel Project, and obtained the surrounding rock pore water pressure and deformation law [6]; Xumei Du et al. analyzed the stability of surrounding rock excavation of the Jiangmen Tunnel from Guangzhou to ZhuHai by using 3D numerical simulation, and obtained effective reinforcement measures [7]; Fenglin Li et al. used the finite difference method to conduct numerical simulation on the SiJiaZhai small spacing tunnel project of Guizhou Panxing Expressway, and obtained the change law of the tunnel excavation displacement field under Grade V surrounding rock [8]; Yanchun Li et al. obtained the distribution characteristics of seepage field, stress field and displacement field of surrounding rock of the water rich fracture zone project of Liangshan Tunnel in Zhangzhou, Fujian Province through simulation analysis [9].

Scholars at home and abroad have summarized the laws of pore water pressure, stress and deformation of surrounding rock of underground projects in different construction strata and different calculation methods, but there is little research on the stability of surrounding rock during excavation through the sudden change area of rock mass grade under the fluid solid coupling effect of shallow tunnel. In this paper, based on the change of seepage field, stress field and displacement field of tunnel surrounding rock, COMSOL Multiphysics software is used to conduct fluid structure coupling simulation analysis on the construction process of shallow tunnel.

2   Fluid Structure Coupling Equation

2.1   Seepage Field Equation

Based on Biot's consolidation theory, the governing differential equation corresponding to fluid structure coupling can be expressed by Darcy's law of fluid motion. We simplify the groundwater seepage mode into laminar flow, and Darcy's law is:

$$q_i = -k \frac{\partial}{\partial x}(p - \rho g x) \tag{1}$$

In the formula, q_i refers to seepage velocity vector, k refers to permeability coefficient of medium, p refers to pore water pressure, ρ refers to liquid density, g refers to gravity acceleration component.

For the convenience of calculation, the rock soil layer can be regarded as a porous elastic medium. According to the principle of seepage mechanics, the seepage field control equation is [10]:

$$\nabla \cdot -\frac{k_m}{u} \cdot (\nabla p - \rho g) + \left(\frac{\alpha - \phi}{k_s} + \frac{\phi}{k_i} \right) \frac{\partial p}{\partial t} + \alpha \frac{\partial \varepsilon}{\partial t} = q_m \qquad (2)$$

In the formula, k_m is the permeability of porous media, α is the Biot coefficient, ϕ is the porosity of porous media, ε is the strain component, q_m is the sink term of the fluid.

The change of volume strain will cause the change of fluid pore pressure, on the contrary, the fluctuation of pore pressure will also cause the occurrence of volume strain [11]. The constitutive equation of porous elastic medium is:

$$\Delta \sigma_{ij} + \alpha \Delta p \delta_{ij} = H_{ij} \left(\sigma_{ij}, \Delta \xi_{i,j} \right) \qquad (3)$$

In the formula, $\Delta \sigma_{ij}$ is the stress variation, Δp is the change of pore water pressure; δ_{ij} is Kronecher factor, H_{ij} is a given function, $\Delta \xi_{i,j}$ is the total strain.

2.2 Stress Field Equation

The control equation of porous elastic material model is:

$$-\nabla \cdot \sigma = \rho g \qquad (4)$$

In the formula, σ is the stress tensor, ρ is the density of liquid, g and is the acceleration of gravity.

The displacement boundary condition is expressed as:

$$u| = u_l \qquad (5)$$

In the formula, u_l is the displacement at the boundary.

The stress boundary condition is expressed as:

$$\sigma_{ij} \cdot n_j| = T_i \qquad (6)$$

In the formula, n_j is the cosine of the angle between the stress and the normal of the projection plane. T_i is the surface force on the boundary.

The displacement and velocity of rock mass particle at the initial time (any selected time) are respectively expressed as:

$$u|_{t=0} = u_i \tag{7}$$

$$\frac{\partial u}{\partial t_i}|_{t=0} = v_i \tag{8}$$

In the formula, u_i is the displacement of rock mass particle when $t = 0$, v_i is the velocity of rock mass particle when $t = 0$.

3 Project Overview

Taking the Er-Jia-Gou Reservoir headrace tunnel in Harbin as an example, the headrace tunnel has a total length of 2410 m, a bottom width of 4.0 m, a tunnel height of 4.0 m, and a tunnel cross section in the shape of a city gate [12]. According to the actual project plan, the buried depth of the tunnel top in some tunnel sections is 15 m, and the surface passing by is partly mountain platform. The lithology of this part of rock is strong and weak coarse-grained granite, and the rock is relatively soft. The rock mass is saturated due to long-term farming and rainwater immersion [13]. According to the construction records, when the headrace tunnel was excavated to 0+628, serious water seepage occurred in the top arch, accompanied by large area of rock collapse.

4 Numerical Simulation Analysis

4.1 Model Establishment and Calculation Scheme

In this paper, COMSOL software is used to simulate and analyze the excavation section of the headrace tunnel, study the stability of surrounding rock in the sudden change area of rock mass grade during shallow tunnel construction, and find out the influence range of the sudden change area. According to the tunnel design, the width of the tunnel bottom is 4.0 m, the tunnel height is 4.0 m, the burial depth (the distance between the tunnel top and the ground) is 15 m, the portal is in the form of a city gate, and the calculation range of the model is set as 40 m × 40 m × 60 m. The tunnel simulation model is shown in Fig. 1, in which 0–60 m, 100–160 m are ordinary areas, and 60–100 m are sudden change areas of rock mass grade. The working condition settings are shown in Table 1, of which working condition 2 is the control group. See Table 2 for physical and mechanical parameters of rock mass.

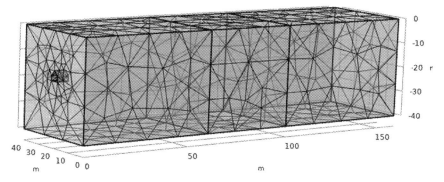

Fig. 1 Simulation model of a 15 m deep tunnel

Table 1 Working condition setting

Working condition	General area	Mutation region
1	Grade III rock mass	Grade IV rock mass
2	Grade IV rock mass	Grade IV rock mass
3	Grade IV rock mass	Grade V rock mass

4.2 Assumptions for Numerical Calculation

(1) The initial pore water pressure before tunnel excavation is equal to the hydrostatic pressure in the rock mass.
(2) Groundwater flow meets Darcy's law before and after excavation.
(3) Rock mass is a homogeneous and isotropic equivalent continuous permeable medium.
(4) The initial stress field of rock mass is calculated according to the dead weight of rock mass.
(5) The influence of support is not considered.

4.3 Boundary Condition

(1) The horizontal displacement is limited around the tunnel model, which is set as roller support constraint, and fixed constraint is set at the bottom.
(2) The upper surface of the model is the ground, which is set as the free boundary together with the excavation section.
(3) The upper surface of the model and the face of the tunnel are in direct contact with the air, and the pore water pressure is set to zero.
(4) To prevent groundwater from flowing around, the model is set with water storage mode and no flow boundary around and at the bottom.

Table 2 Physical and mechanical parameters of rock mass

Material name	Severe /(kN m^{-3})	Elastic modulus/(GPa)	Porosity	Poisson's ratio	Permeability/(m^2)	Cohesion/(MPa)	Internal friction angle/(°)
Grade III rock mass	27	15	0.2	0.28	5×10^{-14}	1.1	42
Grade IV rock mass	23	3.7	0.3	0.31	9.83×10^{-14}	0.42	34
Grade V rock mass	17	0.9	0.5	0.4	2.5×10^{-13}	0.18	26

5   Result Analysis

The tunnel is excavated in full section. In order to facilitate analysis, a monitoring section is set up at 5 m behind the tunnel face to study the pore water pressure, stress and vertical displacement variation of the monitoring section during the construction of the tunnel with a buried depth of 15 m. The COMSOL steady state mode is selected for numerical calculation.

5.1   *Change of Seepage Field*

Pore water pressure data is measured at 5 m from the monitoring section of the tunnel to the left arch waist. This location is selected as the research object to calculate the distribution law of pore water pressure under different working conditions. It can be seen from Fig. 2 that the distribution law of pore water pressure curve under working condition 1 and 3 is basically consistent. At a distance of 20 m from the abrupt change area, the pore water pressure under working condition 1 and 3 has a sudden change, which is greatly reduced after entering the abrupt change area of rock mass grade. It can be seen from the comparison of data under working conditions 1 and 3 that the higher the rock mass grade in the abrupt change area, the greater the amplitude of pore water pressure fluctuation during construction.

Figure 3 shows the dynamic distribution of the pore water pressure of Y–Z monitoring section under different driving distances under working condition 3. The initial pore water pressure of surrounding rock before excavation is layered and increases with the depth from top to bottom. After excavation, the pore water pressure inside

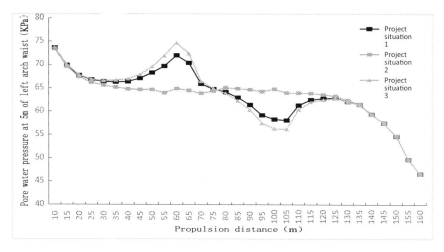

Fig. 2 Distribution of pore water pressure at the research location of the monitoring section with a buried depth of 15 m

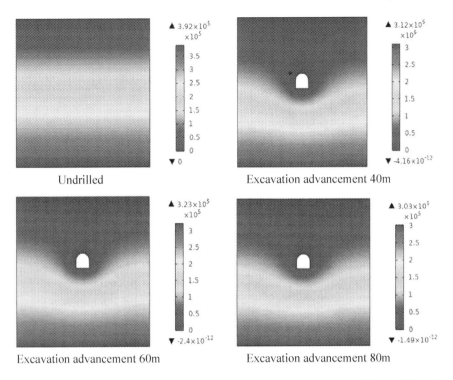

Fig. 3 Dynamic change of pore water pressure of monitoring section under working condition 3

the tunnel is zero, forming a certain pressure difference with the pore water pressure outside the tunnel, causing the balance of the original seepage field to be broken, and finally forming a low pressure area similar to a funnel around the tunnel.

When the tunnel excavation is advanced to the sudden change area, the maximum pore water pressure at the research location of the monitoring section is reduced from 74.7 to 56.2 kPa, a decrease of 24.8%. Under the effect of pressure difference, groundwater is easy to penetrate into the tunnel, causing softening of surrounding rock and stress reduction. Therefore, in the actual construction process, grouting and plugging shall be carried out in time at the leakage location to prevent water inrush and mud leakage. For micropores that cannot be grouted, two layers of waterproof coating can be applied for plugging.

5.2   Change of Stress Field

It can be seen from Fig. 4 that the distribution law of stress curve under working condition 1 and 3 is similar, and the stress of surrounding rock is affected and gradually concentrated at the place 20 m away from the fracture zone when the tunnel

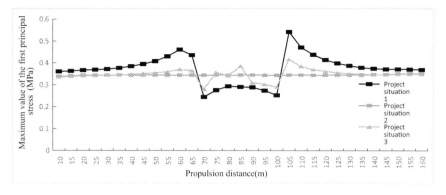

Fig. 4 Distribution of the first principal stress in the monitoring section of the 15 m deep tunnel

is excavated. With the sudden change of rock mass grade, the first principal stress of the tunnel surrounding rock monitoring section under working condition 1 and 3 decreases by 47% and 24% respectively. It can be seen from the data analysis under working conditions 1 and 3 that the higher the rock mass grade in the mutation area, the more frequent the stress mutation.

Figure 5 is the dynamic diagram of the first principal stress of the monitoring section under condition 3. During construction, the initial stress balance is broken and the stress is redistributed. With the advance of excavation, the stress is gradually concentrated. The maximum compressive stress is concentrated at the bottom corner of the tunnel, the stress distribution at the vault and around is small, and the maximum tensile stress is concentrated at the bottom plate of the tunnel. During the excavation of the tunnel, the compressive stress increases to a certain value and then decreases suddenly. This is because at the bottom of the soft rock strength, when the stress level exceeds the bearing range of the surrounding rock, the surrounding rock presents a yield state, resulting in a stress drop. Therefore, in the process of tunnel excavation, stress monitoring and reinforcement measures should be taken in the stress concentration area to prevent the surrounding rock from reaching the stress limit and causing instability and damage.

5.3 Change of Displacement Field

As shown in Fig. 6, during the construction of shallow tunnel, the vertical displacement at the top of the tunnel is greater than that at the bottom, and the distribution law of the displacement curve under working condition 1 and 3 is similar. At a distance of 20 m from the abrupt change area, the vertical displacement of combination mode 1 and 3 is affected and gradually increases. It can be seen from the Fig. 6a that the vertical displacement and settlement of the vault in combination mode 1 increased by

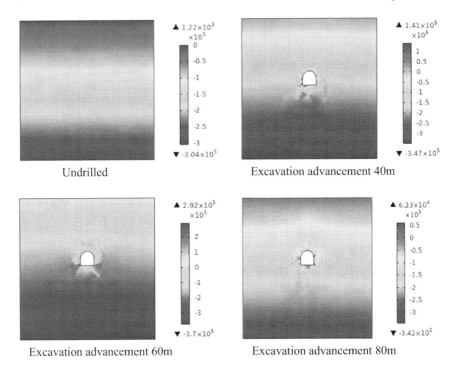

Fig. 5 Dynamic change of the first principal stress of the monitoring section under working condition 3

1.487 mm before and after entering the sudden change area, and the vertical displacement and settlement of the vault in combination mode 3 increased by 2.31 mm. In the Fig. 6b, the vertical displacement and settlement of the arch bottom in combination mode 1 increased by 0.97 mm after entering the sudden change area, and the vertical displacement and settlement of the arch bottom in combination mode 3 increased by 0.93 mm. According to the comparative analysis of the data under working condition 1 and 2, when the rock mass grade in the mutation area is the same, the lower the rock mass grade in the ordinary area, the smaller the vertical displacement. It can be seen from the comparison of data under working conditions 2 and 3 that when the rock mass grade in the general area is the same, the higher the rock mass grade in the mutation area, the greater the vertical displacement.

6 Conclusion

In this paper, the construction process of the diversion tunnel of Er-Jia-Gou Reservoir in Harbin is simulated by fluid structure coupling simulation. Combined with the

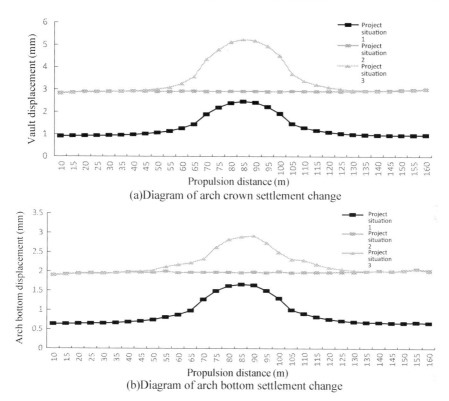

(a)Diagram of arch crown settlement change

(b)Diagram of arch bottom settlement change

Fig. 6 Vertical displacement change of monitoring section of 15 m deep tunnel

change data of pore water pressure, the first principal stress and vertical displacement, the following three conclusions are drawn:

(1) During the construction of the tunnel, the original seepage balance is destroyed, and a funnel like low pressure area is formed around the tunnel, and the stress will be redistributed and gradually concentrated.

(2) When the tunnel construction enters the sudden change area of the rock stratum, the pore water pressure and the first principal stress around the tunnel under working condition 3 decrease by 24.8% and 24% respectively, and the vertical displacement of the vault and arch bottom of the tunnel increase by 2.31 mm and 0.93 mm respectively. The higher the grade of rock mass in the abrupt change area, the greater the fluctuation amplitude of pore water pressure during construction, and the more frequent the sudden change of stress.

(3) When the rock mass grades of ordinary area and sudden change area are different, the pore water pressure, stress and vertical displacement at 20 m away from the sudden change area will be affected. Therefore, in the actual construction process, it is necessary to avoid areas with large changes in rock

mass grade as far as possible, and focus on support and reinforcement within 20 m from the sudden change area to ensure the safety of tunnel construction.

References

1. Weng XJ (2014) Study on inrush of water and mud mechanism and grouting control technology in water-rich fault fracture zone of tunnel. Shandong University, pp 1–121
2. Shiau J, Al-Asadi F (2021) Twin tunnels stability factors Fc, Fs and Fγ. Geotech. Geol Eng 335–345
3. Hu X, Fu W, Wu S et al (2021) Numerical study on the tunnel stability in granular soil using DEM virtual air bag model. Acta Geotech 3285–3300
4. Jinjie Z, Gang C, Penglu G et al (2017) Analysis of ground surface settlements induced by construction of shallow-covered mined tunnel in water-rich and soft-weak strata: a case study of Zizhi Tunnel in Hangzhou. Tunnel Constr 37(2):141–149
5. Liyuan W, Shucai L, Bangshu X et al (2011) Study of solid-fluid coupling model test and numerical analysis of underwater tunnels. Chin J Rock Mech Eng 30(7):1467–1474
6. Kezhong W, Yuqiang T, Weiping L et al (2016) Seepage-stress coupled analysis on the deformation stability of deep diversion tunnels. J Zhejiang Univ Technol 44(2):207–211
7. Xumei D, Xu W, Pengcheng J et al (2013) Stability analysis of surrounding rock of shallow buried tunnel with rich surface water. Railw Eng 12:35–38
8. Fenglin L, Meng W, Xiuyu L et al (2017) Analysis on reasonable clear distance and surrounding rock of shallow-buried small distance tunnel. Subgrade Eng 2:163–166
9. Tingchun L, Lianxun L, Huiling D et al (2016) Water burst mechanism of deep buried tunnel passing through weak water-rich zone. J Central South Univ (Science and Technology) 47(10):3469–3476
10. Lianju Y, Zhenquan L, Shengzhong W et al (2001) Engineering seepage mechanics and its application. China Building Materials Press, Beijing
11. Faben C, Deyong H, Zhuo Y et al (2013) Fluid-structure interaction analysis of rich water tunnel excavation process. J Transp Sci Eng 29(2):56–62
12. Yong Huang F (2021) Caving treatment scheme of diversion tunnel for Erjiagou reservoir. Heilongjiang Sci Technol Water Conservancy 49(1):129–131
13. Jingxi J, Kui Y, Yong H et al (2021) Analysis of self-stabilization capacity of different surrounding rocks of shallow tunnel based on Phase2 software. Heilongjiang Sci Technol Water Conservancy 49(7):14–17

Non-destructive Detection of Grouting Defects Behind Shield Tunnel Wall by Combining Ground Penetrating Radar with Seismic Wave Method

Qing Han, Yihui Huang, Chaochen Li, Peiquan Yang, Xiongjie Deng, Dengyi Wang, Fuan Ma, and Hufeng Shi

Abstract The shield construction is a mature tunnel construction technology. Grouting behind segment wall synchronously during shield construction is an important measure to control stratum deformation, reduce ground settlement and prevent tunnel water seepage. The quality of grouting can significantly affect the quality of tunnel construction. The method of grouting quality judgment in early tunnel construction is inefficient and uncertain. To ensure the safety of tunnel structure, it is an urgent technical innovation to study the non-destructive testing technology for grouting effect of shield segment wall. Nowadays, the general Non-destructive testing methods include ground penetrating radar detection and seismic wave detection. These two methods can meet most of the needs of grouting defect detection behind the wall, but also have their own advantages and disadvantages. Therefore, this study proposes the combination of ground penetrating radar and seismic wave method of shield tunnel wall after grouting defects Non-destructive detection method, first, ground penetrating radar is used for scanning probe shield wall after grouting, and then according to the seismic wave method is used to further verify the results for some suspected injury, while ensure the detection efficiency to improve recognition accuracy. Finally, the defect detection and verification of the proposed method was carried out based on the shield segment model test platform, and this proposed method was successfully applied to the actual project of a metro line in Foshan.

Keywords Tunnel engineering · Shield · Back-wall grouting · Non-destructive detection · Ground penetrating radar · Seismic wave

Q. Han · Y. Huang
CCCC Foshan Investment and Development Co., Ltd, Foshan 528000, China

C. Li · P. Yang (✉) · X. Deng
Guangdong Construction Engineering Quality and Safety Inspection Station Co., Ltd., Guangzhou 510599, China
e-mail: 164157504@qq.com

D. Wang
Department of Geotechnical Engineering, Tongji University, Shanghai 200092, China

F. Ma · H. Shi
Guangxi Nonferrous Survey and Design Institute, Nanning 530031, China

© The Author(s) 2023
S. Wang et al. (eds.), *Proceedings of the 2nd International Conference on Innovative Solutions in Hydropower Engineering and Civil Engineering*, Lecture Notes in Civil Engineering 235, https://doi.org/10.1007/978-981-99-1748-8_8

95

1 Introduction

Shield method is a construction method of tunnel excavation with shield machine. After nearly two hundred years of development, it has become a quite mature construction technology. At present, it has become a main means of tunnel construction in soft soil (saturated soft clay, muddy soft soil, silty clay, saturated sandy silt, silty sand, etc.). In the process of shield, synchronous grouting is needed to reduce the stratum loss. At the same time, the grouting layer can form an effective waterproof layer around the segment peripheral to prevent groundwater from penetrating into the tunnel. However, in the process of grouting, it is difficult to directly observe the grouting effect, quantitatively control the grouting amount and ensure the grouting effect by injecting mortar through the reserved grouting holes in the segments. When the grouting quantity is insufficient, the gap of the pipe wall will not be fully filled, and the cavity will be generated behind the lining wall, resulting in uneven stress on the segments, cracking of the segments and water seepage accidents in the tunnel [1].

Therefore, how to effectively detect the grouting effect behind the shield segment wall is of great significance to ensure the safety and quality of the tunnel. According to the past experience, in the construction process, whether the grouting pressure meets the design requirements and whether the grouting hole overflows or not is often used to judge whether the grouting is full. In the later stage, the grouting quality behind the shield segment wall is mainly verified by observing whether the segment seeps and adopting the drilling method. However, these methods have great limitations, which can't accurately and intuitively reflect the grouting situation behind the wall. In order to ensure the safety of the tunnel structure, it is an urgent technical innovation to study the nondestructive testing technology of the grouting effect behind the shield segment wall, and its technical achievements can be applied in guiding and monitoring the grouting construction behind the shield segment wall. So far, the ground penetrating radar (GPR) method has been widely used to detect the cavity behind the lining wall of mine tunnel. In some places, GPR method has also been applied to the quality detection of grouting behind the shield tunnel segment wall. This method can intuitively reflect the grouting situation behind the segment wall, and can adjust the corresponding parameters according to different structures. It has high detection efficiency and wide coverage, and can quickly survey the grouting defects behind the shield segment wall [2]. However, due to the complexity of segment structure (such as multi-layer steel bar structure, complex and changeable slurry) and the existing technology, there are great difficulties in detecting the density of shield segment grouting, the location of void area and the degree of void. The detection results will be disturbed, which leads to the problem that the defects cannot be identified accurately. With the research of scholars in recent years, seismic wave method has become a potential nondestructive testing method for grouting behind the wall [3]. Seismic wave method is a method to detect underground engineering geological problems by using the characteristics of medium transmitting elastic waves. A small hammer is used as an excitation source to knock on the concrete surface to generate compression waves, and then a receiving sensor placed near the excitation source receives the reflected seismic waves. By analyzing the received data to analyze the defects, this method can effectively avoid some interference. However, the detection efficiency of this method is low, and it can't be widely used. Both methods have their own advantages and disadvantages. Therefore, it is one of the main purposes of this study to combine the two to achieve complementary advantages.

This project relies on a metro line in Foshan, Guangdong Province to carry out scientific research. This project is the backbone of Foshan's east–west route, with a total length of 556 km and 27 stations. It is the east–west trunk line connecting the central group and the express rail passage connecting Foshan and Guangzhou. In the past, through the analysis of the radar signal characteristics of the foundation electrical properties, main interference and grouting defects of shield tunnel in rail transit, it is found that some defects cannot be detected accurately if only the dip radar is used to detect the grouting defects behind the wall. Therefore, it is particularly important to combine the dip radar with seismic waves to ensure the grouting quality of the tunnel.

2 Working Principle of Ground Penetrating Radar and Seismic Wave Method

2.1 Basic Principle of Ground Penetrating Radar Detection

The principle of geological radar detection is that high-frequency electromagnetic wave signals with wide band and short pulse are transmitted to the ground through the transmitting antenna, when the electromagnetic wave encounters the underground interface with electrical differences (i.e. dielectric constant, electrical permeability, etc.) or other target bodies (such as hollows, cavities, metals, etc.), electromagnetic wave phenomena such as reflection and diffraction will occur. The response signal is picked up by the receiving antenna and recorded on the computer. According to the time domain and frequency domain characteristic information of electromagnetic wave, such as waveform, phase, amplitude and frequency spectrum, we can obtain the distribution of different underground electrical bodies and interpret the internal structure of the medium [4]. The schematic diagram of detecting the effect of grouting behind the segment wall by geological radar is shown in Fig. 1.

The nearsightedness expression of electromagnetic wave velocity in medium is:

$$v = \frac{c}{\sqrt{\varepsilon_y}} \tag{1}$$

where, c is the speed of light in vacuum; ε_y is the relative dielectric constant; v is the wave velocity of electromagnetic wave in medium (m/ns).

The formula for calculating the depth of underground interface or target body is:

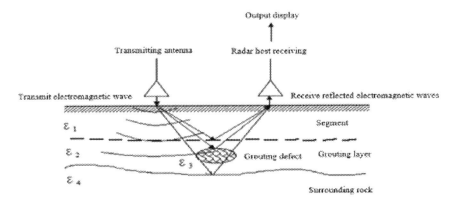

Fig. 1 Schematic diagram of detection of grouting effect behind the wall of ground penetrating radar

$$Z = \frac{v \cdot t}{2} \tag{2}$$

where Z is depth (m); t is the two-way travel time of reflected wave (ns).

According to the propagation characteristics of electromagnetic waves, electromagnetic waves will be reflected and transmitted at different electrical interfaces, and their reflection coefficient R and transmission coefficient T are expressed as follows:

$$R = \frac{\sqrt{\varepsilon_1} - \sqrt{\varepsilon_2}}{\sqrt{\varepsilon_1} + \sqrt{\varepsilon_2}} \tag{3}$$

$$T = \frac{2\sqrt{\varepsilon_2}}{\sqrt{\varepsilon_1} + \sqrt{\varepsilon_2}} \tag{4}$$

where, R is reflection coefficient; T is transmission coefficient; ε_1 is the dielectric constant of the upper layer; ε_2 is the dielectric constant of the lower layer.

It can be seen from the above formula that the reflected energy of electromagnetic waves depends on the difference of dielectric constants of adjacent strata. The greater the difference, the more obvious the reflection, and the easier it is to identify the position of interface and target. Similarly, in practical engineering, the detection effect is influenced by the limited space of shield tunnel, the complex structure of tunnel segments, the radar wave scattering of field obstacles, non-radar electromagnetic fields such as wires and cables, various unfavorable factors of radar wave absorption caused by specific geological surrounding rock media, and the problem that the radar secondary reflected wave signal of shield segment overlaps with the reflected wave signal of grouting layer behind the wall, all of which will lead to problems such as the radar can't distinguish and identify grouting layer thickness and grouting defect signal. Therefore, if the ground penetrating radar is used alone to detect the grouting defects behind the wall, some grouting defects may not be identified.

2.2 Principle of Seismic Wave Method

Seismic wave method is a method to detect underground engineering geological problems by using the characteristics of medium transmitting elastic waves. A small hammer is used as an excitation source to knock on the concrete surface to generate compression waves, and then a receiving sensor placed near the excitation source receives the reflected compression waves. During acquisition, the excitation point and the reception point are always equidistant. When there are bad geological bodies (cavities, unconsolidations, etc.) under the ground, the seismic waves show anomalies from two aspects of kinematics and dynamics. The in-phase axis is staggered, the wave group changes, the amplitude increases or decreases, and even diffraction occurs, and the travel time of elastic waves will also increase or decrease [5]. When displaying the waveform, the software is used to compress (stretch) the seismic wave, and the reflected energy is expressed in different and changeable colors, so as to achieve the effect of visually displaying the shape of underground geological bodies. The schematic diagram of seismic wave method is shown in Fig. 2. In recent years, some scholars have studied the relationship between the reflection law of seismic waves and the vibration, defect size and defect depth of concrete structures [6]. Moreover, the seismic wave method has been successfully applied to the nondestructive inspection of many civil structures, which also does not provide ideas for scholars who study behind-wall grouting. In 2016, Yao et al. began to study the grouting of segment wall thickness by using seismic waves [7], and made further improvements in 2018 [8]. It was found that compared with the ground penetrating radar (GPR) method, the application of seismic waves in the quality detection of segment wall grouting is theoretically better than that of GPR. For example, seismic wave has a better detection ability for unconsolidated grout, because the liquid grout still has a better elastic coefficient difference with the surrounding soil, which can produce obvious reflection. However, in the ground penetrating radar method, many scholars suggest that the grout should be consolidated for a period of time before a better detection result can be obtained. In addition, if the defects in the grouting layer are air-filled, the seismic wave method will have better detection results for the monitoring of the solidified grouting body. The same reason is that the difference of elastic parameters between the air and the grouting layer is greater than the difference of dielectric coefficient, which can produce better reflection. Compared with electromagnetic waves, elastic waves will not be strongly interfered by steel bars, and signals will not be shielded by steel bars, so they have better detection ability for targets behind steel bars.

However, this method also has some limitations in detecting the grouting defects behind the shield wall. For example, the generation of seismic waves depends on knocking, which leads to the low detection efficiency of seismic wave method, and can't achieve scanning detection like ground penetrating radar. The coupling condition of the sensor receiving signals by seismic wave method on the concrete surface is also stricter than that of radar, which further limits the detection efficiency of seismic wave method. At the same time, at present, the impact echo method is mainly based

Fig. 2 Schematic diagram of seismic wave method

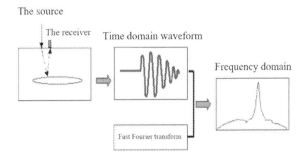

on time–frequency analysis, which can only be used as an auxiliary means of other detection methods, and it is difficult to conduct comprehensive detection independently. Therefore, if the ground penetrating radar and seismic wave are combined, we can complement each other's advantages to achieve better detection results.

3 Example Analysis of Test Platform Detection

In order to select the appropriate radar parameters and be familiar with the image analysis of seismic wave detection defects, we will use the method proposed in this paper to detect the wall thickness grouting defects of test platform of full-scale shield segment model in advance. This platform is a test platform of shield segment ring designed and built by Hu et al. [9] in 2016 with a scale of 1:1. The platform adopts shield segments and synchronous grouting materials of a tunnel in Nanning Rail Transit Line 1 project. The shield segment ring is composed of one uncapped block, two adjacent blocks and three standard blocks, with an outer diameter of 6.0 m, an inner diameter of 5.4 m, a wall thickness of 0.3 m, a ring width of 1.5 m, a concrete strength of C50 and an impermeability grade of P12. By embedding polyvinyl chloride (PVC) pipe in the gap of the building behind the segment wall to simulate the grouting cavity behind the wall, and by not grouting in some areas to simulate the grouting behind the wall. The test platform is shown in Fig. 3.

3.1 Detection Results of Ground Penetrating Radar

Data acquisition of ground penetrating radar (GPR) is carried out for a certain section of shield segment test platform. In order to highlight the contrast, we use the 400 MHz and 900 Hz antennas of American radar for data acquisition. After data processing steps, such as removing DC drift, removing direct wave, adjusting gain, normalizing

Fig. 3 Test platform for post-grouting detection of full-size shield segment

distance, band-pass filtering (100–800 MHz), and moving average, the ground penetrating radar image is formed. The comparison between radar image and grouting defect is shown in Fig. 4.

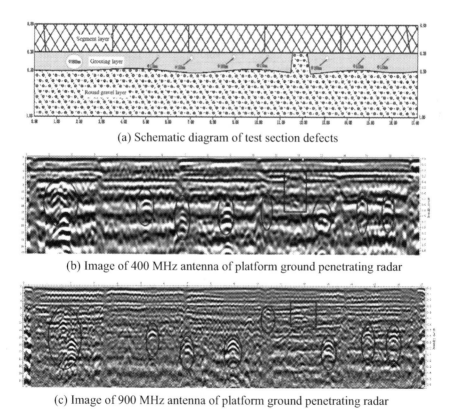

(a) Schematic diagram of test section defects

(b) Image of 400 MHz antenna of platform ground penetrating radar

(c) Image of 900 MHz antenna of platform ground penetrating radar

Fig. 4 Schematic diagram of defects in test section and radar images

Comparing the data images collected by the two groups of radar antennas with the defects of the test platform, it is found that the ground penetrating radar can accurately identify the defects, and the set hole defects show typical hyperbolic strong reflection on the ground penetrating radar image, while the grouting layer defects show the same phase axis and staggered segments on the ground penetrating radar image. Both kinds of radars can clearly identify grouting defects.

3.2   Detection Results of Seismic Wave Method

When seismic wave data acquisition is carried out for a certain section of shield segment test platform, acceleration acquisition is adopted, and the data is played back to the computer for data format conversion. The format data of each single-channel column detector is converted into seismic wave data format, and each single-channel data is spliced into profile data. The comparison between the collected wave train diagram and the defect diagram are shown in Fig. 5.

By analyzing the seismic wave profile data of the test platform, it is found that the grouting defects of the test platform can also be accurately identified by seismic wave method. According to the image analysis, when there are holes or missing grouting layers in the grouting layer, the seismic wave profile is mainly characterized by the discontinuity of in-phase staggered segments, and the seismic wave is reflected at low frequency.

The ground penetrating radar (GPR) method and seismic wave method are used to detect the grouting defects behind the test platform wall. It is found that both methods can accurately and clearly identify the defects, and the set cavity defects show typical hyperbolic strong reflection on the GPR image, and the grouting layer defects show the same phase axis and staggered segments on the GPR image. Both kinds of radars can clearly identify grouting defects. However, when there is a cavity

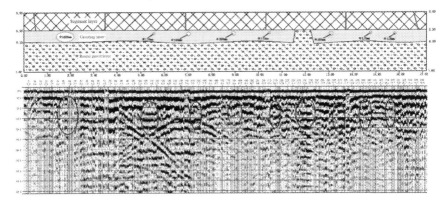

Fig. 5 Comparison between defect schematic diagram of test section and wave train diagram of seismic wave method

in the grouting layer or the grouting layer is missing, the seismic wave profile is mainly characterized by discontinuous in-phase staggered segments, and the seismic wave is reflected at low frequency. From the analysis of test results, it can be seen that these two methods can be applied to practical projects.

4 Case Study

In order to further verify the engineering applicability of the method proposed in this paper, this project relies on a metro line in Foshan, Guangdong Province to carry out scientific research. The 770–870 ring on the left line of section1 and the 305–405 ring on the right line of section2 are selected as the experimental sections of this scientific research project.

Among them, the groundwater in section1 is mainly lateral supplied by river water, and has a close hydraulic connection. It is a weak-medium water-bearing (permeable) layer. The buried depth of the groundwater table in the upper muddy soil, muddy silt and cohesive soil layer is 0.6–3.7 m; The buried depth of confined water in the lower sand and gravel layer is 0.2–4.2 m, and the buried depth of confined water is close to that of phreatic water. The tunnel mainly passes through muddy soil, muddy silty sand, muddy silty soil, silty sand, medium coarse sand, round gravel, pebbles and strongly weathered argillaceous siltstone. The thickness of shield segment used in Section1 is 350 mm; The design thickness of grouting layer behind shield segment wall is 125 mm; Grouting material behind shield segment wall is made of cement, fly ash, bentonite, fine sand and water according to a certain proportion.

The groundwater in the Section2 is mainly lateral supplied by river water, and has a close hydraulic relationship. It is a weak-medium water-bearing (permeable) layer. The stable buried depth of water level is 0.2–6.1 m (elevation—1.24~2.53 m), and the annual variation range is about 1.5 m. The tunnel mainly passes through fine powder sand, fully weathered argillaceous sandstone, strongly weathered argillaceous sandstone, moderately weathered argillaceous sandstone and moderately weathered (gravel-bearing) sandstone. The thickness of shield segment used in Section2 is 300 mm; The design thickness of grouting layer behind shield segment wall is 125 mm; Grouting material behind shield segment wall is made of cement, fly ash, bentonite, fine sand and water according to a certain proportion.

4.1 Setting of On-site Detection Parameters

The actual detection will combine the ground penetrating radar and seismic wave method to realize the complementary advantages of the two methods. In order to achieve a better detection effect, the appropriate detection parameters will be selected according to the research of the test section and the actual geological conditions on site. Among them, the American SIR series radar with relatively stable performance and good measurement effect is selected, and at the same time, the antennas with center frequencies of 200, 400 and 900 MHz are selected for the detection comparison test research. The site drawing is shown in Fig. 6.

As for the instrument selection of seismic wave method, according to the nature and conditions of this task, RSM-EDT(A) column buried depth detector produced by Wuhan Zhongyan Technology Co., Ltd., which has relatively stable performance, good measurement effect and deep measurement depth and supports two data acquisition modes of speed and acceleration, is selected for seismic wave method data acquisition. The site drawing is shown in Fig. 7.

Fig. 6 Field test photos of ground penetrating radar

Fig. 7 Field test photos of seismic wave method

4.2 Detection Results

The on-site detection section is section1 and section2. The total length of the survey lines by the ground penetrating radar method and seismic wave method is 13545.5 m and 44.8 m, respectively. Several representative detection results are analyzed.

Detection Results of Section1. The left line of section1 is located at 7 o'clock azimuth longitudinal survey line, and the ground penetrating radar method is used to detect the scene, and the American SIR 400 and 900 MHz radars are used to collect data. After the data is collected, it is played back to the computer, and after data processing steps such as DC drift removal, direct wave removal, gain adjustment, distance normalization, band-pass filtering and moving average, the ground penetrating radar image is formed, as shown in Figs. 8 and 9. The preset radar wave speed is 0.12 m/ns. Seismic wave method is to collect data from ring 825 to ring 840 by column detector, and speed method is used for data collection. After the data is played back to the computer, the collected velocity signal is differentiated to obtain acceleration signal, and the data format is converted. The format data of each single-channel column detector is converted into seismic wave data format, and the single-channel data is spliced into profile data, as shown in Fig. 10.

It can be seen from the radar images of two frequencies that the radar reflection wave at the lower part of the segment with survey line mileage of 90–91.5 m, 94.5–96 m, 97.5–99 m, 102–103.5 m (corresponding to ring numbers 830, 833, 835, 838) appears strong reflection at low frequency. It is speculated that the grouting layer in this range is not dense, and it may be filled with mud locally. From the wave diagram of seismic data, it can be seen that there are low-frequency reflections and in-phase staggered sections in multiple data in the 828–830 and 833 rings. It is inferred that the grouting in this area is not dense, which corresponds to the anomalies reflected

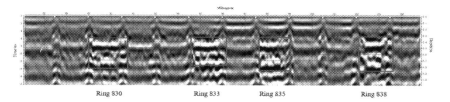

Fig. 8 400 MHz antenna image of 7 o'clock azimuth radar on the left line of Section1

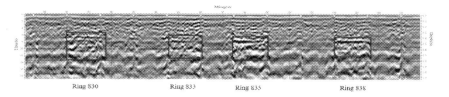

Fig. 9 900 MHz antenna image of 7 o'clock azimuth radar on the left line of Section1

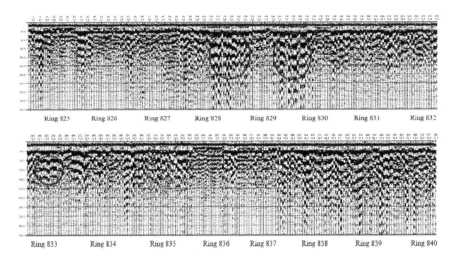

Fig. 10 Normal wave train diagram of seismic wave from ring 825 to ring 840 at 7 o'clock azimuth on the left line of Section1

by the ground penetrating radar method, but fails to correspond one by one. This shows that the velocity signal is used in data acquisition by seismic wave method, and then the acceleration signal is differentiated after playing back the computer, which may result in signal distortion and can't fully reflect the abnormality. The ground penetrating radar can avoid this problem.

Detection Results of Section2. The instrument parameters of ground penetrating radar (GPR) and seismic wave method are consistent with those of section1 on 3 o'clock longitudinal survey line of the right line of section2. The detected GPR image and the normal wave train diagram of seismic wave are shown in Figs. 11, 12 and 13.

From the above image results, it can be seen that when the ground penetrating radar is used, the 400 MHz antenna image shows that the in-phase axis of the radar reflection wave at the lower part of the segment with 70.5–72 m survey line mileage (corresponding to the ring number 352) is staggered and discontinuous. It is speculated that the grouting layer in this range is not dense, but the 900 MHz antenna has no obvious response to this defect due to excessive interference of double-layer steel bars inside the segment. From the wave train diagram of seismic data, it can be seen

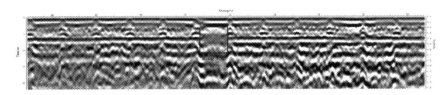

Fig. 11 400 MHz antenna image of 3 o'clock azimuth radar on the right line of Section2

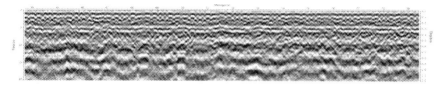

Fig. 12 900 MHz antenna image of 3 o'clock azimuth radar on the right line of Section2

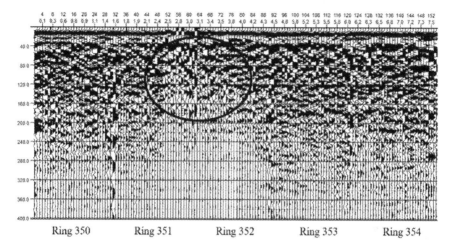

| Ring 350 | Ring 351 | Ring 352 | Ring 353 | Ring 354 |

Fig. 13 Normal wave train diagram of seismic waves from ring 350 to ring 354 at 3 o'clock on the right line of section2

that there are low-frequency reflections and in-phase staggered sections in many data in the 351 and 352 rings, and the seismic wave data in the 351 and 352 rings in the isoline diagram and spectrum analysis diagram are obviously low-frequency, so it can be inferred that grouting is not dense in this area. From the results of this section, it can be shown that the electromagnetic wave of the ground penetrating radar will be interfered by steel bars, and it is easy to produce misjudgment, and the seismic wave method can make up for this defect. If the signal results detected by the two radars are inconsistent, the seismic wave method can be used for verification.

5 Conclusions and Suggestions

Non-destructive testing is a very suitable method for detecting the quality of grouting behind the shield tunnel wall. In this paper, based on a metro line in Foshan, we study two methods for detecting the quality of grouting behind the shield wall: ground penetrating radar method and seismic wave method. By comparing the advantages, characteristics and limitations of each method between the test section and the actual project, we have summarized several conclusions of applying the ground penetrating radar method and seismic wave method to the grouting detection behind the wall.

The ground penetrating radar can accurately identify the grouting defects behind the shield tunnel wall, and can adjust the corresponding parameters according to different structures, with high detection efficiency and wide coverage, and can quickly make a general survey of the grouting defects behind the shield segment wall. However, the resolution of identification is greatly affected by the difference of dielectric constants of adjacent strata, and the electromagnetic wave is easily interfered by shield materials, resulting in misjudgment of identification.

Seismic wave method can also accurately identify the grouting defects behind the shield tunnel wall. At the same time, seismic waves will not be strongly interfered by steel bars, signals will not be shielded by steel bars, and there is still a good detection ability for targets behind steel bars. At the same time, in some media, seismic waves can produce better reflection than electromagnetic waves. However, this method has stricter requirements for sensors and signal processing, and its detection efficiency is low, so it can't achieve scanning detection.

Combining the advantages and disadvantages of the two methods and the research results of the test section, it is found that by using the two methods comprehensively, we firstly use the ground penetrating radar method to scan the shield tunnel, and then use the seismic wave method to further detect the suspicious defect section according to the identification results, which can accurately identify various defects of grouting behind the segment wall.

References

1. Xiang L, Wang C, Hao S et al (2020) Experimental study and application of ground penetrating radar in shield tunnel detection. Urban Rapid Rail Transit 3:12–17
2. Huang H, Liu Y, Xie X (2003) Application of GPR to grouting distribution behind segment in shield tunnel. Rock Soil Mech S2:353–356
3. Fang X, Xue Y (2019) Experimental analysis of refined detection of void behind tunnel lining using impact-echo method. Tunnel Constr 8:1284–1292
4. Wang D, Luo X, Zhang Z et al (2009) Detecting ground settlement of shield tunneling in soft soil by ground penetrating radar. J Hefei Univ Technol 10:97–100
5. Jiang Y, Wu J, Ma Y et al (2020) Application of impact echo acoustic method in quality testing of railway tunnel lining. Railw Eng 060(005):6–10
6. Wang J, Chang T, Chen B et al (2010) Evaluation of resonant frequencies of solid circular rods with impact-echo method. J Nondestr Eval 29(2):111–121
7. Yao F, Chen G, Su J (2016) Experimental research and numerical simulation on grouting quality of shield tunnel based on impact echo method. Shock Vib 1025276
8. Yao F, Chen G, Abula A (2018) Research on signal processing of segment-grout defect in tunnel based on impact-echo method. Constr Build Mater 187:280–289
9. Hu S, Xu G, Tang F et al (2016) Optimization research on gpr detecting of grouting behind shield tunnel s in water-soaked sand and pebble stratum. In: 2016 international conference on intelligent transportation, big data & smart city (ICITBS), IEEE, pp 141–146

Service Performance Evaluation of Rubber Floating Slab Track for Metro in Operation

Qiuyi Li and Wei Luo

Abstract Guangzhou metro line 1 is one of the first to adopt floating slab track in the Chinese mainland, it has been in operation for more than 20 years up to now. In order to obtain the vibration reduction performance of the floating slab track after long service, systematic field test and laboratory test are carried out. The results show that: (1) The rubber bearing of floating slab track keeps good appearance. (2) The measurement results of shore hardness and elongation at break show that, the rubber bearing has a certain degree of hardening after long service. Other mechanical properties of the rubber bearing still meet the design requirements. (3) The first natural frequency of the floating slab track is 33.9 Hz, its vibration reduction effect keeps good, the test result is 12.9 dB. The effective damping frequency band of the floating slab track is above 35 Hz. (4) The vertical dynamic displacement amplitude of rail and track bed is lower than the limit value given in relevant specifications. (5) The test results show that after 20 years of service, the floating slab track is in good condition, it's still qualified to maintain normal service.

Keywords Metro · Floating slab track · Vibration reduction · Service performance · Experimental study

1 Introduction

In 1997, the floating slab track was introduced into China (mainland) and first used in Guangzhou metro line 1 [1–4]. Up to now, its service time has been more than 20 years, the operation of 6 type A trains. The floating slab track of Guangzhou metro line 1 uses rubber bearings as vibration isolation elements. Considering the extreme environment (wet, ozone, etc.) in underground line tunnels, the aging, deformation and damage that may occur in the long-term service of rubber materials are directly

Q. Li (✉) · W. Luo
China Railway Siyuan Survey and Design Group Co. Ltd., Wuhan 430063, China
e-mail: 813352696@qq.com

Railway Service Safety Key Laboratory of Hubei Province, Wuhan 430063, China

© The Author(s) 2023
S. Wang et al. (eds.), *Proceedings of the 2nd International Conference on Innovative Solutions in Hydropower Engineering and Civil Engineering*, Lecture Notes in Civil Engineering 235, https://doi.org/10.1007/978-981-99-1748-8_9

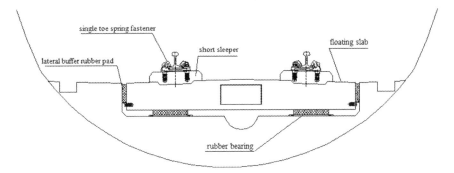

Fig. 1 Typical cross section of floating slab track of Guangzhou metro line 1

related to the safe service and vibration reduction performance of the floating slab track [5–7]. It is necessary to evaluate the performance of the floating slab track comprehensively and systematically after long service.

The research team to which the author belongs has carried out an experimental study on the floating slab track of Guangzhou metro line 1.

2   Design Overview of Floating Slab Track

A floating slab track of 1110 m is laid on the main line of Guangzhou metro line 1, which is a single rubber bearing floating slab track. Each floating slab is 2.95 m in length, 2.8 m in width, 0.335 m in thickness at the center and 0.3 m at the edge, and weighs about 6.5 t. The bottom of the standard floating slab is prefabricated with 4 circular grooves, and 4 rubber bearings are installed. The rubber bearings are natural rubber products, with a diameter of 400 mm and a thickness of 75 mm. The design static stiffness of the rubber bearing is 12–16 kN/mm, and the lateral static stiffness of the lateral buffer rubber pad is 3.2–4.2 kN/mm [2]. The single toe spring fastener is adopted for the section of floating slab track. The typical cross section of the track is shown in Fig. 1.

3   Field Test and Analysis

3.1   Appearance Detection of Rubber Bearings

Electronic video endoscope was used to detect the appearance of rubber bearings. As shown in Fig. 2, the electronic video industrial endoscope consists of a display, a joystick, a cable and a probe, among which the probe is generally composed of two parts: a light source and a miniature camera. The endoscope works as follows:

Fig. 2 Electronic video endoscope for industrial use

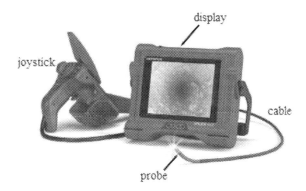

A tiny camera takes pictures of the scene under the light source, and then converts the images into digital signals that are transmitted by cable to the display screen [8].

By using the sampling method of "separate one and pump one" (separate one floating slab and pump one floating slab for detection), 50% of the floating slab track of line 1 was extracted and its rubber bearings were tested. The appearance images of the rubber bearing obtained on site are shown in Figs. 3 and 4. The main conclusions of appearance detection are as follows:

(1) No obvious cracks were observed;
(2) No obvious damaged blocks were found;
(3) There was no obvious void or joint between the rubber bearing and the bed slab, and between the tunnel base;
(4) There is no obvious inclination and deviation of the rubber bearing;
(5) During the test, the rubber bearing is not wet and soaked in water;
(6) Surface cracking and peeling of some rubber bearings, accounting for about 11% of the total amount detected.

Fig. 3 Appearance detection image 1 of rubber bearing

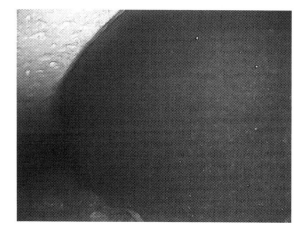

Fig. 4 Appearance detection image 2 of rubber bearing

3.2 Laboratory Test of Rubber Bearings

Three rubber bearing samples were obtained from line 1 in the way of "take one for one", and then the mechanical property parameters of rubber bearings were tested in the laboratory. The test items included: appearance size, Shore hardness, tensile strength, elongation at break, permanent deformation under constant compression, static stiffness and dynamic stiffness. Figure 5 shows the rubber bearing sample, and the rubber bearing stiffness testing device is shown in Fig. 6.

The laboratory test results of the mechanical properties of the rubber bearing are shown in Table 1. The difference between the size of the bearing and the design size after permanent deformation is less than 1 mm, and the tensile strength and static stiffness still meet the design requirements. The Shaw hardness value of the rubber bearing material is about 15% higher than the designed upper limit value, and the elongation at break is about 9.5% lower than the designed lower limit value, indicating that the rubber bearing material has a certain degree of hardening after a long time of service.

Fig. 5 Rubber bearing sample

Fig. 6 Rubber bearing
stiffness detection device

Table 1 Laboratory test results of mechanical properties of rubber bearings

Test items		Test results	Design requirements
Appearance size	Diameter/mm	400.00~400.56	400
	Thickness /mm	74.20~74.82	75, After 1 × 107 cycles of load, the change value is no more than 1 mm
Shore hardness/shore A		52	40 ± 5
Tensile strength/MPa		22.8	≥ 20
Elongation at break/%		543	≥ 600
Constant compression permanent deformation		24%	–
Stiffness	Static stiffness/kN/mm	15.22	10~16
	Static and static stiffness ratio	1.343~1.442	–

3.3 *Natural Vibration Characteristics of Floating Slab Track*

In this study, the field test of the vibration characteristics of the floating slab track was completed. The block length of line 1 floating slab track is 2.95 m. Three rows of measuring points are arranged, as shown in Fig. 7, with 6 measuring points in each row and 1 vertical acceleration sensor installed in each measuring point. The rigid force hammer is used as the excitation equipment, as shown in Fig. 8, and the measuring points 3, 4, 15 and 16 are used as the four hammering points.

The random subspace method is used to identify the modes and modal parameters of the floating slab track [9]. The natural vibration frequency and damping test results corresponding to modes 1–7 of line 1 floating slab track are shown in Table 2. The first order natural frequency is 33.9 Hz, and the corresponding structural damping is 6.866%. Figure 10 shows the mode shape correlation matrix of modes 1–7. The main diagonal elements of the matrix are all 1, and the values of the other

Fig. 7 Layout diagram of measuring points for testing natural vibration characteristics of floating slab track

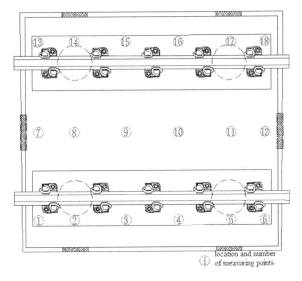

Fig. 8 Field test of free vibration characteristics of floating slab gauge bed

Table 2 Natural frequencies and damping of floating slab track of order 1–7

Order number	Frequency/Hz	Damping/%
1	33.903	6.866
2	46.235	6.924
3	114.103	3.151
4	143.840	3.734
5	185.530	0.977
6	253.801	1.695
7	263.984	1.397

elements are very small, which indicates that the mode shapes of modes 1–7 obtained through identification have good orthogonality, and the identification results have high reliability.

3.4 Damping Effect of Floating Slab Track

Layout of Test Section and Measuring Point. Combined with the data of lines and running vehicles, this study selected two sections as shown in Table 3 to carry out field tests. The acceleration measurement point used to evaluate the vibration reduction effect of the floating slab track is arranged at the side wall of the tunnel 1.5 m away from the top surface of the rail, as shown in Fig. 9. The acceleration sensor used has been verified as qualified product by the third party testing institution, with a range of 10 g and a resolution of 0.0004 m/s^2.

The Measurement Value. In the Standards of *Technical Guidelines for Environmental Impact Assessment for Urban Rail Transit (HJ453-2018), Standard for Environmental Vibration in Urban Areas (GB10070-88)*, and *Measurement Method for Environmental Vibration in Urban Areas (GB10071-88)*, the measurement value is the maximum value of VL_z vibration level in the process of train passing, which is shown as VL_{zmax} in Fig. 11.

The frequency range of the analysis was 1–80 Hz, and the weighting factor was the whole-body Z-weighting factor stipulated by *ISO2631/1-1985*, as shown in Table 4.

Table 3 Profiles of each test section

Section number	The speed of the car	Straight/curve	Type of orbit	Note
Section 1	46 km/h	A straight line	Ordinary orbit	They are a group of contrast cross sections
Section 2	48 km/h	A straight line	Floating slab track	

Fig. 9 Vertical acceleration
sensor on tunnel side wall
installed on site

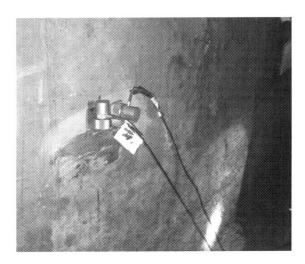

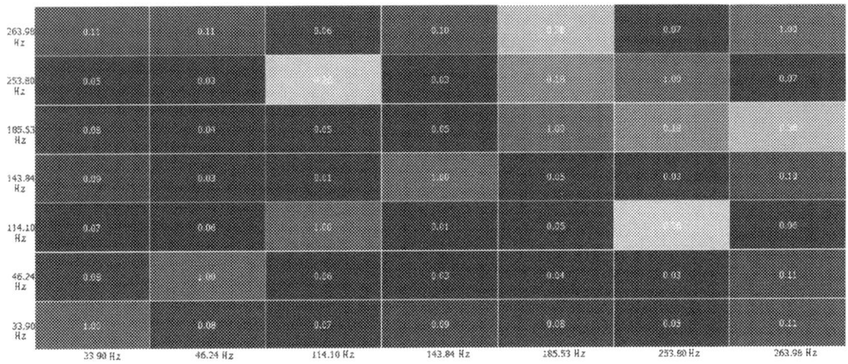

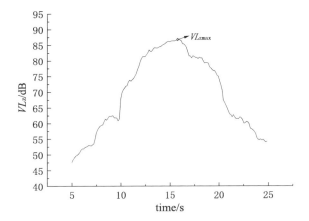

Fig. 10 Mode correlation matrix

Fig. 11 Typical time history
curve of VL_Z vibration level
during train passing

Table 4 Z-weighting factors stipulated in *ISO2631/1-1985*

Central/Hz	1	1.25	1.6	2	2.5	3.15	4	5	6.3	8
Weighting factor	0.5	0.56	0.63	0.71	0.80	0.90	1.0	1.0	1.0	1.0
Central/Hz	10	12.5	16	20	25	31.5	40	50	63	80
Weighting factor	0.8	0.63	0.5	0.4	0.315	0.25	0.2	0.16	0.125	0.1

Aiming at the vibration acceleration data of a passing event, it is divided into several 1 s length data paragraphs (which can consider certain overlap coefficient). The method of calculating the VL_z vibration level of each paragraph data is detailed as follows.

Fourier transform is performed on the data of time history of vibration acceleration to find out the vibration components in the frequency band corresponding to each central frequency, and inverse Fourier transform is performed on the vibration components identified. Then, the effective value of acceleration a_w corresponding to each central frequency can be obtained according to Eq. (1).

$$a_w = \left[\frac{1}{T} \int_0^T a_w^2(t)dt \right]^{\frac{1}{2}} \tag{1}$$

where, $a_w(t)$ is the acceleration data whose frequency is in the frequency band corresponding to a certain central frequency, and is the time function; T is the length of measurement time.

Further, the calculation formula of VLz vibration level is as follows:

$$VL_z = 20 \log_{10} \left\{ \frac{\left[\sum_{j=1}^n (W_j \times a_{wj})^2 \right]^{\frac{1}{2}}}{a_0} \right\} \tag{2}$$

where: W_j is the weighting factor corresponding to the Jth central frequency; a_{wj} is the effective value of acceleration in the frequency band corresponding to the Jth central frequency. a_0 is the base acceleration, $a_0 = 10^{-6} \text{m/s}^2$

The damping effect Δ of floating slab track is calculated by the following formula:

$$\Delta = VL_{z\,\text{max}\,1} - VL_{z\,\text{max}\,2} \tag{3}$$

In the formula, $VL_{z\text{max}1}$ is the maximum value of VL_z vibration level in the passing process of trains in the ordinary track section, and $VL_{z\text{max}2}$ is the maximum value of VL_z vibration level in the passing process of trains in the floating slab track section.

Test Results. When the train passes through, the frequency domain distribution comparison of vertical acceleration level on the tunnel side walls of Sections 1 and 2 is shown in Fig. 12. Compared with the ordinary track, the vertical acceleration

Fig. 12 Frequency domain distribution curve of vertical acceleration level of tunnel sidewall of Sections 1 and 2

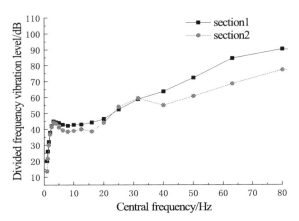

Table 5 VL_{zmax} calculation values of tunnel sidewall of Sections 1 and 2

Project	Section 1 VL_{zmax1}	Section 2 VL_{zmax2}	Damping effect Δ
VL_{zmax}/dB	76.2	63.3	12.9

level of the tunnel side walls is significantly reduced by the floating slab track. The corresponding frequency is 63 Hz.

VL_{zmax} calculation values of tunnel sidewall of Sections 1 and 2 are shown in Table 5. The test result of damping effect of floating ballast bed is 12.9 dB.

3.5 Dynamic Displacement of Floating Slab Track

At the location of Section 2 in Table 3, the floating slab track is selected to carry out dynamic displacement test. The measuring points are arranged as shown in Fig. 14, where the rail displacement is the relative displacement between the rail and the track bed, and the track bed displacement is the relative displacement between the track bed and the tunnel backfill layer. The displacement sensor is verified as qualified product by the third party testing institution, the measuring range is ± 10 mm, and the accuracy can reach 0.01 mm. The typical time-history curve of the dynamic displacement of the floating plate track bed at the time when the train passes through is shown in Fig. 13. The peaks of the dynamic displacement curve reflect the impact of the wheel when it passes through the test section. The statistical values of dynamic displacement amplitude of rail and track bed are shown in Table 6. The average vertical dynamic displacement amplitude of floating plate is 1.32 mm at the end of the slab and 1.21 mm in the middle of the slab, which is less than the limit value of 3 mm given in Technical Specification for Floating slab track (CJJ/T 191-2012). The average vertical dynamic displacement amplitude of the rail is 3.5 mm at the end of

Fig. 13 Typical time-history curve of dynamic displacement of floating slab track (vertical)

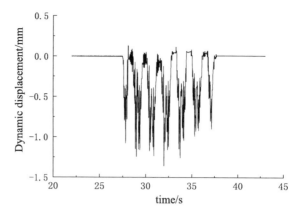

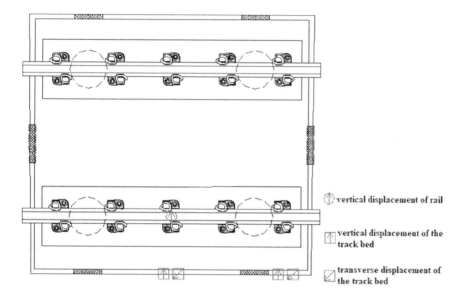

Fig. 14 Plane layout of dynamic displacement measuring points of floating slab track

the slab and 2.5 mm in the middle of the slab, which is less than the limit value of 4 mm given in Technical Specification for Floating slab track (CJJ/T 191–2012).

4 Conclusion

Combined with the laboratory and field tests, the current performance of the floating slab track of Guangzhou metro line 1 was systematically studied. The main conclusions were drawn as follows:

Table 6 Statistical table of dynamic displacement amplitude of rail and track bed (unit: mm)

Sample number	Vertical relative displacement of rail		Vertical displacement of floating slab		Transverse displacement of floating slab		Vertical absolute displacement of rail	
	The end of sab	The middle of sab	The end of sab	The middle of sab	The end of sab	The middle of sab	The end of sab	The middle of sab
1	−2.1	−1.1	−1.1	−1.0	−0.9	−1.1	−3.2	−2.1
2	−2.2	−1.1	−1.3	−1.1	−0.9	−0.9	−3.5	−2.2
3	−2.1	−1.1	−1.2	−1.2	−1.0	−1.1	−3.3	−2.2
4	−2.0	−1.2	−1.2	−1.2	−1.1	−1.0	−3.2	−2.4
5	−2.2	−1.3	−1.4	−1.3	−1.1	−1.1	−3.6	−2.6
6	−2.4	−1.2	−1.5	−1.3	−0.9	−0.8	−3.9	−2.5
7	−2.1	−1.1	−1.4	−1.2	−1.0	−0.8	−3.4	−2.3
8	−2.2	−1.2	−1.3	−1.1	−0.9	−0.9	−3.4	−2.2
9	−2.2	−1.2	−1.3	−1.2	−1.0	−1.0	−3.5	−2.3
10	−2.3	−1.3	−1.3	−1.2	−0.9	−0.9	−3.5	−2.6
11	−2.3	−1.3	−1.3	−1.1	−1.0	−0.9	−3.5	−2.4
12	−2.2	−1.3	−1.3	−1.2	−1.1	−1.1	−3.6	−2.5
13	−2.4	−1.4	−1.5	−1.3	−1.2	−1.2	−3.8	−2.7
14	−2.3	−1.4	−1.4	−1.3	−1.2	−1.2	−3.7	−2.7
15	−2.2	−1.3	−1.4	−1.2	−1.0	−1.0	−3.6	−2.5
16	−2.2	−1.3	−1.2	−1.2	−1.0	−1.1	−3.4	−2.5
17	−2.3	−1.2	−1.4	−1.3	−1.0	−0.9	−3.7	−2.5
18	2.6	−1.6	−1.7	−1.5	−0.8	−0.8	−4.3	−3.2
19	−2.3	−1.4	−1.3	−1.1	−0.9	−0.8	−3.5	−2.5
20	−2.0	−1.1	−1.1	−1.0	−0.9	−0.8	−3.1	−2.1
Mean value	−2.2	−1.2	−1.3	−1.2	−1.0	−1.0	−3.5	−2.5

(1) The rubber bearing of the floating slab track is generally in good appearance except for a few surface cracks and peeling.

(2) The mechanical properties of the rubber bearing of the floating slab track still meet the design requirements, except that the test results of Shore Hardness and tensile elongation reflect a certain degree of hardening.

(3) The first natural frequency test result of the floating slab track is 33.9 Hz, and the vibration reduction effect is maintained well, which is 12.9 dB. Above 35 Hz is the effective vibration reduction frequency band of the floating slab track.

(4) When the train passes, the vertical displacement amplitude of the rail and track in the section of floating slab track is lower than the relevant limit given in the *Technical Specification for Floating Slab Track*.

(5) The test results of various indexes show that, after 20 years of service, the state of the floating slab track is generally good, and it has the conditions to maintain normal service.

References

1. Yao JC, Yang YQ, Sun N (2003) Development of floating slab track structure. China Railway 7:20–22
2. Yan H, Yao L (2002) Design of floating slab track. J Railway Eng Soc 4:12–15
3. Wang D (2005) Analysis on vibration characteristics of floating slab track of Guangzhou Metro Line 2. Southwest Jiaotong University
4. Xu ZQ, Yao JC, Yang YT (2003) Rubber support floating slab track structure dynamic calculation and analysis. Railway Stand Des 8:11–13
5. Zhu YZ (2020) Study on aging characteristics of rubber under prestress. Harbin Institute of Technology University
6. Dong ZH, Zhang JQ, Wei H (2020) Shear property of aging ordinary slab rubber bearing research on energy. Eng Mech 37(S1):208–216
7. Jin QQ, Zheng XT, Wu DM (2020) Natural rubber aging time on rubber springs Influence of mechanical properties. Plastics 4:131–134
8. Han FD (2020) Application of endoscope detection in internal inspection of power plant boiler. China Spec Equip Saf 3:44–48
9. Tao J (2009) Structural modal analysis and damage recognition based on random subspace method. Chongqing University

Study on the Influence of the Arrangement of Thermal Insulation Floor on the Thermal Insulation and Mechanical Properties of Hollow Slab

Yuchen Liu

Abstract In order to study the influence of the arrangement of thermal insulation floor on the thermal insulation and mechanical properties of hollow slab, ABAQUS is used to establish the model of thermal analysis and mechanical behavior of hollow slab. By investigating distribution of temperature, distribution of heat flux, damage and deformation of floor, deformation of mid-span deflection and other characteristics of the floor section, it is concluded that although the transmission of heat can be effectively obstructed by the thermal insulation slab, the heat will be transferred to the interior of the floor through the gap between the thermal insulation slabs. The arrangement of thermal insulation slab is not the main factor which affect the thermal insulation properties of the floor with the same coverage area. Different arrangement of the thermal insulation slab has a certain impact on the mechanical performance of floor. It is recommended to arrange the thermal insulation slab in equal sections to fully improve the contact area between steel bar and concrete, which can effectively provide the bearing capacity of the floor.

Keywords Thermal insulation properties · Mechanical performance · The arrangement of thermal insulation floor · Hollow slab · Finite element simulation

1 Introduction

The floor system is an important part in the whole system of prefabricated structure, in addition to being the main stress component, the floor also undertakes the function of space separation, insulation and so on. At present, some enterprises and colleges at home and abroad have carried out researches on the mechanical and thermal performance of floor, and achieved certain results [1].

In mechanical performance, most of the researches focus on the flexural rigidity of floor. For example, Lu et al. [2]. explored the influence of different factors on the

Y. Liu (✉)
Infrastructure Development Office, Wuhan University of Technology, Wuhan 430070, Hubei, China
e-mail: 630751260@qq.com

© The Author(s) 2023
S. Wang et al. (eds.), *Proceedings of the 2nd International Conference on Innovative Solutions in Hydropower Engineering and Civil Engineering*, Lecture Notes in Civil Engineering 235, https://doi.org/10.1007/978-981-99-1748-8_10

flexural rigidity of composite sandwich slabs. Studies have shown that increasing the thickness of the concrete pavement, decreasing shear span, the attachment of the shear screws can increase the post-rigidity (before the slip observed) of the composite sandwich slabs. Also, they proposed a modified calculation formula of flexural rigidity. Tian et al. [3]. put forward a kind of large-span composite slab, the results show that mid-span deflection and crack width of the large-span composite slab specimen under serviceability limit state meet the requirements of the relative codes, which shows the obvious bidirectional force transmission characteristics, that can be designed as two-way slab. Yang et al. [4] proposed a type of composite slab with additional steel trusses, and carried out static test and numerical analysis on its bottom slab, which includes that the new type of composite slab can improve the bearing capacity, increase the bending stiffness and enhance the crack resistance effectively. In thermal performance such as thermal insulation and fire prevention, Kovalow et al. [5] used the ANSYS software to study the thermal performance of the hollow reinforced concrete floor, and compared it with the results of experimental study. Basing on this, an approach is proposed that allows to take into account all types of heat transfer by specifying cavities as a solid body with an equivalent coefficient of thermal conductivity. Zhao et al. [6] studied the thermal performance of composite slabs comprised of closed profiled steel decking and recycled aggregate concrete (RAC). Results indicated that the temperatures within composite slabs declined with the increasing coarse recycled aggregate (CRA) content. Finally, simplified formulas were developed to predict the cross-section temperature in RAC composite slabs.

It can be seen that although there are researches on the thermal performance of floor at home and abroad, there are few on the thermal performance of fully prefabricated floor. In particular, there is barely research on the thermal insulation performance of the fully prefabricated floor, nearly there is no more systematic and comprehensive report on it so far. Therefore, this paper intends to use finite element software to model and calculate a fully prefabricated floor which are applicated in engineering, and analyze its impact on the thermal insulation and mechanical properties of hollow slab by changing the arrangement and spacing of thermal insulation slab, aiming to provide a reference for the engineering application and design.

2 Establishment of Finite Element Model

2.1 Fully Prefabricated Floor Structure

In order to study the influence of the arrangement of thermal insulation slab on its thermal insulation and mechanical performance, we change the number, size, spacing and other parameters of its thermal insulation slab on the premise that the rate of thermal insulation coverage of the fully prefabricated hollow slab is the same (48.7%). As shown in Table 1, in which board 0 is the control board. The laying mode is shown in Fig. 1.

Table 1 Different forms of cavity floor

Cavity form	Size/mm	Number	Longitudinal spacing/mm	Transverse spacing/mm
0	750 × 500 × 50	10	100	100
1	750 × 250 × 50	20	44 × 4, 48, 44 × 4	100
2	750 × 300 × 50 + 750 × 350 × 50	16	60 × 3, 40, 60 × 3	100
3	375 × 500 × 50	20	100	33, 34, 33
4	500 × 500 × 50	15	100	50, 50
5	750 × (450, 500, 600, 500, 450 mm) × 50	10	100	100
6	750*(300, 500, 900, 500, 300 mm) × 50	10	100	100
7	750 × (150, 500, 1200, 500, 150 mm) × 50	10	100	100

2.2 Finite Element Mesh

In the model of the thermal analysis model in ABAQUS, eight node linear heat transfer hexahedron element DC3D8 is used, and two node transfer connection element DCID2 is used for reinforcement analyzing. In the model of rigidity of structural, eight node linear hexahedral element C3D8R is selected for concrete analyzing, and two node linear three-dimensional truss element T3D2 is used for reinforcement analyzing. The grid element division of the fully prefabricated hollow slab is shown in Fig. 2.

2.3 Properties of Materials

For the thermal parameters of reinforcement and concrete in the temperature field in this model, such as specific heat capacity, thermal expansion coefficient, thermal conductivity and density, can be determined by referring to reference [7]. In the model of stress analysis, the model of plastic damage in ABAQUS is adopted for concrete, the curves in specification [8] are adopted in the stress–strain model of concrete which are under compression and tension, in which the strength grade of concrete is C30; The stress–strain curve of reinforcement adopts the model of linear strengthening, and the strength grade of reinforcement is HRB335. Settings of floor contact constraints are shown in Fig. 3.

The floor is composed of the upper panel, the lower panel and the insulation material filled in the hollow. The thermal insulation materials are made of extruded

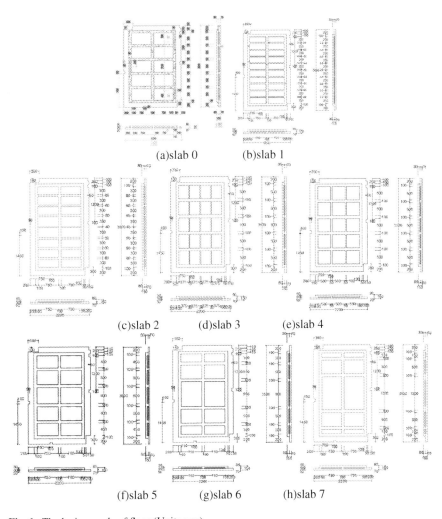

(a)slab 0   (b)slab 1

(c)slab 2   (d)slab 3   (e)slab 4

(f)slab 5   (g)slab 6   (h)slab 7

Fig. 1 The laying mode of floor (Unit: mm)

Fig. 2 Fully prefabricated
hollow slab

Fig. 3 Settings of floor
contact constraints

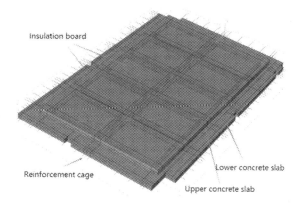

Insulation board

Reinforcement cage

Lower concrete slab

Upper concrete slab

Table 2 Temperature of the physical parameters of materials

Materials	Dry density (kg/m^3)	Thermal conductivity/W(m · K)	Specific heat capacity/J/(kg · K)
Concrete	2500	1.74	920
Steel bar	7800	58.2	480
Steel	7850	16.2	550
XPS board	35	0.030	1.38

polystyrene foam (XPS). The thermal and physical parameters of the materials are
shown in Table 2.

2.4   Contact and Boundary Conditions

In the model of thermal insulation analysis of floor, the concrete floor and the thermal
insulation slab in it are connected by a surface-to-surface of tie, and concrete slab
and the reinforcement fabric in it are connected by a set-to-surface of tie, which can
effectively transfer the temperature of the joint. The initial temperature of the floor is
uniformly set as 20 °C, the temperature of the upper surface of the floor is uniformly
set as 35 °C, and the temperature of the lower surface of the floor is uniformly set
as 5 °C. The convective heat transfer coefficient of the outer surface of the wall is
23.3 W/(m·K), and that of the inner surface of the wall is 8.7 W/(m·K).

In the model of stress analysis of floor, the concrete slab and the thermal insulation
slab inside are connected by a surface-to-surface of tie, while the concrete slab and the
reinforcement fabric in it are connected by an embedded connection. The boundary
conditions of the floor are simply supported at both ends. One end of the constrained
edge of the floor in the model is UI = U2 = U3 = URI = UR2 = UR3 = 0, while the

Fig. 4 Boundary conditions
of floors

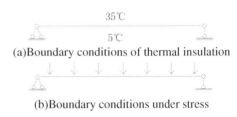

(a)Boundary conditions of thermal insulation

(b)Boundary conditions under stress

other is UI = U2 = URI = UR2 = 0. When uniform load is applied to the surface of the floor, the boundary conditions are shown in Fig. 4.

3 Results Obtained Through Simulation

3.1 Analysis of Thermal Insulation Performance

(1) Temperature response of floor

With different arrangement of thermal insulation slab, heat nephogram of cross-section of floor is shown in Fig. 5. It can be seen from the figure: (1) The color of the upper surface of floor is deeper, that is, the temperature is higher but less than 35 °C. The color of the lower surface is lighter, that is, the temperature is lower but more than 5 °C. That is because of the heat is transferred from the higher side to the lower, thus there is a certain temperature gradient. (2) The temperature on the surface of the hollow slab is mainly concentrated at the place where the thermal insulation slab is arranged. There is little difference in color in the middle layer of the floor, that is, the temperatures are basically similar. From which we can indicate that although the transmission of heat can be fully obstructed by the thermal insulation slab, the heat will be transferred to the interior of the floor through the gap between the thermal insulation slabs. (3) Although the reinforcement fabric runs through the whole floor, the thermal bridge effect of it is not obvious.

From the heat nephogram of the reinforcement in floor shows in Fig. 6, the temperature distribution of reinforcement is similar to that of floor. It can be seen that, firstly, the temperature of the upper reinforcement fabric in hollow slab is higher, which of the lower reinforcement fabric is lower. Secondly, the temperature of the reinforcement at the gap of the thermal insulation slab is lower than that of which in the thermal insulation slab, which is due to the effective transfer of the temperature of the floor from the above to the interior.

(2) The heat flux of floor

Figure 7 shows the nephogram of heat flux of floor, from which we can see the heat flux of floor presents a phenomenon similar to the thenephogram of floor temperature, that is the distribution of heat flux of floor presents well blocks around thermal

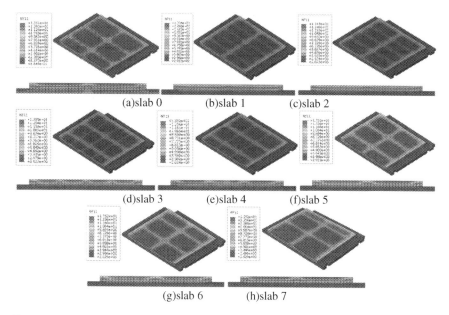

Fig. 5 Heat nephogram of cross-section of floor

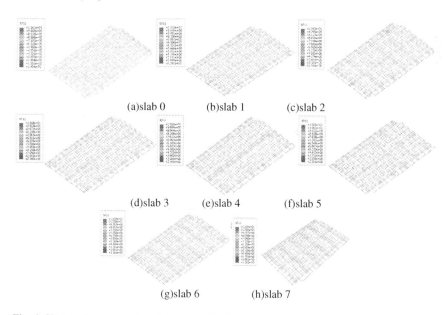

Fig. 6 Heat nephogram of the reinforcement in the floor

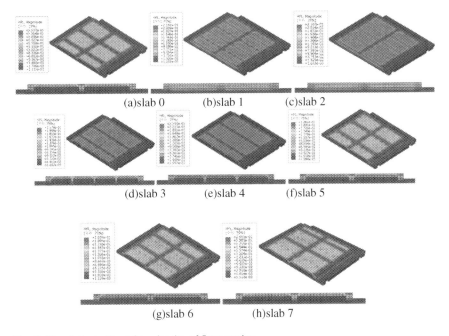

Fig. 7 Cloud chart of heat flow density of floor section

insulation slab, while the difference is that the heat flux at the gap between the thermal insulation slab is deeper, which means that the heat is effectively transferred here. It also indicates that the thermal insulation slab can obviously obstruct the tranfer of heat in cross sections of the floor. However, the thermal bridge effect will be formed along the gap. Meanwhile, the vertical reinforcement in the section also plays the role of thermal bridge, but is not obvious.

(3) Comparison of heat transfer coefficients

Table 3 shows the results of finite element numerical simulation of fully prefabricated hollow slab. Taking slab 0 as a reference, the ratio of heat transfer coefficients from slab 0 to slab 7 is about 1:1.04:1.03:1.03:1.02:1.03:1.04:1.04, and the difference of heat transfer coefficients between each slab is uniformly less than 5%. It is obviously that the arrangement of the thermal insulation slab is not the main factor which affect the thermal insulation properties of the floor with the same coverage area.

3.2 Analysis of Mechanical Performance

(1) Damage and deformation of floor

Table 3 Numerical results of finite element simulation

Number	Average temperature of inner surface (°C)	Average temperature of outer surface (°C)	Average heat flux W/m²	Heat transfer coefficient W/(m²·K)	Ratio
Slab 0	12.1539	3.30430	162.091	18.31610	1.00000
Slab 1	12.0938	3.46031	163.941	18.98896	1.03674
Slab 2	12.1123	3.45644	163.245	18.85948	1.02967
Slab 3	12.1327	3.44963	163.030	18.77562	1.02509
Slab 4	12.1391	3.44832	163.053	18.76161	1.02432
Slab 5	12.0884	3.44608	163.615	18.93184	1.03362
Slab 6	12.0645	3.45469	163.912	19.03782	1.03940
Slab 7	12.0641	3.45480	163.832	19.02965	1.03896

The above research shows that the arrangement of thermal insulation slabs has little effect on the thermal insulation performance, but may affect the mechanical performance of floor. Therefore, the static loading of the floor with different arrangement of thermal insulation slab is simulated and the uniform load is applied on the upper surface. The deform0ation form of the floor is shown in Fig. 8.

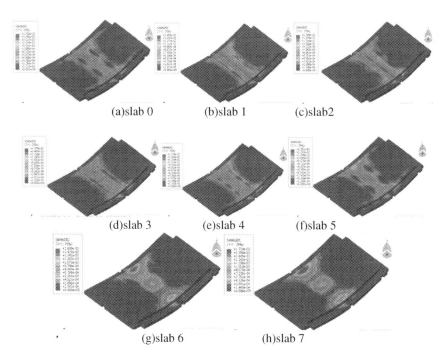

(a)slab 0   (b)slab 1   (c)slab2

(d)slab 3   (e)slab 4   (f)slab 5

(g)slab 6   (h)slab 7

Fig. 8 Cloud chart of floor damage and deformation

As can be seen from Fig. 8, the deformation forms of the floor with different arrangement of thermal insulation slabs are basically the same. Mainly showing that the deformation of mid-span deflection is too large, and the plastic damage of the floor is mainly distributed in the mid-span section, which also indicates that cracks are firstly developed here. However, the plastic damage distribution of the floor is influenced by different arrangement of thermal insulation slab. With more thermal insulation slabs, and the more uniform the size is, the distribution of plastic damage of the floor is more uniform. For example, the distribution of plastic damage in the mid span section of slab 1, slab 2, slab 3, and slab 4 is more uniform than that of slab 0, slab 5, slab 6, and slab 7. Among them, the uniformity of the plastic damage distribution of slab 4 is the best, and that of slab 7 is the worst.

(2) Stress of the reinforcement in floor

Figure 9 shows the stress and deformation nephograph of the reinforcement fabric in hollow slab with different arrangement of thermal insulation slab, which shows the longitudinal reinforcement of board 7 basically reaches the yield strength, while that of board 4 still has stresses. It can be indicated that the mechanical behavior of the floor with the arrangement of thermal insulation slab of 4 is the best, of which the coordination of the joint deformation between the reinforcement and the floor is better. This is due to the fact that in the working condition of slab 4, the reinforcement have full access to the concrete and the force is uniform, while in the working condition of slab 7, the reinforcement have no access to the concrete, thus they cannot bear the force together.

(3) Change in the deflection of the floor

Extract the curve of the mid span deflection of the floor with the change of load, as shown in Fig. 10a, it can be seen that curves in this figure. Basically present a similar pattern. With the increasing of the load imposed on the floor, the mid span deflection of the floor also increases, and the rate of increase gradually accelerates. Taking the corresponding loads of when the mid-span deflection of the floor is 1/20 of the span (175 mm), among them, the load corresponding to plate 4 is the maximum (252.7 KN), and the plate 7 is the minimum (241.3 KN), which is basically consistent with the results of the previous analysis.

In order to further explore the magnitude of the influence, when the uniform load applied to the floor is $0.1N /mm^2$, the curves of full deflection of the floor along the longitudinal are as Fig. 10b, in which the deflection of board 4 is the smallest and the mid-span deflection is 223.1 mm, and the deflection of board 7 is the largest and the mid-span deflection is 283.6 mm. The difference between them is about 27%. Therefore, it is recommended that the thermal insulation slabs be evenly divided as possible when with the same area, so as to fully improve the contact between reinforcement and concrete which can effectively provide the bearing capacity of the floor.

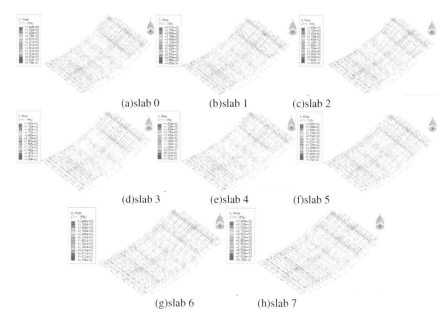

(a)slab 0   (b)slab 1   (c)slab 2

(d)slab 3   (e)slab 4   (f)slab 5

(g)slab 6   (h)slab 7

Fig. 9 Cloud chart of floor reinforcement stress

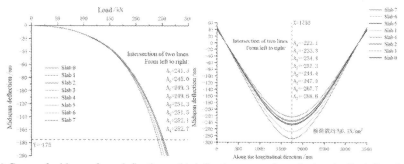

(a) Curve of mid span from deflection   (b) deflection curve along the longitudinal direction

Fig. 10 Deflection change of floor slab

4   Conclusion

Based on the simulating carried by finite element of mechanical performance of the fully prefabricated hollow slab, the following results are obtained:

The range of temperature is larger at the surface of the floor surface is large, but smaller within the floor, which means that although the transmission of heat can be effectively obstructed by the thermal insulation slab, the heat will be transferred to the interior of the floor through the gap between the thermal insulation slabs.

The ratio of the heat transfer coefficient of hollows slabs with different arrangement of thermal insulation panels is about 1: 1.04: 1.03: 1.03: 1.02: 1.03: 1.04: 1.04, it can be seen that the difference of heat transfer coefficient between each board is uniformly less than 5%. That is, the arrangement of thermal insulation slab is not the main factor which affects the thermal insulation properties of the floor with the same coverage area.

Different arrangement of the thermal insulation slab has a certain impact on the mechanical performance of floor. It is recommended to arrange the thermal insulation slab in equal sections to fully improve the contact area between steel bar and concrete, which can effectively provide the bearing capacity of the floor.

References

1. Liu Y, Li ZW, Yang SZ et al (2019) Some recent advances in research of composite slabs of prefabricated building components and measures. China Concr Cement Prod 01:61–68 (In Chinese)
2. Lu YF, Cao PZ, Wu K et al (2021) Experimental research on flexural rigidity of re-entrant trough profiled steel sheeting-concrete composite sandwich slabs with function of heat insulation. Building Struct 51(04):109–113 (In Chinese)
3. Tian LM, Kou YF, Hao JP et al (2017) Study on flexural behavior of bamboo composite slabs with sprayed thermal insulation material. J Huazhong Univ Sci Technol (Nat Sci Ed) 45(11):41–45 (In Chinese)
4. Yang XY, Wang YZ, Liu YG et al (2021) Experimental study and numerical simulation on mechanical properties of the bottom plate in the assembled composite slab with additional steel trusses. Adv Civ Eng
5. Kovalov A, Otrosh Y et al (2022) Research of fire resistance of fire protected reinforced concrete structures. Mater Sci Forum 6604
6. Zhao H, Zhao JZ, Wang R et al (2022) Thermal behavior of composite slabs with closed profiled steel decking and recycled aggregate concrete in fire. Fire Saf J 132
7. Ding J, Li GQ, Sakumoto Y (2014) Parametric studies on fire resistance of fire-resistant steel members. J Constr Steel Res 60(7):1007–1027
8. North China Electric Power Design Institute DL/T 5085-2021 code for design of steel concrete composite structures. China Planning Press, Beijing (2021)

Analysis of Slurry Ratio of Rotary Digging Pile in Deep Sand Layer

Wanjun Zang and Jiang Wen

Abstract Slurry ratio is a crucial link in the construction of bored pile, which directly determines the quality of bored pile. In order to determine the key performance parameters of the slurry required to form piles in the deep sand layer, relying on Huizhou north station engineering, an orthogonal test of slurry proportioning was designed and carried out, and SPSS statistical analysis software was used to carry out bivariate correlation analysis and multivariate stepwise analysis of the test results, combined with the slurry performance index test regression equation and using MATLAB software optimization processing, slurry optimal mix ratio and verify, the research results show that: orthogonal test screening value, software calculation value, test value is not different. Conclusion: The results show that bentonite and CMC have significant influence on slurry indexes, while Na_2CO_3 and PHP can adjust slurry performance to meet the slurry use standard; the optimal mix ratio is 148 g bentonite, Na_2CO_3 5.2 g, CMC 3.5 g, PHP 0.05 g; the slurry ratio test analysis and treatment, and the optimization mix ratio is feasible and reasonable, class I pile proportion more than 98% to meet the actual engineering requirements.

Keywords Mud · Orthogonal test · Correlation analysis · Regression analysis · Mix ratio

1 Introduction

With the continuous development of urban construction, high-rise residential buildings and large buildings are increasing, and the pile foundation is being promoted because of its large bearing capacity. The superstructure transfers the load to the surrounding soil through the pile foundation, effectively reducing the settlement

W. Zang (✉) · J. Wen
The School of Civil Engineering, Fujian University of Technology, Fujian 350118, China
e-mail: zangwanjun@fjut.edu.cn

The Key Laboratory of Underground Engineering, Colleges and Universities in Fujian Province, Fujian 350118, China

© The Author(s) 2023
S. Wang et al. (eds.), *Proceedings of the 2nd International Conference on Innovative Solutions in Hydropower Engineering and Civil Engineering*, Lecture Notes in Civil Engineering 235, https://doi.org/10.1007/978-981-99-1748-8_11

of the pile body and meeting the seismic requirements of the building. Circular excavation bored pile has the advantages of strong adaptability, no noise and high bearing capacity, and is widely used in pile foundation construction, but in practical engineering, geology conditions are complex and changeable, and quality problems emerge in endlessly. The bearing capacity of pile foundation plays a decisive role in engineering safety, and the slurry quality directly determines the quality of bored pile foundation [1].

Slurry mainly plays the role of protecting the hole wall, slag discharge and hole cleaning in the construction. At present, the research on the relationship between hole forming (pile) and slurry has made some results [2–6]. However, there is not much research on the design and analysis of slurry mix ratio, especially the optimization of different soil layers. Wang X P determined the proportion of the main component of the pile wall protection slurry suitable for the ultra-thick sand layer through the ratio test [7]; Wang J et al. determined the optimal quality ratio suitable for the volcanic ash deposition areas through the slurry performance test [8]; Li et al. adopted the uniform design test method for the mix ratio test, and obtained the optimal slurry mix ratio through analysis and optimization [9]. Most of the above scholars determine the slurry mix ratio through the analysis of the test results, but there are very few studies to determine the optimal mix ratio through further optimization and comparative verification.

Slurry preparation is a crucial part of the construction of bored pile. For different geological conditions, different proportions of slurry should be prepared. Combined with the station project of Huizhou North Railway Station, the range of bentonite and additives was determined according to the actual formation conditions, and then the orthogonal test was designed, and SPSS and MATLAB were analyzed to determine the optimal slurry ratio.

2   Project Overview

Huizhou North Station Railway is a line side flat station with elevated waiting room, including station room, platform canopy, underground contact passage, subway section, elevated ramp bridge, etc., with a total construction area of 49,998 m^2. The foundation form is bored cast-in-place pile, the foundation design grade is grade A, and the design grade of the building pile foundation is grade A. The revealed strata in the proposed site are mainly Quaternary strata, and the conditions of each stratum are shown in Table 1.

As can be seen in Table 1, the sand layer of the site is deep and thick, and the rotary excavation into the hole in the deep sand layer can easily cause engineering problems such as poor hole wall integrity, fast sand precipitation, excessive sediment and buried drilling. According to the distribution of on-site strata, refer to the Technical Specification for Highway Bridge and Culvert Construction (JTG/T 3650-2020) and relevant documents [10–12], The basic performance parameters of the slurry are determined as shown in Table 2.

Table 1 Stratigraphic distribution

Order number	Formation name	Depth/m	Volumetric weight/(kN m^{-3})	Cohesive forces/kPa	The angle of internal friction/($^\circ$)
1	Plain fill	1.00~4.80	18.8	13	12
2	Silty clay	1.20~10.90	18.7	20	15
3	Silt	0.60~2.90	19.4	—	10
4	Fine sand	1.00~4.60	19.2	—	18
5	Medium sand	1.00~7.50	19.2	—	23
6	Thick sand	1.20~24.20	19.2	—	30
7	Sandstone	0.60~29.38	23.0	35	25

Table 2 Basic performance parameters of slurry

Proportion /(g/cm^{-3})	Viscosity /s	Plasticity viscosity/(mPa · s)	Yield point/Pa	Water loss/(mL · (30 min)$^{-1}$)	Filter cake thickness/mm
1.08–1.15	22–30	8–13	1.5–3.0	≤ 15	0.5–2.0

3 Trial

3.1 Test Materials

The raw material of the slurry is calcium bentonite with large mesh and strong stability, and the admixture is selected from Carboxymethyl Cellulose (CMC) which can increase the slurry viscosity and form a film on the soil surface to prevent the hole wall from peeling; Sodium Carbonate (Na_2CO_3), Can control slurry PH, increase the thickness of hydration film, improve slurry stability, reduce the pollution of slurry by calcium ions and groundwater; Partially Hydrolyzed Polyacrylamide (PHP), can make the slurry colloid form a chemical film in fine sand, coarse sand and gravel soil, close the hole wall, and keep the hole wall stable.

Table 3 Measurement indicators and related instruments

Project	Instrument
Proportion	N B-1 slurry gravity meter
Viscosity	N C-1006
Plastic viscosity	Z NN-D6 6-speed rotation viscometer
Yield point	
Water loss	Z NS-2 slurry for water loss determinator
Filter cake thickness	

Table 4 Factor levels table

Horizontal	Factor			
	a	b	c	d
1	128	4.4	2.7	0.05
2	148	4.8	3.3	0.10
3	168	5.2	3.9	0.15
4	188	5.6	4.5	0.20

3.2 Test Instrument

The test and measurement content and the required instruments are shown in Table 3.

3.3 Test Design

Orthogonal tests were determined by orthogonal tables, and the results were analyzed statistically. Among them, the same number of tests is performed at the level of either or both factors, which makes them very representative according to the orthogonal table and reduces the number of trials. The orthogonal test method in the ratio design can avoid unnecessary calculation and optimize the design process. Using four levels and four factors orthogonal test, the factors and level distribution are shown in Table 4. Among these, a is bentonite, b is Na_2CO_3, c is CMC and d is PHP.

16 trials were designed according to the orthogonal test table as shown in Table 5.

3.4 Test Method

According to the orthogonal test table, the bentonite, Na_2CO_3, CMC, PHP in addition to 1000 mL water in turn, using JJ-1 electric mixer to stir and mix at a speed of 850

Table 5 Orthogonal test table

Number	a	b	c	d
1	2	1	2	2
2	4	3	1	3
3	1	1	1	1
4	3	4	4	1
5	1	3	4	2
6	3	3	2	4
7	2	4	1	4
8	1	2	3	4
9	4	1	4	4
10	2	3	3	1
11	3	2	1	2
12	4	2	2	1
13	4	4	3	2
14	2	2	4	3
15	1	4	2	3
16	3	1	3	3

r/min, hydrate and expand for a certain period of time at room temperature, and test the data with reference to the measurement method given in the relevant specifications.

4 Analysis of the Test Results

4.1 Test Results

According to the 16 sets of test ratio shown in Table 4, the performance test of each formula slurry was tested separately, and the calculated mean value of the measured test data is shown in Table 6. Among them, $\varphi300$ and $\varphi600$ are the readings of the rotating viscosimeter at 300 r/min and 600 r/min, respectively, and the plastic viscosity and yield point are calculated from the two sets of data.

$$\eta = \varphi_{600} - \varphi_{300} \tag{1}$$

$$\tau_d = 0.511(\varphi_{300} - \eta) \tag{2}$$

Table 6 Slurry test data table of each group

Test number	Proportion/(g/cm^{-3})	Viscosity/s	φ_{300}/(°)	φ_{600}/(°)	Plasticity viscosity/(mPa·s)	Yield point/Pa	Water loss/[mL (30 min)$^{-1}$]	Filter cake thickness/mm
1	1.12	21.5	14.5	25.4	10.9	1.8	10.8	1.8
2	1.14	23.8	23.5	38.4	14.9	4.3	14.4	2.5
3	1.10	19.3	12.0	21.2	9.2	1.4	12.3	2.0
4	1.11	25.1	24.4	39.0	14.6	4.9	12.2	0.9
5	1.09	24.2	22.1	36.0	13.9	4.1	9.9	1.1
6	1.11	22.6	17.1	29.8	12.7	2.2	12.3	2.3
7	1.10	22.9	14.8	27.2	12.4	1.2	14.6	2.6
8	1.10	21.8	17.3	29.2	11.9	2.7	11.1	2.1
9	1.15	29.2	28.4	45.6	17.2	5.6	10.0	1.2
10	1.11	23.7	19.5	33.0	13.0	3.0	10.0	1.1
11	1.11	20.7	12.6	21.6	9.0	1.8	14.6	2.2
12	1.13	24.9	23.8	38.2	14.4	4.7	12.3	1.5
13	1.13	26.9	24.4	39.2	14.8	4.8	12.9	1.3
14	1.12	27.2	25.2	40.8	15.6	4.8	9.1	1.1
15	1.08	20.6	14.9	26.2	11.3	1.8	14.4	2.0
16	1.12	25.3	26.1	41.8	15.7	5.2	9.1	1.6

Note The P H value of this test is located between 8 and 10

4.2 Correlation Analysis

Correlation analysis is one of the very mature basic theories in statistics. The interdependence between various phenomena can be manifested as functional relations or correlations among variables [13].

SPSS analysis software was used to carry out bivariate correlation analysis on the test results [14], analyze the correlation degree between each factor and the performance of the slurry, and obtain the correlation coefficient. In this paper, the four factors: bentonite, Na_2CO_3, CMC, PHP, slurry six indexes: specific gravity, viscosity, plastic viscosity, yield point, water loss, filter cake thickness were statistically analyzed, and the correlation coefficient of the factors and each performance of slurry was obtained, as shown in Table 7.

In correlation analysis, positive correlation: correlation coefficient $r > 0$; negative correlation: $r < 0$; high correlation: $|r| 0.8$; moderate correlation: $0.8 > |r| 0.5$; low correlation: $0.5 > |r| 0.3$; weak correlation: $|r| < 0.3$.

As shown from Table 7:

(1) Positive correlation with bentonite: specific gravity, viscosity, plastic viscosity, yield point, water loss, and negative correlation with bentonite: slurry cake thickness. Among them, bentonite is weakly correlated with water loss and slurry cake thickness, moderately related to viscosity, plastic viscosity and yield point, and highly related to specific gravity, which shows that bentonite can significantly increase the proportion of slurry and increase the strength of the internal gel mesh structure when slurry flows. https://baike.baidu.com/item/% E7%BD%91%E7%8A%B6%E7%BB%93%E6%9E%84/5106492.

(2) Positive correlation with Na_2CO_3: water loss, slurry cake thickness, and negative correlation with Na_2CO_3: specific gravity, viscosity, plastic viscosity, and yield point. Among them, Na_2CO_3 is weakly correlated with viscosity, plastic viscosity, yield point and slurry cake thickness, is lowly related to specific gravity and moderately related to water loss, indicating that Na_2CO_3 energy decomposes slurry particles, which can reduce the rate of Na_2CO_3 and improve the rate of slurry stability.

(3) Positive correlation with CMC: specific gravity, viscosity, plastic viscosity, yield point, and negative correlation with CMC: water loss, filter cake thickness.

Table 7 Correlation coefficient between various factors and various slurry performance

Factor	Project					
	Proportion	Viscosity	Plastic viscosity	Yield point	Water loss	Filter cake thickness
Bentonite	0.856	0.595	0.549	0.588	0.139	−0.089
Na_2CO_3	−0.349	−0.191	−0.101	−0.206	0.519	0.037
CMC	0.127	0.803	0.665	0.697	−0.750	−0.861
PHP	0.063	0.278	0.302	−0.062	0.035	0.464

Among them, CMC is weakly related to the specific gravity of slurry, moderately related to plastic viscosity, yield point, and water loss, and highly related to viscosity and filter cake thickness, indicating that It shows that CMC can significantly increase the viscosity of the slurry, enhance the internal friction between the suspended particles of the slurry and between the slurry particles and the liquid phase, improve the slag carrying capacity of the slurry, and reduce the water loss.

(4) Positive correlation with PHP: slurry specific gravity, viscosity, plastic viscosity, water loss, filter cake thickness, and negative correlation with PHP: yield point. Among them, PHP is weakly correlated with slurry viscosity, specific gravity, yield point and water loss, and is weakly correlated with filter cake thickness and plastic viscosity, indicating that PHP mainly plays the role of flocculation. In this test, the lifting effect was not obvious, perhaps because the slurry PH value was not increased to above 10, and the PHP was not fully effective.

4.3   Regression Analysis

As a screening method, the stepwise regression introduces the regression equations one by one according to the effect of each performance indicator, while the non-significant factors are always not introduced [15].

The independent variables were bentonite, Na_2CO_3, CMC and PHP, and the specific gravity, viscosity, plastic viscosity, yield point, water loss and slurry skin thickness of the slurry were the dependent variables. To determine the regression equation between the respective variables and the dependent variables, the multiple stepwise regression analysis was conducted using SPSS software, and the results are shown in Table 8.

Comparing Tables 7 and 8, the magnitude of the significance of the influence of the independent variable on the dependent variable in the regression equation coincides with the correlation analysis, e.g., The regression equation of the specific gravity of slurry is composed of x_1 (bentonite) and x_2 (Na_2CO_3), which is consistent with the

Table 8 Regression equations for each factor

Factor	Regression equation
Proportion	$y_1 = 1.076 + 0.001x_1 - 0.014x_2$
Viscosity	$y_2 = 3.084 + 0.069x_1 + 2.712x_3$
Plastic viscosity	$y_3 = -3.658 + 0.056x_1 + 2.250x_3$
Yield point	$y_4 = -8.423 + 0.039x_1 + 1.554x_3$
Water loss	$y_5 = 8.495 + 2.200x_2 - 2.117x_3$
Filter cake thickness	$y_6 = 3.625 - 0.688x_3 + 4.45x_4$

Note 1 x_1 is bentonite, x_2 is Na_2CO_3, x_3 is CMC, and x_4 is PHP
Note 2 y_1 is the specific gravity, y_2 is the viscosity, y_3 is the plastic viscosity, y_4 is the yield point, y_5 is the water loss, and y_6 is the thickness of the filter cake

Table 9 Analysis of variance of the regression equations

Dependent variable	R^2	F_1	P_1	F_2	P_2
Proportion	0.815	18.111	0.000	2.265	0.158
Viscosity	0.854	2.910	0.078	4.206	0.032
Plastic viscosity	0.854	2.743	0.148	3.522	0.042
Yield point	0.831	2.466	0.112	3.997	0.035
Water loss	0.832	2.851	0.092	5.700	0.012
Filter cake thickness	0.957	11.524	0.001	2.854	0.098

large correlation coefficient between bentonite, soda ash and slurry proportion in the correlation analysis. Therefore, the correlation analysis and regression analysis are suitable for this test analysis and meet the actual requirements.

4.4 Regression Equation Test

To determine the fitting effect of the regression equation and the slurry performance index, the regression equation test is required. Given the significance level $\alpha = 5\%$, a one-way ANOVA was performed on the dependent variable versus the independent variable composing the regression equation. For example, in the specific gravity analysis, F_1, P_1, F_2, and P_2 in the table are the results of the variance analysis of specific gravity and x_1 (bentonite) and x_2 (Na_2CO_3). The cut-off value F is obtained through the F distribution table $F0.05(3,12) = 3.49$, the F value obtained in the Table is compared with the F value found in Table 9, it can be seen that the value of F_1 and F_2 is less than 3.49 but close to the critical value, indicating that the respective variable had a significant influence on the dependent variable. And the coefficient of determination R^2 is close to 1, indicating that the regression equation fits well and meets the requirements of this experiment.

5 Mix Ratio Optimization

Further, slurry specific gravity, viscosity, plastic viscosity, yield point, water loss, filter cake thickness six indicators are the standard to measure the performance of slurry. To determine the optimal slurry mix ratio, the function is optimized and the following objective function is established:

$$Z = y_n \tag{3}$$

In formula: n = 1~6. The constraints on y and x in the objective function are:

$$\begin{cases} 1.08 \le y_1 \le 1.15 \\ 22 \le y_2 \le 30 \\ 8 \le y_3 \le 13 \\ 1.5 \le y_4 \le 3.0 \\ y_5 \le 15 \\ 0.5 \le y_6 \le 2.0 \end{cases} \tag{4}$$

$$\begin{cases} 128 \le x_1 \le 188 \\ 4.4 \le x_2 \le 5.6 \\ 2.7 \le x_3 \le 4.5 \\ 0.05 \le x_4 \le 0.20 \end{cases} \tag{5}$$

According to the function relationship of y and x determined by the regression equation, the objective function was optimized by MATLAB software. After calculation, after calculation, when the material dosage is 148 g of bentonite, 5.2 g of Na_2CO_3, 3.5 g of CMC, and 0.05 g of PHP, the function has an optimal solution, and each index is tested. The screening results of orthogonal tests (148 g of bentonite, 5.2 g of Na_2CO_3, CMC 3.9 g, and PHP 0.05 g), software calculation results and test results are shown in Table 10.

According to the table, the orthogonal test screening value, software calculation value and test value do not differ much. Taking the test value as a reference, the smallest difference is the specific gravity: the deviation of the screening value is 1.77%, and the deviation of the calculated value is 0.88%; the largest difference is the thickness of the slurry: the deviation of the screening value is 21.43%, and the deviation of the calculated value is 7.14%. All of the above meet the requirements, indicating that the orthogonal test is the method of finding the principal contradiction and selecting the better scheme in this test. The optimized treatment results of MATLAB software meet the requirements of various slurry indexes, so this method is applicable for the optimization analysis of slurry mix ratio.

The slurry ratio is specifically applied to Huizhou North Railway Station project and detected by low strain and acoustic wave transmission method: 1181 Class I piles, accounting for 98.504% of the tested piles; 18 class piles, accounting for 1.495% of the tested piles; There are no III and IV piles. The proportion of pile foundation integrity testing type I pile has reached more than 98%, which has ensured the project quality and project progress, achieved good results, and accumulated certain engineering experience.

6   Conclusion

(1) Orthogonal test screening value, software calculation value and test value are not much different. With the test value as the reference, the smallest difference is the proportion: The deviation of the screening value is 1.77%, and the deviation

Table 10 Orthogonal test screening values, software calculation values and test values

Project	Proportion/(g/cm^{-3})	Viscosity /s	Plasticity viscosity /(mPa · s)	Yield point/Pa	Water loss/[mL·(30 min)$^{-1}$]	Filter cake thickness/mm
Screening value	1.11	23.7	13.0	3.0	10.0	1.1
Calculated value	1.14	22.0	11.8	2.4	10.3	1.3
Test value	1.13	22.2	12.3	2.6	12.6	1.4

of the calculated value is 0.88%; the biggest difference is the thickness of the slurry skin: the deviation of the screening value is 21.43%, and the deviation of the calculated value is 14.29%. All the above meet the requirements, indicating that the orthogonal test, as a method of finding the main contradiction from multiple factors and multiple levels and selecting better schemes, has achieved good results in this test.

(2) Bentonite and CMC have a significant impact on various slurry indexes, while Na_2CO_3 and PHP can be adjusted as additives to meet the standard of slurry use. The optimal mix ratios were 148 g of bentonite, 5.2 g of Na_2CO_3, 3.5 g of CMC, and 0.05 g of PHP.

(3) Using SPSS statistical software for correlation analysis and regression analysis, and the optimized treatment through MATLAB software is suitable for slurry ratio test analysis and treatment, and the optimization mix ratio sought is feasible and reasonable, which can meet the actual engineering needs.

Acknowledgements This project is supported by Key Technology and Application Research Project of New Construction of Huizhou North Station of Ganzhou-Tangxia Section of Ganzhou-Shenzhen Railway (GY-H-21017).

References

1. Zhang XW, Guan YJ, Zhou JH (2005) Application of PHP slurry to drilling of overlength and extra-large-diameter bored piles. Chin J Rock Mech Eng 24(14):2571–2575
2. Chi XW, Yao ZW, Lin C (2012) Analysis of bored piles slurry mixing ratio in thick sand layer. J Wuhan Univ (Eng Sci) 45(04):477–480
3. Liu WB (2021) Application of the large-diameter and extra-long rotary bored pile in the thick sand layer area. IOP Conf Ser Earth Environ Sci 651(3):032–098
4. Wan ZH, Dai GL, Gong WM (2019) Field study on post-grouting effects of cast-in-place bored piles in extra-thick fine sand layers. Acta Geotech 14(5):1357–1377
5. Zhu DW, Zhao CF, Zhao C et al (2015) The influence of slurry gravity on the variation of the pore diameter in cast-in-situ bored pile. J Hydraul Eng 46(S1):349–353
6. Chen YQ, Lei JS, Xu L et al (2019) Model test study on influence of mud cake on friction performance of pouring pile. J Railway Sci Eng 16(07):1660–1665
7. Wang XP (2015) The application of rotary drilling and reamed pile construction in super-thick sand layer. Shijiazhuang Railway University, Shijiazhuang
8. Wang J, Ren WF, Wang QL (2007) Character of boring slurry in Pozzuolana sediment area. J Chang'an Univ (Nat Sci Ed) 27(06):67–71
9. Li SH, Liu HQ (2018) Experimental study on optimum mixture ratio of slurry based on uniform design. Silic Bull 37(07):2280–2284+2296
10. Jiang P, Zhang JQ, Zhu ZP (2019) Research on the control of the proportion of mud in the construction of bored piles. Highway 64(07):145–148
11. Wu YH, Zhou Z, Chen WQ et al (2020) Research on construction technology of cast-in-situ bored pile under complex geological conditions. IOP Conf Ser Earth Environ Sci 510(5):052–089
12. Dong YT, Niu XB, Dong K et al (2021) Study on slurry ratio of sludge-water balance shield in fine silty sand formation. Hans J Civ Eng 10(04):319–328

13. Li ZJ, Wang XL, Shi DP (2011) Analysis of the correlation among indexes of sulfide ores in oxidation process at ambient temperature. Sci Technol Herald 29(36):28–32
14. Cao LY, Zhang XX, Zhao YL et al (2014) Study on the relationship between soil nutrient and yield analysis based on bivariate correlation between. Adv Mater Res 3349(998-999):1466–1469
15. Hu SW (2017) Analysis of influencing factors on reliability of depth submarine-launched system based on multivariate stepwise regression. Acta Armamentarii 38(05):986–994

Numerical Analysis of Bearing Capacity of Basic Stress Unit Frame of Socket-Type Wheel Buckle Formwork Support Frame Body

Heng Liu, Weiwen Xu, Hongtao Xu, and Xuemiao Xiang

Abstract Socket-and-socket wheel-buckle steel pipe formwork support frame is a new formwork support system, with fixed component specifications and fast erection speed, which can significantly improve the construction efficiency. Its application performance is different from that of fastener-type, bowl-buckle-type and plate-buckle-type, and there are not enough experimental and numerical analysis results to consult. Therefore, based on the bearing capacity test of the basic bearing units of socket-type wheel-buckle steel pipe formwork support, the finite element software ANSYS is used to conduct numerical analysis of 12 groups of basic bearing units, and the influence degree of vertical bearing capacity of different specifications of unit supports is studied by stud spacing and wedge tightness of cross bar plug. It is pointed out that the stiffness of formwork support in the vertical and horizontal directions should not be too different in the construction process.

Keywords Formwork support frame · Numerical analysis · Bearing capacity test

1 Introduction

In recent years, with the continuous development of civil engineering in China, various tall and complex structures have emerged. In the pursuit of efficiency today, people begin to seek new formwork support systems to solve the disadvantages of slow construction efficiency such as fastener-type, bowl-buckle-type and pan-buckle-type formwork support frames. Therefore, various formwork support systems with new structural forms have emerged as the times require. Xu et al. [1, 2] introduced a new type of plug-in formwork support frame, which realized the advantages of axial force transmission of vertical bar, good connection performance, high bearing capacity and strong applicability. Qiang [3] has studied the practical application of bolt-type formwork support and the semi-rigid connection of bolt, welding lug, welding seam and node, etc. The results show that the bolt will self-lock in the

H. Liu (✉) · W. Xu · H. Xu · X. Xiang
China Construction Seventh Engineering Division. Corp. Ltd., Zhengzhou 450000, China
e-mail: 70354911@qq.com

© The Author(s) 2023
S. Wang et al. (eds.), *Proceedings of the 2nd International Conference on Innovative Solutions in Hydropower Engineering and Civil Engineering*, Lecture Notes in Civil Engineering 235, https://doi.org/10.1007/978-981-99-1748-8_12

normal work of formwork support, and its connection performance is good, and it is suggested that the connection stiffness k of node can be taken as a low value. Through analysis, it is considered that it is safe and reasonable to take K = 20 kN·m/rad. Huang [4] carried out horizontal force loading tests of two types of bolt-type plane frames under four wedge tightness conditions and two groups of basic unit load-bearing performance tests, and studied the joint characteristics and load-bearing performance of unit frames. He pointed out that when bolt-type formwork supports are used, the members are mainly subjected to axial force, and the bending moment is usually less than 0.2 KN·m. Therefore, the semi-rigid connection stiffness of the joint that affects the structural performance is the initial connection stiffness of the joint, but the initial tangent stiffness of the joint. Similarly, socket-and-socket wheel-buckle steel pipe formwork support frame is gradually applied to various construction sites, and it has similarities with plug-in type, plug-in type, plug-in type and other frame bodies in bearing mechanism. However, its structural form is simpler [5], and the outer diameter of steel pipes is 48 mm. The difference between it and other forms of steel pipe formwork supports lies in the vertical and cross bar connection joints, and the connection joints are shown in Fig. 1. There is no bracing hole at the vertical pole bearing plate, so this form of formwork support needs to use the bar of fastener rack and rotating fastener to set up bracing. Its installation and disassembly are convenient and quick, and its stability is easy to ensure, which can shorten the construction period and improve the construction efficiency. However, its cross bar is disconnected at the node, so it can't be applied to the side formwork engineering of vertical members. Therefore, the load directly borne by the cross bar is mainly a small amount of materials piled up or the passage of personnel and equipment, and its load is small, which won't cause plastic deformation of the bearing plate.

The node size is shown in Fig. 2. The inner diameter of the bearing plate is 48 mm and the outer diameter is 108 mm. The plug hole is 18 mm long and 12 mm wide, and four jacks are evenly distributed on the circumference, and the distance between adjacent holes is 90; The bearing plate thickness is 20 mm. The end of the cross bar has an arc joint of 16 mm long, 3.5 mm wide and 8 mm thick. A plug of 10 mm wide

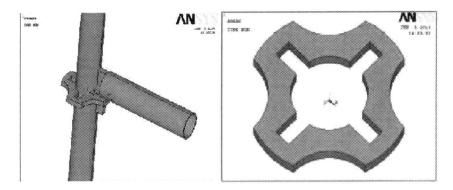

Fig. 1 Detail structure of socket-type wheel-buckle steel pipe formwork support frame node

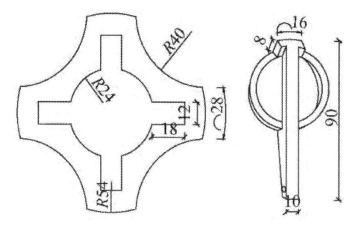

Fig. 2 Detail dimensions of socket-type wheel-buckle steel pipe scaffold node

and 90 mm long is connected to the joint, and the tail of the plug is tapered with a φ6 mm pin hole.

In this paper, the finite element verification is carried out for the experimental study of the basic stress unit frame of the socket-and-socket type steel pipe formwork support frame in reference [6]. The difference of joint stiffness is directly caused by the unsatisfactory contact surface during the erection of the frame body, so it is expressed by the wedge tightness of the cross bar plug in different degrees. On this basis, the variable parameter analysis is carried out with the wedge tightness of the cross bar plug and the vertical and horizontal spacing of the vertical poles as variables. During the verification process, the cross section of the pipe rack is still considered as 48 × 2.75; However, the normal wedge tightness of the cross bar plug corresponds to 7~10 times during erection in the formwork support project, and the minimum wedge tightness of the cross bar plug corresponds to 0~3 times during erection in the formwork support project.

2   Finite Element Analysis Assumption

Combined with the erection and boundary conditions of formwork support system in engineering application, the finite element software ANSYS is used for numerical analysis to make the following assumptions.

(1) The components in the formwork support frame are mutually constrained, and there is no slip at the bottom of the vertical pole, while no other treatment is done at the top of the vertical pole, so it is assumed that the bottom of the vertical pole is hinged and the top is free.

(2) When the middle stud is extended, there should be a sleeve to limit the translational and rotational freedom of each other, while the stud and the crossbar

transmit shear force and bending moment through the friction at the node, so it is assumed that the stud is rigidly connected, and the junction between the stud and the crossbar is semi-rigidly connected.

(3) When checking the ultimate bearing capacity, the material properties are still based on the reference [5], and the results obtained from the unidirectional material tensile test are the actual material properties.

3 Numerical Analysis of Bearing Capacity of Different Specifications of Frame Body

3.1 Definition of Material Parameters

According to the setting of different structural analysis units in the finite element software ANSYS, Beam188 element with shear deformation is used to simulate the vertical bar, and Combin39 element is used to simulate the stiffness of the vertical bar node in the numerical analysis of formwork support. The material parameters are obtained by uniaxial tensile test: $E = 2.21 \times 105$ MPa, fy $= 359$ MPa; The definition of bending moment-turning angle relationship of spring element under two working conditions of normal wedging degree and minimum wedging degree is shown in Table 1 [7].

3.2 Establishment of Finite Element Model

Taking the vertical and horizontal spacing of vertical poles and the wedge tightness of nodes as variables, the influence degree of vertical bearing capacity of socket formwork support frame is studied. As the vertical pole bearing plate spacing and cross bar size are fixed, finite element models of formwork support frames with different specifications are established, and the frame body specifications are shown in Table 2. The finite element model of each unit frame is established, as shown in Fig. 3. In the analysis, it is assumed that each vertical pole carries load synchronously, so the vertical translational degrees of freedom of the top of four vertical poles in the unit frame are coupled.

3.3 Analysis of Bearing Capacity Results

Non-linear buckling analysis is carried out on each basic stress unit frame model, and the ultimate bearing capacity of different specifications of frame bodies is obtained, which is summarized in Table 3. Comparative analysis is made according to the

Table 1 Bending moment-turning angle relationship of 1Combin39 unit

Corner/rad		−0.03	−0.025	−0.02	−0.015	−0.01	−0.005	−0.002
Bending moment/(N·mm^{-2})	Smallest	−525,000	−472,500	−456,750	−430,500	−399,000	−357,000	−309,750
	Normal	−750,000	−675,000	−652,500	−615,000	−570,000	−510,000	−442,500
Corner/rad		−0.0017	−0.0014	−0.001	0.001	0.0014	0.0017	0.002
Bending moment/(N·mm^{-2})	Smallest	−246,750	−184,800	−63,000	399,000	184,800	246,750	309,750
	Normal	−352,500	−264,000	−90,000	9000	264,000	352,500	442,500
Corner/rad		0.005	0.01	0.015	0.02	0.025	0.03	
Bending moment/(N·mm^{-2})	Smallest	357,000	399,000	430,500	456,750	472,500	525,000	
	Normal	510,000	570,000	615,000	625,000	675,000	750,000	

Table 2 Summary of unit rack specifications

Number	1	2	3	4	5	6	7	8	9	10	11	12
L1	0.6	0.6	1.2	1.2	0.9	0.9	0.9	0.9	0.9	0.9	1.2	1.2
L2	0.6	0.6	0.6	0.6	0.6	0.6	0.9	0.9	1.2	1.2	1.2	1.2
h	1.2	1.2	1.2	1.2	1.2	1.2	1.2	1.2	1.2	1.2	1.2	1.2
R	R1	R2	R1	R2	R1	R2	R1	R2	R1	R2	R1	R2
a	0.36	0.36	0.36	0.36	0.36	0.36	0.36	0.36	0.36	0.36	0.36	0.36
b	0.2	0.2	0.2	0.2	0.2	0.2	0.2	0.2	0.2	0.2	0.2	0.2

Note L1-longitudinal distance; L2-horizontal distance; H-step distance; R-wedge tightness of nodes; R1-normal; R2-minimum; A-height of sweeping pole; B-height of free end

Fig. 3 Schematic diagram of unit frame model

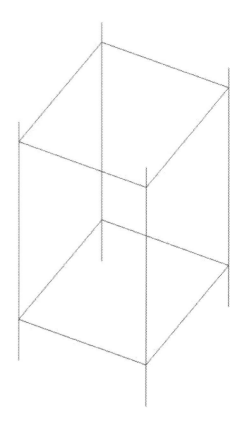

difference of bearing capacity caused by the change of vertical pole spacing and wedge tightness of cross bar plug in each unit frame.

As can be seen from Table 3, the unit frame 1 with small vertical pole spacing and large wedge tightness of cross bar plug has the largest bearing capacity, whereas the unit frame 12 with large vertical pole spacing and small wedge tightness of

Table 3 Nonlinear ultimate bearing capacity of each unit frame

Unit number	1	2	3	4	5	6
Ultimate bearing capacity/kN	342.724	302.970	254.549	227.410	337.308	290.376
Unit number	7	8	9	10	11	12
Ultimate bearing capacity/kN	292.638	263.840	262.171	229.807	233.366	202.032
Contrast formula	(1–2)/2	(3–4)/4	(5–6)/6	(9–10)/10	(11–12)/12	
Comparative value/%	13.12	11.93	16.16	14.08	15.51	
Contrast formula	(3–1)/1	(5–1)/1	(7–1)/1	(9–1)/1	(11–1)/1	
Comparative value/%	−25.73	−1.58	−14.61	−23.50	−31.91	

cross bar plug has the smallest bearing capacity, and the bearing capacity changes with the change of vertical pole spacing. According to the difference of vertical and horizontal spacing of vertical poles, the bearing capacity of unit frame is better when the stiffness difference in two directions is small; However, the bearing capacity is low when the stiffness difference between two directions is large. For example, the ultimate bearing capacity of unit frame 3 and unit frame 4 with 600 mm horizontal distance and 1200 mm vertical distance is slightly lower than that of unit frame 9 and unit frame 10 with 900 mm horizontal distance and 1200 mm vertical distance, which indicates that the deformation of unit frame develops earlier when the stiffness difference between two directions is large. In practical application, attention should be paid to the vertical and horizontal stiffness of frame body. When the stiffness difference between the two directions is large, additional pulling and connecting measures should be added.

In the numerical analysis, the ultimate bearing capacity of cross-bar plugs with different specifications under normal wedging degree is 11.93~16.16% higher than that under minimum wedging degree, and the difference is fluctuating and irregular. The nonlinear load–displacement curve of each unit frame is shown in Fig. 4. From Fig. 4, it can be seen that the smaller the vertical pole spacing of the unit frame is, the faster the bearing capacity decreases when it reaches the ultimate bearing state. On the contrary, when the vertical pole spacing is large, although the bearing capacity is relatively small, the bearing capacity can continue when the ultimate bearing capacity is reached. In order to visually compare the ultimate bearing capacity of different specifications of unit frames, the summary is shown in Fig. 5. It can be seen from Fig. 5 that, regardless of the joint stiffness, the bearing capacity will be greatly reduced when the vertical and horizontal spacing of the unit frame exceeds 900 mm. Under the normal joint wedge tightness, the bearing capacity of the unit frame decreases obviously when the vertical pole spacing increases; Under the minimum joint wedge tightness, the bearing capacity of the unit frame decreases slightly less than that of the normal wedge tightness when the vertical pole spacing is increased; When the vertical pole spacing is small, the ultimate bearing capacity of the unit frame with normal wedge tightness is much higher than that of the unit frame with minimum wedge tightness, but the difference between them is small when the vertical pole spacing of the unit frame is large.

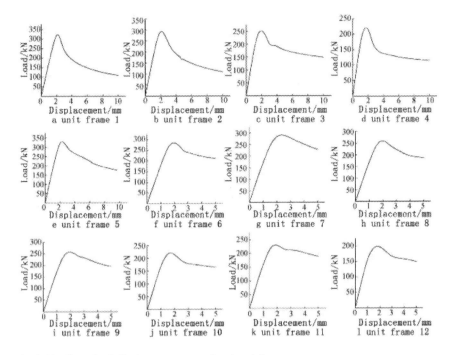

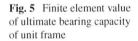

Fig. 4 Nonlinear load–displacement curve of each unit frame

Fig. 5 Finite element value
of ultimate bearing capacity
of unit frame

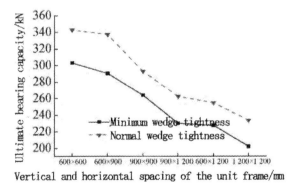

4   Conclusion

In this paper, the general finite element software ANSYS is used to verify the vertical
ultimate bearing capacity of the basic stress unit frame of the socket-type wheel-
buckle steel tube formwork support frame, and the following conclusions are drawn
from the analysis results.

(1) The wedge tightness of the cross bar plug will affect the vertical bearing capacity of the socket-type wheel-buckle steel pipe formwork support frame. Under the support of reliable data, the bearing capacity of the frame with large wedge tightness is more than 10% higher than that of the frame with small wedge tightness.

(2) In the socket-type wheel-buckle steel pipe formwork support frame, if the vertical and horizontal spacing of the vertical poles exceeds 900 mm in one direction, its bearing capacity will be much lower than that when the vertical and horizontal spacing of the vertical poles are both < 900 mm, so excessive erection spacing of the vertical poles should be avoided in application.

(3) The bearing capacity of socket-type wheel-buckle steel pipe formwork support frame is related to the vertical and horizontal stiffness of the frame body. Excessive difference in vertical and horizontal stiffness will cause the frame body to lose stability earlier. Therefore, during the construction of long and narrow concrete members, the weaker side of the frame body should be connected with the existing structure.

References

1. Jiang X, Zhang QL, Wang HJ (2007) Research on numerical model of new plug-in scaffold and improved design of joints. Constr Technol 36(S1):202–206
2. Jiang X, Zhang QL, Gu MJ et al (2008) Experimental and numerical simulation study of new plug-in scaffold. J Civ Eng 41(7):55–60
3. Huang Q (2006) Study on the performance of bolted steel pipe scaffold joints. Chongqing University, Chongqing
4. Huang H (2006) Experimental study on semi-rigid joints and basic stress units of bolt-type steel pipe scaffold. Chongqing University, Chongqing
5. Wang QL, Xue DZ, Wang T et al (2010) Buckling steel pipe scaffold for cast-in-place reinforced concrete slab construction: China, 201020178718.7
6. Guo Y (2014) Study on the bearing capacity of socket steel pipe formwork support frame. Xi'an: Xi'an University of Architecture and Technology
7. Hu CM (2008) Experimental and theoretical research on steel structure connected by fasteners. Xi'an: Xi'an University of Architecture and Technology

Investigation of Web Hole Effects on Capacities of Cold-Formed Steel Channel Members

Ngoc Hieu Pham

Abstract Cold-formed steel structures have been widely applied in structural buildings with advantages in manufacturing, transportation and assembly. Holes can be pre-punched in the sectional members to allow technical pipes to go throughout such as electricity, water or ventilation. This affects the capacities of these such members which have been considered in the design standards in America or Australia/New Zealand. The paper, therefore, investigates the effects of web holes on the capacities of cold-formed steel channel members under compression or bending. Their capacities can be determined according to the American Specification AISI S100-16. The investigated results are the base for analysing the effects of web hole dimensions on the behaviors and capacities of cold-formed steel channel members. It was found that the capacity reductions were obtained for compressive members with the increase in hole sizes, but the flexural capacities were noticeable increase with the increase in the hole heights.

Keywords Effects · Capacities · Web holes · Cold-formed steel channel members

1 Introduction

Cold-formed steel members have been progressively applied in structural buildings due to their advantages compared to traditional steel structures [1]. Channel sections are the common products worldwide for many decades [2]. Holes are pre-punched in the webs of such members to allow the technical pipes such as water and electricity to go throughout with the variation of hole shapes. The presence of holes has been demonstrated to reduce the capacities of these cold-formed steel members and has been considered in the American and Australian/New Zealand design standards ([3, 4]). According to these standards, a new method has been introduced in the design of cold-formed steel members namely the Direct Strength Method (DSM) which

N. H. Pham (✉)
Faculty of Civil Engineering, Hanoi Architectural University, Hanoi, Vietnam
e-mail: hieupn@hau.edu.vn

© The Author(s) 2023
S. Wang et al. (eds.), *Proceedings of the 2nd International Conference on Innovative Solutions in Hydropower Engineering and Civil Engineering*, Lecture Notes in Civil Engineering 235, https://doi.org/10.1007/978-981-99-1748-8_13

has been illustrated to be innovative compared to the traditional design method—the Effective Width Method [5]. The DSM will be used for the investigation in this paper.

The DSM allows the designer to directly predict the capacities of cold-formed steel members based on the determination of elastic buckling loads. These buckling loads can be provided by utilising several software programs for buckling analysis of cold-formed steel sections such as CUFSM [6] or THIN-WALL-2 [7]. The application of these software programs in the design is reported in the works of Pham [8] or Pham and Vu [9]. For cold-formed steel sections with perforations, the determination of elastic buckling loads was studied and subsequently proposed by Moen and Schafer ([10–16]). Their research results were the base for the development of a module software program CUFSM funded by the American Iron and Steel Institute ([17, 18]). This module software program will be introduced and applied in the buckling analysis of cold-formed steel sections with perforations in this paper.

The paper presents the application of the DSM in the determination of capacities of cold-formed steel channel members with perforations according to the American Specification AISI S100-16 [3]. The rectangular hole shapes are considered and are arranged evenly with the variation of the hole sizes. The hole heights vary from 0.2 to 0.8 times of the investigated sectional depths whereas their lengths are from 0.5 to 2.0 times of the depths. The material properties are regulated in the American Specification [3]. The boundary conditions are varied to obtain different buckling modes of the investigated cold-formed steel members under compression or bending. The investigated results will be used to analyse the influence of hole sizes on the capacities of cold-formed steel channel members with perforations under compression or bending.

2 Determination of Capacities of Cold-Formed Steel Members Under Compression or Bending According to the American Specification AISI S100-16

The provisions for the design of cold-formed steel members are presented in Chapter E for compression and Chapter F for bending according to the American Specification AISI S100-16 [3]. The DSM for the design of cold-formed steel members is presented in this paper.

2.1 Member in Compression

The nominal axial strength (P_n) is the least of three following strength values including global buckling strength (P_{ne}), local buckling strength (P_{nl}), and distortional buckling strength (P_{nd}).

Global Buckling Strength (P_{ne})

$$P_{ne} = \left(0.658^{\lambda_c^2}\right) P_y \quad \text{if } \lambda_C \leq 1.5 \qquad (1)$$

$$P_{ne} = \left(\frac{0.877}{\lambda_c^2}\right) P_y \quad \text{if } \lambda_C > 1.5 \qquad (2)$$

where $\lambda_c = \sqrt{P_y/P_{cre}}$;

P_y Is the yield strength of the gross section;

P_{cre} Is the elastic global buckling strength, is taken as the smaller of the following values.

$$P_{ey} = \frac{\pi^2 E I_y}{(K_y L)^2} \qquad (3)$$

$$P_{exz} = \frac{1}{2\beta} \left[(P_{ex} + P_t) - \sqrt{(P_{ex} + P_t)^2 - 4\beta P_{ex} P_t} \right]$$

$$P_{ex} = \frac{\pi^2 E I_x}{(K_x L)^2}; \ P_t = \frac{1}{r_o^2}\left(GJ + \frac{\pi^2 E C_w}{(K_t L)^2}\right); \ r_o = \sqrt{x_o^2 + y_o^2 + \frac{I_x + I_y}{A_g}} \qquad (4)$$

The sectional properties (I_x, I_y, J, A_g, x_o, y_o, r_o) are determined on the basis of the gross-section. These properties for perforated sections can be determined using the "weighted average" approach as presented in table 2.3.2-1 of the specification [3] based on the ratio between the segment length of the gross-section and the net section, as follows:

$$I_{avg} = \frac{I_g L_g + I_{net} L_{net}}{L}; \ J_{avg} = \frac{J_g L_g + J_{net} L_{net}}{L}$$

$$r_{o,avg} = \sqrt{x_{o,avg}^2 + y_{o,avg}^2 + \frac{I_{x,avg} + I_{y,avg}}{A_{avg}}}; \ A_{avg} = \frac{A_g L_g + A_{net} L_{net}}{L}$$

$$x_{o,avg} = \frac{x_{o,g} L_g + x_{o,net} L_{net}}{L}; \ y_{o,avg} = \frac{y_{o,g} L_g + y_{o,net} L_{net}}{L}$$

$C_{w,net}$ is the net warping constant assuming the hole height h_{hole*} as determined in Eq. (5), where h_{hole} is the actual hole height and D is the sectional depth

$$h_{hole*} = h_{hole} + \frac{1}{2}(H - h_{hole})\left(\frac{h_{hole}}{H}\right)^{0.2} \qquad (5)$$

Local Buckling Strength (P_{nl})

$$P_{nl} = \begin{cases} P_{ne} & for \; \lambda_1 \leq 0.776 \\ \left[1 - 0.15\left(\frac{P_{crl}}{P_{ne}}\right)^{0.4}\right]\left(\frac{P_{crl}}{P_{ne}}\right)^{0.4} P_y & for \; \lambda_1 > 0.776 \end{cases} \tag{6}$$

where λ_l is the slenderness factor for local buckling, $\lambda_l = \sqrt{P_{ne}/P_{crl}}$;

P_{crl} is the elastic local buckling load of the gross section or perforated section that can be determined using elastic buckling analyses.

Distortional Buckling Strength (P_{nd})

$$P_{nd} = \begin{cases} P_y & for \; \lambda_d \leq 0.561 \\ \left[1 - 0.25\left(\frac{P_{crd}}{P_y}\right)^{0.6}\right]\left(\frac{P_{crd}}{P_y}\right)^{0.6} P_y & for \; \lambda_d > 0.561 \end{cases} \tag{7}$$

where λ_d is the slenderness factor for distortional buckling, $\lambda_d = \sqrt{P_y/P_{crd}}$;

P_{crd} is the elastic distortional buckling load of the gross section or perforated section that can be determined using elastic buckling analyses. For the perforated section, if $\lambda_d \leq \lambda_{d2}$, where λ_{d2} is determined as in Eq. (10) then:

$$P_{nd} = \begin{cases} P_{ynet} & for \; \lambda_d \leq \lambda_{d1} \\ P_{ynet} - \left(\frac{P_{ynet} - P_{d2}}{\lambda_{d2} - \lambda_{d1}}\right)(\lambda_d - \lambda_{d1}) & for \; \lambda_{d1} < \lambda_d \leq \lambda_{d2} \end{cases} \tag{8}$$

P_{ynet} Is the axial yield strengths of the net section;
$\lambda_{d1}, \lambda_{d2}$ Are the slenderness factors of distortional buckling;
P_{d2} Is the nominal axial strength of distortional buckling at λ_{d2}.

$$\lambda_{d1} = 0.561\left(\frac{P_{ynet}}{P_y}\right) \tag{9}$$

$$\lambda_{d2} = 0.561\left[14\left(\frac{P_y}{P_{ynet}}\right)^{0.4} - 13\right] \tag{10}$$

$$P_{d2} = \left[1 - 0.25\left(\frac{1}{\lambda_{d2}}\right)^{1.2}\right]\left(\frac{1}{\lambda_{d2}}\right)^{1.2} P_y \tag{11}$$

2.2 Member in Flexure

The nominal moment of a beam (M_n) is the least of three values including global buckling moment (M_{ne}), local buckling moment (M_{nl}), and distortional buckling moment (M_{nd}).

Global Buckling Moment (M_{ne})

$$M_{ne} = M_y \quad \text{if } M_{cre} \geq 2.78 M_y \tag{12}$$

$$M_{ne} = \frac{10}{9}\left(1 - \frac{10 M_y}{36 M_{cre}}\right) M_y \quad \text{if } 0.56 F_y < M_{cre} < 2.78\, M_y \tag{13}$$

$$M_{ne} = M_{cre} \quad \text{if } M_{cre} \leq 0.56 M_y \tag{14}$$

where M_y is the yield moment of the gross sections; M_{cre} is the elastic global buckling moment that can be determined as follows:

$$M_{cre} = \frac{\pi}{K_y L}\sqrt{E I_y \left(GJ + \frac{\pi^2 E C_w}{(K_t L)^2}\right)}$$

The sectional properties are defined and determined as presented in Sect. 2.1 for the gross section and the perforated section.

Local Buckling Moment (M_{nl})

$$M_{nl} = \begin{cases} M_{ne} & for\ \lambda_l \leq 0.776 \\ \left[1 - 0.15\left(\frac{M_{crl}}{M_{ne}}\right)^{0.4}\right]\left(\frac{M_{crl}}{M_{ne}}\right)^{0.4} M_y & for\ \lambda_l > 0.776 \end{cases} \tag{15}$$

where λ_l is the slenderness factor for local buckling, $\lambda_l = \sqrt{M_{ne}/M_{crl}}$;
M_{crl} is the elastic local buckling moment of the gross section or perforated section that can be determined using elastic buckling analyses.

Distortional Buckling Moment (M_{nd})

$$M_{nd} = \begin{cases} M_y & for\ \lambda_d \leq 0.673 \\ \left[1 - 0.22\left(\frac{M_{crd}}{M_y}\right)^{0.5}\right]\left(\frac{M_{crd}}{M_y}\right)^{0.5} M_y & for\ \lambda_d > 0.673 \end{cases} \tag{16}$$

where λ_d is the slenderness factor for distortional buckling, $\lambda_d = \sqrt{M_y/M_{crd}}$;
M_{crd} is the elastic distortional buckling moment of the gross section or perforated section that can be determined using elastic buckling analyses. For the perforated section, if $\lambda_d \leq \lambda_{d2}$, where λ_{d2} is determined as in Eq. (19) then:

$$M_{nd} = \begin{cases} M_{ynet} & for\ \lambda_d \leq \lambda_{d1} \\ M_{ynet} - \left(\frac{M_{ynet} - M_{d2}}{\lambda_{d2} - \lambda_{d1}}\right)(\lambda_d - \lambda_{d1}) & for\ \lambda_{d1} < \lambda_d \leq \lambda_{d2} \end{cases} \tag{17}$$

M_{ynet} Is the yield moment of the net section;
λ_{d1} and λ_{d2} Are the slenderness factors of distortional buckling;

M_{d2} Is the nominal moment of distortional buckling at λ_{d2}.

$$\lambda_{d1} = 0.673 \left(\frac{M_{ynet}}{M_y} \right)^3 \qquad (18)$$

$$\lambda_{d2} = 0.673 \left[1.7 \left(\frac{M_y}{M_{ynet}} \right)^{2.7} - 0.7 \right] \qquad (19)$$

$$M_{d2} = \left[1 - 0.22 \left(\frac{1}{\lambda_{d2}} \right) \right] \left(\frac{1}{\lambda_{d2}} \right) M_y \qquad (20)$$

3   Elastic Buckling Analyses for Cold-Formed Steel Channel Members with Perforations

Elastic buckling analysis is a compulsory step to apply the DSM in the determination of capacities of cold-formed steel members. Section C20015 taken from the commercial sections is selected for the investigation, as illustrated in Fig. 1, where "C" indicates the channel section; the nominal dimensions include the depth D = 203 mm, the width B = 76 mm, the lip length L = 19.5 mm, and the thickness t = 1.5 mm. Rectangular hole shape is considered in this investigation with the hole heights varying from 0.2 to 0.8 times of the sectional depth (D), and the hole lengths varying from 0.5 to 2.0 times of the depth (D).

3.1   Elastic Sectional Buckling Analyses

The elastic sectional buckling analyses are carried out with the support of the module software program CUFSM [18]. This program requires simple input and directly provides output results including the local buckling and distortional buckling loads of both gross-section and perforated sections, as illustrated in Fig. 2. It was found that elastic local buckling loads only depend on the hole heights whereas they are hole lengths for distortional buckling loads [10]. The material properties regulated in the American Specification [3] include the strength F_y = 345 MPa, and Young's modulus E = 203,400 MPa. The elastic buckling loads, therefore, are determined and reported in Table 1.

Table 1 shows that the elastic loads are seen as an increasing trend for local buckling modes, and the opposite trend is obtained for distortional buckling modes when the hole sizes increase. The distortional buckling loads of perforated sections are always less than those of the gross section. Meanwhile, the local buckling loads of

Fig. 1 Nomenclature

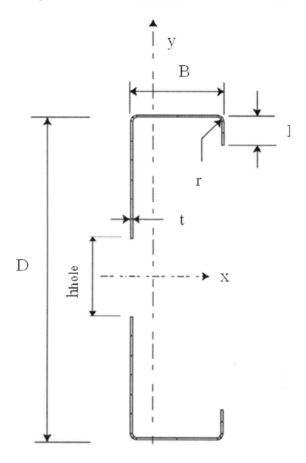

the net section are even higher than those of this gross section (see the hole heights of 0.5D and 0.8D for compression, or 0.8D for bending). This means that local buckling modes will occur at the net section for small hole heights and at the gross section areas between holes for large hole heights. The reason for this has been explained in previous studies [19]. Therefore, the elastic local buckling loads (P_{crl}, M_{crl}) of the perforated sections can be taken as the smaller of these load values of the gross section and the net section for the design.

3.2 Global Buckling Analyses

The C20015 section is chosen for the investigation with the length of 2500 mm. There are 05 symmetrical and even holes in the web of the investigated section as shown in Fig. 3. The variations of hole sizes have been presented above.

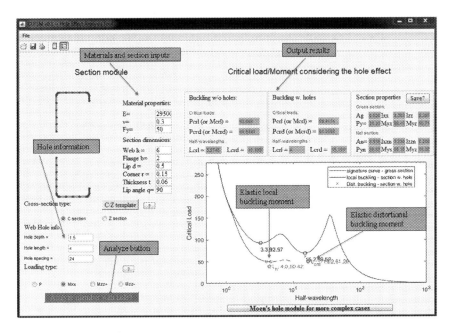

Fig. 2 The CUFSM software program

Table 1 Elastic buckling loads of C20015 section

Hole dimensions		Local buckling		Distortional buckling	
		Compression (kN)	Flexure (kNm)	Compression (kN)	Flexure (kNm)
No hole		33.01	10.49	76.67	10.31
h_{hole}	0.2D	31.93	5.65		
	0.5D	66.47	8.01		
	0.8D	114	10.98		
L_{hole}	0.5D			71.55	9.71
	D			66.29	9.04
	1.5D			60.85	8.32
	2D			55.18	7.53

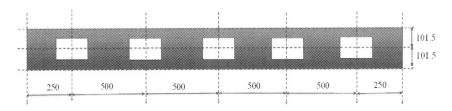

Fig. 3 The web hole locations of the specimen C20015

The specimen will be investigated under compression and bending. For compression, two different boundary conditions are applied to obtain two different global buckling modes. The first configuration allows the specimen to freely rotate about the strong axis (x-x) to obtain the flexural–torsional buckling mode (see Fig. 4) whereas the free rotation is for the weak axis (y-y) in the second configuration to get the flexural buckling mode (see Fig. 5). Warping displacements are restrained at two ends for both two configurations. The effective lengths, therefore, can be taken as follows: $L_x = L$, $L_y = L_z = 0.5L$ for the first configuration; $L_y = 0.5 L$, $L_x = L_z = 0.5L$ for the second configuration, where L is the specimen length. For bending, the specimen can freely rotate about both the strong axis (x-x) and the weak axis (y-y), and free warping displacements are applied at two ends as illustrated in Fig. 6. The effective lengths in all axes are equal to the specimen length (L). The elastic global buckling loads are determined as presented in Sect. 2 and are given in Table 2.

Table 2 shows that the buckling loads of flexural–torsional buckling modes under compression are the most significant impacts due to the appearance of web holes with the 50% capacity reduction compared to those of the gross section specimen, whereas they are about 30% reduction for the other buckling modes.

Both elastic sectional and global buckling loads calculated in this section will be used for the determination of the capacities of the investigated specimen with perforations in Sect. 4.

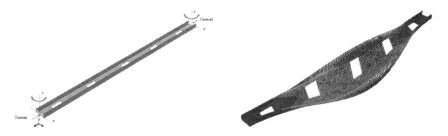

Fig. 4 The first configuration for compression and flexural–torsional buckling mode

Fig. 5 The second configuration for compression and flexural buckling mode

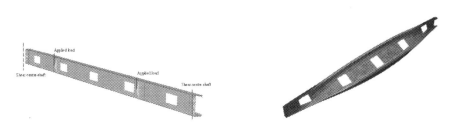

Fig. 6 The configuration for bending and flexural–torsional buckling mode

Table 2 Elastic global buckling loads of the C20015 specimen under compression or bending

h_{hole}/D	L_{hole}/D	Compression (kN)				Bending (kNm)	
		F-T	$\Delta\%$	F	$\Delta\%$	F-T	$\Delta\%$
0	0	389.32	100	137.98	100	12.61	100
0.2	0.5	370.2	95.09	136.06	98.61	12.27	97.30
	1	364.04	93.51	134.15	97.22	12.18	96.59
	1.5	357.89	91.93	132.24	95.84	12.09	95.88
	2	351.75	90.35	130.32	94.45	12	95.16
0.5	0.5	329.13	84.54	132.16	95.78	11.49	91.12
	1	314.06	80.67	126.35	91.57	11.23	89.06
	1.5	299.21	76.85	120.53	87.35	10.96	86.92
	2	284.53	73.08	114.71	83.14	10.68	84.69
0.8	0.5	266.49	68.45	126.11	91.40	10.19	80.81
	1	244.78	62.87	114.24	82.79	9.68	76.76
	1.5	224.18	57.58	102.36	74.18	9.15	72.56
	2	204.46	52.52	90.49	65.58	8.59	68.12

F, F-T stand for flexural and flexural–torsional buckling modes; $\Delta\%$ stands for the deviations between the elastic global buckling loads of the perforated section members and the gross section members

4   Investigation of Capacities of Cold-Formed Steel Channel Members with Perforations Under Compression or Bending

The obtained elastic buckling loads in Sect. 3 are used to directly determine the capacities of C20015 specimen with its length of 2500 mm and the variation of end boundary conditions by applying the Direct Strength Method design as presented in Sect. 2. There are three component strength values including global buckling strengths (P_{ne}, M_{ne}), local buckling strengths (P_{nl}, M_{nl}), and distortional buckling strengths (P_{nd}, M_{nd}) as listed in Tables 3, 4 and 5, respectively. The member capacities of the investigated specimens are the least of the above strength values as given in Table 6 and illustrated in Fig. 7.

Table 3 Global buckling strengths of the C20015 specimen under compression or bending

Configurations	Hole dimensions		L_{hole}			
			0.5D	D	1.5D	2D
Config. 1 under compression (kN)	No holes		158.200			
	h_{hole}	0.2D	156.712	156.128	155.526	154.905
		0.5D	152.449	150.641	148.706	146.623
		0.8D	143.797	139.935	135.703	131.009
Config. 2 under compression (kN)	No holes		108.037			
	h_{hole}	0.2D	107.139	106.224	105.291	104.339
		0.5D	105.253	102.296	99.148	95.793
		0.8D	102.171	95.509	87.897	79.147
Bending (kNm)	No holes		9.831			
	h_{hole}	0.2D	9.728	9.699	9.670	9.640
		0.5D	9.465	9.367	9.262	9.149
		0.8D	8.935	8.689	8.403	8.063

Table 4 Local buckling strengths of the C20015 specimen under compression or bending

Configurations	Hole dimensions		L_{hole}			
			0.5D	D	1.5D	2D
Config. 1 under compression (kN)	No holes		77.732			
	h_{hole}	0.2D	76.352	76.171	75.985	75.792
		0.5D	75.924	75.351	74.734	74.067
		0.8D	73.155	71.898	70.505	68.941
Config. 2 under compression (kN)	No holes		60.945			
	h_{hole}	0.2D	59.914	59.586	59.250	58.906
		0.5D	59.932	58.846	57.676	56.413
		0.8D	58.801	56.306	53.372	49.875
Bending (kNm)	No holes		8.525			
	h_{hole}	0.2D	6.883	6.870	6.856	6.842
		0.5D	7.611	7.559	7.502	7.441
		0.8D	8.122	7.970	7.791	7.575

In terms of global buckling strengths (see Table 3), the effects of web holes are insignificant with less than 5% reductions for small hole heights of 0.2D in comparison with those of the gross section members, but they become noticeable with more than 20% reductions for large hole heights of 0.8D.

In terms of distortional buckling strengths (see Table 5), the hole impacts are unchanged for the hole heights of 0.2D and 0.5D, and have minor changes for the hole heights of 0.8D, whereas the effects of hole lengths become noticeable. The

Table 5 Distortional buckling strengths of the C20015 specimen under compression or bending

Configurations	Hole dimensions		L_{hole}			
			0.5D	D	1.5D	2D
Config. 1 under compression (kN)	No holes		95.098			
	h_{hole}	0.2D	92.321	88.815	85.003	80.805
		0.5D	92.321	88.815	85.003	80.805
		0.8D	80.304	78.738	76.909	74.723
Config. 2 under compression (kN)	No holes		95.098			
	h_{hole}	0.2D	92.321	88.815	85.003	80.805
		0.5D	92.321	88.815	85.003	80.805
		0.8D	80.304	78.738	76.909	74.723
Bending (kNm)	No holes		8.908			
	h_{hole}	0.2D	8.768	8.533	8.263	7.946
		0.5D	8.768	8.533	8.263	7.946
		0.8D	8.430	8.309	8.162	7.946

Table 6 Member buckling strengths of the C20015 specimen under compression or bending

Configurations	Hole dimensions		L_{hole}			
			0.5D	D	1.5D	2D
Config. 1 under compression (kN)	No holes		77.732			
	h_{hole}	0.2D	76.352	76.171	75.985	75.792
		0.5D	75.924	75.351	74.734	74.067
		0.8D	73.155	71.898	70.505	68.941
Config. 2 under compression (kN)	No holes		60.945			
	h_{hole}	0.2D	59.914	59.586	59.250	58.906
		0.5D	59.932	58.846	57.676	56.413
		0.8D	58.801	56.306	53.372	49.875
Bending (kNm)	No holes		8.525			
	h_{hole}	0.2D	6.883	6.870	6.856	6.842
		0.5D	7.611	7.559	7.502	7.441
		0.8D	8.122	7.970	7.791	7.575

reductions of distortional buckling strengths are about 20% for compression and about 10% for bending in comparison with those of gross section members.

In terms of member capacities, it is found that the member failure modes are governed by local buckling modes for both cases due to the small thickness of the investigated specimen (see Tables 3 and 6). For compression, they are seen as downward trends for both two configurations if the hole dimensions increase. The capacity reductions are insignificant for small and intermediate hole heights ($h_{hole} = 0.2D$ and

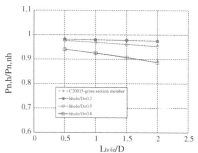

a) Configuration 1 for compression with flexural- torsional buckling modes

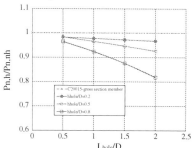

b) Configuration 2 for compression with flexural buckling modes.

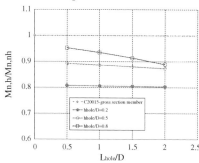

c) Bending configuration with flexural-torsional buckling modes.

Fig. 7 The capacities of C20015 section members

0.5D), but they are significant for large hole heights ($h_{hole} = 0.8$D), especially in the second configuration. For bending, it is found that the impacts of hole lengths are negligible, as seen in the minor deviations of member capacities with the variation of hole lengths. The novel point herein is the member capacities become higher for larger hole heights as seen in Fig. 7c. As seen in Table 1 for sectional buckling analyses, the local buckling moments of net sections are less than those of the gross section, which means that local buckling modes governed the member failures occurring at the net sections. Therefore, the member capacities are seen as an upward trend with the increase in the hole heights due to the increasing trend of elastic local buckling moments as discussed in Sect. 3.

5 Conclusions

The paper investigated the effects of web hole dimensions on the capacities of cold-formed steel channel members under compression or bending. Variations of boundary conditions were used for the investigation to obtain different global buckling modes. The Direct Strength Method applied for the investigation was regulated in the American specification AISI S100-16. The CUFSM software program was used to support the elastic buckling analyses of the investigated section. The obtained strengths were the base for the analysis of the member behaviors. The several remarks are given as follows:

(1) For global buckling strengths, the impacts of web holes are negligible for small hole heights, but become significant for large hole heights.
(2) For distortional buckling strengths, the influence of web holes remains unchanged for hole heights of 0.2D and 0.5D, and has minor changes for hole heights of 0.8D compared to those of the rest of hole heights, whereas the impacts of hole lengths are found to be more significant.
(3) Member buckling failures are governed by local buckling modes due to the small thickness of the investigated section.
(4) For member buckling strengths, the compressive capacities undergo decreasing trend if the hole sizes increase, whereas the flexural capacities are found to significantly increase with the increase in the hole heights although they have negligible reductions due to the effects of hole lengths.
(5) These remarks provide the base understanding of the behavior and strength of cold-formed steel channel members due to the effects of the web holes.

References

1. Yu WW, Laboube RA, Chen H (2020) Cold-formed steel design. Wiley, New York
2. Hancock GJ, Pham CH (2016) New section shapes using high-strength steels in cold-formed steel structures in Australia. Elsevier Ltd
3. American Iron and Steel Institute: North American Specification for the Design of Cold-Formed Steel Structural Members (2016)
4. AS/NZS 4600-2018: Australian/New Zealand Standard—Cold-formed steel structures. The Council of Standards Australia (2018)
5. Schafer BW, Peköz T (1998) Direct strength prediction of cold-formed members using numerical elastic buckling solutions. In: Fourteenth international specialty conference on cold-formed steel structures
6. Li Z, Schafer BW (2010) Buckling analysis of cold-formed steel members with general boundary conditions using CUFSM: conventional and constrained finite strip methods. In:

Twentieth international specialty conference on cold-formed steel structures. Saint Louis, Missouri, USA

7. Nguyen VV, Hancock GJ, Pham CH (2015) Development of the thin-wall-2 for buckling analysis of thin-walled sections under generalised loading. In: Proceeding of 8th international conference on advances in steel structures

8. Pham NH (2022) Investigation of sectional capacities of cold-formed steel SupaCee sections. In: Proceedings of the 8th international conference on civil engineering, vol. 213. ICCE 2021. Springer Singapore, pp 82–94

9. Pham NH, Vu QA (2021) Effects of stiffeners on the capacities of cold-formed steel channel members. Steel Constr 14(4):270–278

10. Moen CD, Schafer BW (2009) Elastic buckling of cold-formed steel columns and beams with holes. Eng Struct 31(12):2812–2824

11. Moen CD (2008) Direct strength design for cold-formed steel members with perforations. Ph.D. thesis. Johns Hopkins University. Baltimore

12. Moen CD, Schafer BW (2008) Experiments on cold-formed steel columns with holes. Thin-Walled Struct 46(10):1164–1182

13. Moen CD, Schafer BW (2006) Impact of holes on the elastic buckling of cold- formed steel columns. In: International specialty conference on cold-formed steel structures. pp 269–283

14. Moen CD, Schafer BW (2010) Extending direct strength design to cold-formed steel beams with holes. In: 20th international specialty conference on cold-formed steel structures—Recent research and developments in cold-formed steel design and construction. pp 171–183

15. Cai J, Moen CD (2016) Elastic buckling analysis of thin-walled structural members with rectangular holes using generalized beam theory. Thin-Walled Struct 107:274–286

16. Moen CD, Schafer BW (2009) Elastic buckling of thin plates with holes in compression or bending. Thin-Walled Struct 47(12):1597–1607

17. American Iron and Steel Institute (2021) Development of CUFSM hole module and design tables for the cold-formed steel cross-sections with typical web holes in AISI D100. Research report. RP21-01

18. American Iron and Steel Institute (2021) Development of CUFSM hole module and design tables for the cold-formed steel cross-sections with typical web holes in AISI D100. Research report. RP21-02

19. Pham NH (2022) Elastic buckling loads of cold-formed steel channel sections with perforations. Civil Eng Res J 13(1):9

A Method for Determining the Pile Location of Pile Based on the Point Safety Factor Distribution of Reinforced Slope

Shengjun Hou, Gaojin Zhao, Yongfeng Yang, Fengjiao Fu, and Qilin Li

Abstract Anti-slide pile is one of the supporting structures commonly used in land-slide treatment, while the determination of pile location is empirical. A highway landslide in Yunnan Province was selected as a study case, this paper proposes a method to determine the anti-slide pile location based on the point safety factor distribution of sliding surface. The study found that the local sliding surface has a large value of point safety factor in the anti-slide section. With increase of the proportion of the anti-slide section, the anti-sliding ability of the slide surface can be fully utilized, and the reinforcement effect of the anti-slide pile will be great. Using the point safety factor to determine the pile location is a quantitative method, which enriches the design theory of landslide support structure.

Keywords Point safety factor · Anti-slide piles · Optimal reinforcement position · Landslide soil

1 Introduction

Within a certain thickness of soil layer from the ground surface to a certain depth, there will be full and strong weathering soil layers in the shallow layer. The surface rock and soil mass is generally loose in structure and poor in mechanical properties [1, 2] and the rock-mass deformation modulus is also decreased relative to Young's modulus [3]. In order to ensure the stability of the highway [4] and the railway track [5], the surface rock and soil mass needs to be removed before subgrade filling [6]. When the thickness of the surface rock and soil mass is not large, the removal method can be adopted [7, 8]. Otherwise, the foundation needs to be reinforced. For example, pile and borehole pressure grouting technique were adopted to reinforce the silty clay

S. Hou · G. Zhao · Y. Yang · F. Fu
China Power Construction Honghe Prefecture Jiangeyuan Expressway Co., Ltd., Mengzi 661100, Yunnan, China

Q. Li (✉)
School of Civil Engineering, Southwest Jiaotong University, Chengdu 610031, Sichuan, China
e-mail: 2021210074@my.swjtu.edu.cn

© The Author(s) 2023
S. Wang et al. (eds.), *Proceedings of the 2nd International Conference on Innovative Solutions in Hydropower Engineering and Civil Engineering*, Lecture Notes in Civil Engineering 235, https://doi.org/10.1007/978-981-99-1748-8_14

overlying the shallow-buried tunnel section, according to Zhong et al. [9], and then fill the subgrade of highway [10, 11]. If the surface rock and soil mass is not treated properly, the weak interlayer will form between the subgrade fill and the lower stable rock mass. Zhao et al. [12–14] proved that the existence of the weak interlayer will affect the integrity and creep failure model of the rock mass. During the operation of the project, the groundwater will seepage along the weak interlayer, further softening the rock and soil mass, which will lead to the instability of the subgrade fill, and then a landslide will occur, afterwards the weak interlayer becomes the sliding belt in the landslide [14–16].

As a commonly used landslide control and reinforcement measure, anti-slide piles are flexible in design, simple in construction and good in anti-sliding effect [17–19]. The selection of the anti-slide pile location is very important. The appropriate reinforcement location can not only give full play to the performance of the anti-slide pile [20], but also save the project cost, ensure the safety of the project, and achieve the optimal economic and technical ratio [21]. Therefore, the choice of its layout location is a very important issue in the scheme design [22]. For this reason, many researchers have conducted research and analysis on this issue using different research methods. Nian et al. [23] used the comparative analysis method to find that the optimal reinforcement location of the anti-slide pile is located in the middle of the slope; Tan et al. [24] thought of using energy analysis method to determine the best location for re-reinforcing slope with piles should be located in the lower slope. Li et al. [25] proposed a method based on the stress level distribution characteristics of the sliding zone and draw the following conclusions, that anti-slide piles should be placed in areas with higher stress levels in the sliding zone. Pan et al. [26] used the finite element software ABAQUS to analyze the stability of soil slope, the result shows that when the ratio of the distance from the pile to the toe to the horizontal distance of the toe is 0.6, the safety factor is the largest. Fang et al. [27] used the finite element strength reduction method to study the multi-slip belt slope, the anti-slide piles are always set in the middle and lower part of the slip zone. In addition, based on the relationship between slope deformation and slope stability. Yang et al. [28] proposed a method to determine the optimal reinforcement location of anti-sliding piles based on the slope deformation field, and found that the anti-sliding piles set up in places with large slope displacements have better reinforcement effects.

Although relevant research has been able to provide some reference opinions for engineering design, due to the diversity of landslide types, the occurrence of landslides and the location of landslides are quite different, resulting in large differences in the optimal reinforcement positions of anti-slide piles for different types of landslides. At present, the choice of the location of the landslide anti-slide pile is still based on experience in most cases, and there is no unified quantitative method. As a quantitative index that can describe the sliding mechanism of the landslide, the safety factor of the landslide point can describe the stability of different parts of the landslide, and it is widely used in the research of landslide stability [29]. However, there needs more research on the application of point safety factor in the design of landslide support structures. Taking a typical roadbed slope as an example, this paper discusses the guiding role of point safety factor distribution on the setting of anti-sliding piles for landslides, and provides suggestions for engineering design.

2 The Point Safety Factor

Generally, the instability sliding of a landslide begins locally, and the process from stabilization to instability is a process of asymptotic development, which can be described by sliding mechanism. After summarizing a large number of landslide cases, the engineering community has reduced the sliding mechanism of landslides to three categories: traction, translational, and pushing. The sliding mechanism classification describes the asymptotic failure process of landslide from local instability to overall sliding from a qualitative point of view, but this qualitative description method is relatively rough and is not easy to apply in engineering design. In order to quantitatively describe the sliding mechanism of landslides, Yang [29] proposed the concept of the point safety factor at landslide and derived the corresponding calculation formulas.

Using numerical methods to obtain the stress state, sliding surface normal vector and sliding direction of sliding belt elements, according to the theory of elasticity, the normal stress of the sliding belt element and the shear stress in the sliding direction can be obtained, and the point safety factor of the sliding belt element can be defined by Eq. 1,

$$F_E = \frac{\tau_u}{\tau} = \frac{c + \sigma_n \tan \phi}{\tau} \tag{1}$$

In formula, c is the cohesion of the slip belt material, ϕ is the internal friction angle of the slip belt material.

A weighted average of the area of the slip belt unit yields the overall safety factor for landslides in Eq. 2,

$$F_{3d} = \frac{\sum\limits_{i=1}^{ne} F_E^i S_E^i}{\sum\limits_{i=1}^{ne} S_E^i} \tag{2}$$

In formula, ne is the total number of slip belt units, F_E^i is the safety factor of unit i, S_E^i is the area of the slip belt represented by unit i.

For slope engineering, the potential sliding surface needs to be identified first. Yang et al. [30] believed that the strength reduction method was used to make the slope reach the limit equilibrium state, and the displacement contour of the potential sliding zone represents the potential sliding surface. According to this, the slope stress field can also be decomposed to obtain the normal stress and shear stress on the displacement isosurface, calculating the safety factor of the slope point and the overall safety factor according to formulas (1) and (2).

3 Layout of Anti-slide Pile for Landslide of the Subgrade Fill

3.1 General Situation of Roadbed Landslide Project

A highway in Yunnan Province was constructed and opened to traffic in December 2017, and the K32+780~K32+860 subgrade was built above the slope. The length of the filling slope is 80 m, the width is 12.75 m, the maximum filling height is 32.44 m, and the maximum center height is 15.45 m, which is located at the K32+840 section. On July 12, 2021, due to continuous rainfall, the subgrade suddenly underwent landslide, and the highway surface formed a pull crack. Since then, the landslide has continued to develop, and by August 20, the width of the highway crack has reached 24 cm, the hind edge of the landslide has been wrong by 53 cm, and the subgrade slope has been seriously damaged and needs to be remediated, as shown in Fig. 1.

After on-site investigation and analysis, it was determined that the cause of the subgrade landslide was that the shallow loose soil on the original ground was not thoroughly cleaned during the subgrade filling construction, and a weak interlayer with an average thickness of about 1.0 m was formed between the subgrade filling and the lower rock layer. The groundwater formed by rainfall infiltrates downward along the weak interlayer, softening the weak interlayer and forming a landslide.

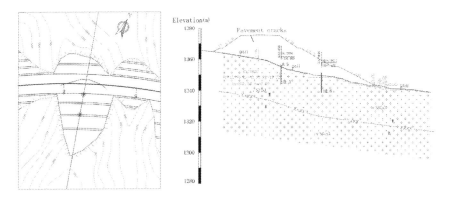

Fig. 1 Subgrade plan (left) and subgrade profile (right)

3.2 Computational Model and Parameters

Firstly, the landslide is numerically calculated and the distribution of point safety factor on the sliding surface is obtained [31]. According to the typical section of subgrade slope shown in Fig. 1, a numerical calculation model of subgrade slope is established. In order to reflect the control effect of the original highway on the stability of the subgrade slope, the overburden layer with thickness of 1.0 m was built. Accordingly, the calculation model consists of three materials, subgrade filling, original ground cover and underlying rock foundation. The unit width of the model is longitudinally stretched to form a quasi-three-dimensional calculation model, so as to obtain the spatial stress state of the slope body. The models are all discretized with hexahedral elements, and the element size is less than 0.4 m to improve the calculation accuracy. The divided model calculation grid is shown in Fig. 2.

The finite difference software FLAC3D 6.0 is used for numerical calculation, and the yield criterion is the Mohr–Coulomb yield criterion. The calculation parameters are shown in Table 1.

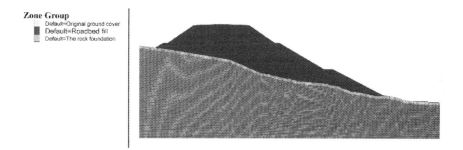

Fig. 2 Numerical computation model networks

Table 1 Rock mass calculation parameters

Rock mass categories	Volumetric weight/kN/m³	Cohesive forces/kPa	The angle of internal friction/(°)	Elastic Modulus/10^6 Pa	Poisson's ratio
Subgrade filling	19.2	25.5	28.2	125.5	0.30
Barrier covers	18.8	18.6	15.6	65.0	0.34
Rock foundation	21.5	300.0	32.5	1000.0	0.25

3.3 Computational Model and Parameters

The displacement field of slope limit state before reinforcement is calculated as shown in Fig. 3. Among them, the range of 0.05–0.6 m contour line better shows the range and location of the potential instability slip zone of the slope body, which is consistent with the actual deformation cracks on site, as shown in Fig. 4. The potential slip zone consists of two sections, the front section develops along the overburden, and the rear section develops in the subgrade fill.

The slope stress field in its natural state is introduced, and the safety factor of the slip belt point is calculated according to Eq. (1), and its distribution is shown in Fig. 4. Equation 2 calculates the overall safety factor of the slope is 1.217. It can be clearly seen from the distribution of the safety factor of the sliding belt points that the landslide presents an obvious composite sliding mechanism of front traction, rear pushing, and middle anti-sliding, which is consistent with the current development characteristics of landslides.

According to the distribution characteristics of the point safety factor on the sliding surface, the sliding surface can be divided into three sections with the design safety factor of 1.30 as the standard. Those are, the section less than 1.30 in the rear fill is the

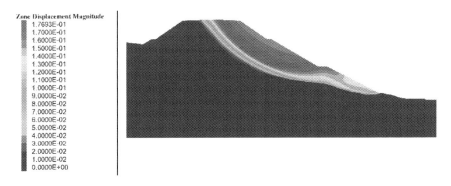

Fig. 3 Displacement field in the ultimate state

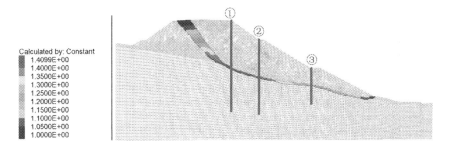

Fig. 4 Point safety factor distribution

Fig. 5 Anti-slide piles placed on the shoulder of the road (position①) calculate the results of the model

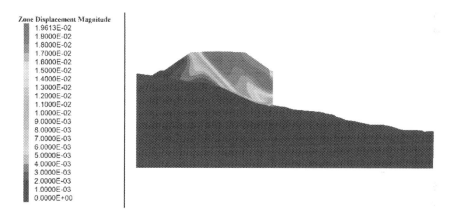

Fig. 6 Anti-slide pile calculation model results placed on a primary platform (position②)

Fig. 7 Anti-slide pile calculation model results placed on a secondary platform (position③)

main sliding section, the section with the leading edge less than 1.30 is the traction section, and the section with the middle section greater than 1.30 is the anti-slip section. It can be seen that in the subgrade fill, the main sliding section has nearly penetrated the entire filling body, which is consistent with the poor slope stability. However, the central anti-slip section still exists, and the length is about 1/2 of the entire slip surface, which plays a good anti-slip effect, so that the slope of the roadbed has not yet occurred overall instability.

3.4 Reinforcement Location of Anti-slide Piles

According to the calculation results of the point safety factor, considering the stability of different positions of the slope, it is proposed to carry out the calculation and analysis of the pile in three positions. The three pile positions are located at the road shoulder (location①), the first-level platform (location②) and the second-level platform (location③), as shown in Fig. 4. In order to constrain the displacement of the pile head, a row of anchor cables is arranged at the pile head, and the form of a pile and an anchor support structure is adopted. The diameter of the anti-slip pile is 1.5 m × 2.0 m, the anchor cable tonnage is 500 kN, the angle between the anchor cable and the horizontal surface is 20°, and the anchorage length in rock is 8 m. The pile length is determined according to the different thickness of the slider at each pile position, and is determined according to the method that the length of the anchoring section and the free section is the same. In addition, in order to ensure the safety of the project, the anti-slip ability of the slider before the pile is not considered in the calculation, that is, the anti-slip pile is considered as cantilever pile.

3.5 Analysis of Reinforcement Effect of Different Pile Positions

The reinforcement effect evaluation method of anti-slip pile is as follows: (1) Establish a calculation model according to Figs. 5, 6 and 7, assign the corresponding model parameters, and calculate to the initial balance of the model. (2) The displacement of the model is cleared to zero, and the shear strength parameter is reduced according to formula (3). At the same time, a uniform load of 30 kPa is applied to the road surface to simulate the long-term traffic load. Recalculate formula (3) to balance, analyze the displacement field of the fill slope, and judge the quality of the pile position.

$$\begin{cases} c_e = \frac{c}{k} \\ \varphi_e = \arctan(\tan \varphi / k) \end{cases} \tag{3}$$

In formula: C_e—cohesion of the reduced belt material.

φ_e—internal friction angle of the reduced belt material. The reduction coefficient k is taken as 1.3.

It can be seen from the calculation results shown in Figs. 5, 6 and 7 that due to the reinforcement of the anti-slide piles, the stability of the embankment slope can be guaranteed, and the overall instability and damage will not occur. However, there will be a certain degree of deformation of the subgrade, which is reflected in the vertical settlement and the horizontal displacement along the direction of the free surface, which is comprehensively reflected as the combined displacement of the filled subgrade. Therefore, in this paper, the combined displacement of the filled roadbed is used as the evaluation index of the reinforcement effect of the anti-slide pile. The calculation results are shown in Table 2.

It can be seen that different pile positions have obvious effects on the reinforcement effect of the landslide.

(1) Set anti-slide piles at the road shoulder, the cantilever section is long, about 20 m, and the maximum combined displacement is 10.2 mm; when the anti-sliding piles are set on the first-level slope, the maximum combined displacement is 8.2 mm; when the second-level slopes are set with anti-sliding piles, the cantilever section is shorter, but the maximum combined displacement reaches 15.2 mm.

(2) The anti-sliding pile is set at the shoulder of the road (location①). Although the pile is located in the anti-sliding section of the sliding surface, the proportion of the anti-sliding section is only 1/3 of the sliding surface behind the pile, and most of the sliding surface points have a safety factor, less than 1.30. After the

Table 2 Comparison of supporting effect

Pile position	Cantilever pile length (m)	Anchor cable length (m)	Maximum combined displacement of roadbed (mm)	Sliding point safety factor characteristics
① Shoulder	19.4	45	10.2	Located in the anti-slip section, the anti-slip section accounts for about 1/3 of the rear sliding surface of the pile
② Level 1 platform	15.2	33	8.2	Located in the anti-slip section, the anti-slip section accounts for about 3/5
③ Level 2 platform	6.5	25	15.2	Located in the leading edge traction section, the anti-slip section accounts for about 1/2

shear strength parameter of the sliding belt is reduced, the landslide is further unstable, which increases the landslide thrust of the anti-slide pile. In addition, the cantilever section of the anti-sliding pile is long and the lateral displacement is large, so the combined displacement of the sliding body behind the pile is obvious.

(3) The anti-sliding pile is set at the first-level platform (location②), the pile is located in the anti-sliding section of the sliding surface, the proportion of the anti-sliding section is 3/5 of the sliding surface behind the pile, and the shear strength parameter of the sliding surface is folded. After the reduction, the point safety factor of the anti-slip section is still greater than 1.0, and it still has anti-slip capability. The anti-sliding ability of the anti-sliding pile and the anti-sliding section of the sliding surface is fully exerted, so the displacement of the slope body is relatively small.

(4) The anti-sliding pile is set at the secondary platform (location③), although the pile length is very short (6.5 m), the potential instability section of the sliding surface is long. After the shear strength parameters of the sliding surface are reduced, the main sliding section of the trailing edge and the traction section of the leading edge will be unstable at the same time. Reinforcement effect of anti-slide piles.

(5) In general, the reinforcement effect of anti-sliding piles at different positions depends on two aspects, one is the length of the pile, and the other is the proportion of the anti-sliding section behind the pile. The longer the pile length is, the more obvious the deformation of the sliding body behind the pile is. The larger the proportion of the anti-slip section, the more the anti-slip ability of the sliding belt can be fully exerted, and the better the reinforcement effect is. The proportion of the anti-sliding section of the sliding surface behind the pile is more important. In actual engineering, this can be the main basis to determine the pile location.

4   Conclusion

By applying the calculation method of the point safety factor of landslides and side slopes, the distribution of the point safety factor of the sliding surface in the landslide can be obtained, and the anti-sliding section of the landslide can be determined accordingly. The proportions of the anti-sliding sections of the three pile positions are 1/3, 3/5 and 1/2, respectively, and the maximum displacement of the sliding body after the shear strength parameters of the sliding surface are reduced are 10.2 mm, 8.2 mm and 15.2 mm, respectively.

The proportion of the anti-sliding section of the sliding surface behind the pile is an important factor affecting the reinforcement effect of the anti-sliding pile. In actual engineering, the selection of anti-sliding pile positions should be based on a combination of various engineering factors, make full use of the anti-sliding effect of the anti-sliding section of the sliding surface as much as possible, and increase the proportion of the anti-sliding section of the sliding surface behind the pile. The method in this paper provides a quantifiable standard for the selection of anti-slide pile positions and enriches the design theory of landslide support structures.

References

1. Zhang ZY, Wang SQ, Wang LS (1994) Principles of engineering geological analysis—2nd Edition. Geological Press
2. Lisjak A, Grasselli G (2014) A review of discrete modeling techniques for fracturing processes in discontinuous rock masses. J Rock Mech Geo Tech Eng 6(4):301–314
3. Richard A, Schultz (1996) Relative scale and the strength and deformability of rock masses. J Struct Geol 18(9):1139–1149
4. Xu T, Zhou Z, Yan R et al (2020) Real-time monitoring method for layered compaction quality of loess subgrade based on hydraulic compactor reinforcement. Sensors
5. Lisa N, Wheeler W, Take A, Neil A (2017) Performance assessment of peat rail subgrade before and after mass stabilization. Can Geotech J 54(5):674–689
6. Weng S, Ma SD (2001) Effect of the geotextile-reinforced sand cushion on the deformation and stability of the foundation under a embankmen. Rock Soil Mech
7. Liu HL, Zhao MH (2016) Review of ground improvement technical and its application in China. Chin Civil Eng J 49(01):96–115
8. Hsein Juang C, Jie Z, Shen MF, Hu JZ (2019) Probabilistic methods for unified treatment of geotechnical and geological uncertainties in a geotechnical analysis. Eng Geol 249:148–161
9. Zhong ZL, Chao L, Liu XR et al (2021) Analysis of ground surface settlement induced by the construction of mechanized twin tunnels in soil-rock mass mixed ground. Tunnelling and Underground Space Technology
10. Zheng YR, Lu X, Li XZ, Feng YX (2000) Research on theory and technology of improving soft clay with DCM. Chin J Geotech Eng 21–25
11. Zhao MH, Zou XJ, Patrick X, Zou (2017) Disintegration characteristics of red sandstone and its filling methods for highway roadbed and embankment. J Mater Civ Eng 404–410
12. Zhao N, Zhang YB, Miao HB, Meng LX (2022) Study on the creep and fracture evolution mechanism of rock mass with weak interlayers. Adv Mater Sci Eng
13. Duffaut P (1981) Structural weaknesses in rocks and rock masses tentative classification and behaviour ISRM international symposium
14. Zhong Z, Yong R, Tang H et al (2021) Experimental studies on the interaction mechanism oflandslide stabilizing piles and sandwich-type bedrock. Landslides 18:1369–1386
15. Cheng Q, Zhou DP, Feng ZJ (2009) Research on shear creep property of typical weak intercalation in redbed soft rock. Chin J Rock Mech Eng 28(S1):3176–3180
16. Ding XL, Fu J, Liu J et al (2005) Study on creep behavior of alternatively distributed soft and hard rock layers and slope stability analysis. Chin J Rock Mech Eng 12–20
17. Gao Q, Wang H, Wan Z, Liu H, Huang R, Wang H (2021) Research on stability analysis of soft foundation slope and its anti-slide pile support technology. In: 2021 4th international symposium on traffic transportation and civil architecture (ISTTCA)
18. Zhang Q, Hu J, Du Y et al (2021) A laboratory and field-monitoring experiment on the ability of anti-slide piles to prevent buckling failures in bedding slopes. Environ Earth Sci 80:44
19. Zhao XW (2015) Analysis of soil and rock slope stability influence by anti-slide piles position. Electron J Geotech Eng 20(11):4527–4534
20. Liu X, Cai G, Liu L et al (2020) Investigation of internal force of anti-slide pile on landslides considering the actual distribution of soil resistance acting on anti-slide piles. Nat Hazards 102:1369–1392

21. Qiao S, Xu P, Teng J et al (2020) Numerical study of optimal parameters on the high filling embankment landslide reinforced by the portal anti-slide pile. KSCE J Civ Eng 24(5):1460–1475

22. Yang GH, Zhang YX, Zhang YC et al (2011) Optimal site of anti-landslide piles based on deformation field of slopes. Chin J Geotech Eng 8–13

23. Nian TK, Xu HY, Liu HS (2012) Several issues in three-dimensional numerical analysis of slopes reinforced with anti-slide piles. Rock Soil Mech 33(08):2521–2526

24. Tan HH, Zhao LH, Li L, Luo Q et al (2011) Energy analysis method for pre-reinforcing slopes with anti-slide piles. Rock Soil Mech 190–197

25. Li XZ, Li T, Pan D et al (2021) Optimal reinforcement location of antislide piles based on stress level of sliding zone. J Eng Geol 29(3):640–646

26. Pan J, Wang Z, Dong T et al (2017) Analysis on the best location and the pile distance of anti-slide pile of reinforced soil slope. In: IOP conference series: earth and environmental science. IOP Publishing 61(1)

27. Fang L, Mei L, Huang X et al (2018) The optimization of the anti-slide piles location in slope control project. AIP Conf Proc 1973(1)

28. Yang GH, Zhong ZH, Zhang YC et al (2012) Identification of landslide type and determination of optimal reinforcement site based on stress field and displacement field. Chin J Rock Mech Eng 31(09):1879–1887

29. Yang T, You Y, Qin YT (2010) Application of point safety factor to study on spatial sliding mechanism of landslide. J Southwest Jiaotong Univ 45(05):794–799

30. Yang T, Zhou DP, Ma HM et al (2010) Point safety factor method for stability analysis of landslide. Rock Soil Mech 31(3):971–975

31. Yang T, Ma HM, Dai J et al (2011) Application condition of point safety factor method for stability analysis of landslide. J Southwest Jiaotong Univ 46(6):966–972

Study on Performance of Pervious Concrete Modified by Nano-Silicon + Polypropylene Fiber Composite

Jingsong Shan, Chengfa Song, Shengbo Zhou, TongJun Duan, Shuai Zheng, and Bo Zhang

Abstract In order to improve the comprehensive performance of pervious concrete, nano-silicon and polypropylene fiber were added to pervious concrete to study the change of performance of pervious concrete. Firstly, the effect of single doped nano-silicon on the properties of cement slurry and pervious concrete was studied, and the optimal water-binder ratio and nano-silicon content were determined. Based on this, mixed polypropylene fiber with different proportions of length of 18 mm to determine the reasonable amount of polypropylene fiber. The results showed that the compressive strength of pervious concrete was the highest when the nano-silicon content was 0.5% and the water-binder ratio was 0.32. Based on this ratio, the maximum compressive strength can be obtained by adding 1.0 kg/m^3 polypropylene fiber, and the compressive strength of 7d and 28d increased by 29.9% and 42.2%, respectively. Adding 1.5 kg/m^3 polypropylene fiber was the most beneficial to improve the freezing resistance of pervious concrete. For example, after 300 freeze–thaw cycles, the compressive strength residual rate was 62%. That's much higher than the 40 percent that was found when nano silicon was mixed alone.

Keywords Nano-silicon · Polypropylene fiber · Compressive strength · Void fraction · Freezing resistance

J. Shan · C. Song (✉)
Shandong University of Science and Technology Qingdao, Qingdao 266590, China
e-mail: 747082124@qq.com

S. Zhou
Jimo District Housing and Urban-Rural Development Bureau Qingdao, Qingdao 266590, China

T. Duan · S. Zheng
Shandong Road and Bridge Group Co., Ltd., Jinan 250014, China

B. Zhang
Tianjin University College of Construction Engineering Tianjin, Tianjin 300072, China

© The Author(s) 2023
S. Wang et al. (eds.), *Proceedings of the 2nd International Conference on Innovative Solutions in Hydropower Engineering and Civil Engineering*, Lecture Notes in Civil Engineering 235, https://doi.org/10.1007/978-981-99-1748-8_15

1 Introduction

Pervious concrete is a kind of multi-air mixed material, which is widely used in sponge city construction because of its advantages such as permeability, air permeability and noise absorption. Due to its many connected voids, its strength is low, which limits its application in the field of practical engineering. Therefore, many scholars began to add admixture materials such as fibers and nanomaterials [1, 2] into pervious concrete to achieve the purpose of improving the performance of pervious concrete.

Nano-silicon is an inorganic material with small particle size and has strong pozzolanic activity, nucleation and micro-aggregate filling effect, which is conducive to improving macro mechanical properties and durability of pervious concrete [3, 4]. For example, Tarangini [5] added nano-silicon with 3% cement mass to pervious concrete to study the durability of pervious concrete. The results showed that compared with ordinary pervious concrete, the freeze–thaw resistance of pervious concrete modified by nano-silicon had been effectively improved, and the durability coefficient was up to more than 70%. Fiber can improve the performance of pervious concrete to a certain extent [6–9]. Under the action of external forces, fibers will be in different stress states such as stretching, pulling out and breaking, thus consuming part of the energy transferred under the action of external forces, restraining the expansion of cracks, and making concrete structures have certain flexibility. For example, Banthia [10] studied the relationship between shrinkage cracks and polypropylene fiber size, and the study showed that the incorporation of fine fibers was more effective in suppressing cracks than that of coarse fibers, while the incorporation of long fibers was more effective in suppressing cracks than that of short fibers. In order to comprehensively utilize admixture materials to improve the performance of pervious concrete, many scholars began to study the influence of admixture materials on the performance of pervious concrete. Ali [11] added volcanic ash and fiber into the permeable concrete structure, and the results showed that the mechanical properties of the concrete structure were effectively improved and basically met the requirements of practical application without changing the permeability.

To sum up, when different admixtures are added to the pervious concrete structure, the comprehensive performance of pervious concrete will be significantly improved if the advantages of each additive material can be brought into play. Therefore, nano silicon and polypropylene fiber were added to pervious concrete in this paper to explore the effect of composite additive on the performance of pervious concrete.

2 Test Raw Materials and Forming Methods

(1) Test raw materials

The cement used for the permeable concrete was ordinary Portland cement P42.5R, the water was local tap water, the water reducing agent was high efficiency poly-carboxylate concrete water reducing agent, the aggregate was basalt aggregate, the gradation is 4.75~9.5 mm (80%): 9.5~13.2 mm, (20%). The nano silicon is Degussa A380 vapor phase nano silica, and the fiber is 18 mm polypropylene fiber.

(2) Test block forming method

The test block of pervious concrete was compacted by Marshall automatic compactor. The weight of the instrument drop hammer was 10.21 kg. According to practical experience, the striking times were set to 100 times.

3 Effect of Single Doped Nano Silicon on Cement Slurry and Pervious Concrete

3.1 Effect of Nano-Silicon on Water Consumption of Standard Consistency of Cement Paste

Take the cement weight of 500 g for the test, added nano silicon at the same time, reduced the same weight of cement. The concrete cement paste test mix was divided into eight groups, as shown in Table 1.

It can be seen that with the increase of nano-silicon content, the water consumption of the standard consistency of the cement paste increases significantly. When the nano-silicon content was 3%, the water consumption of the cement paste was 192 mL, 47.7% higher than the water consumption of the cement paste without the nano-silicon content. Therefore, from the perspective of cement slurry to reach the standard consistency, there should be different water-binder ratio under different nano-silicon content.

Table 1 Experimental results

Serial number	Cement/g	Nano silicon/%	Water consumption/g	Water-binder ratio
1	500	0	129.6	0.26
2	497.5	0.5	141.6	0.28
3	495	1	154.6	0.31
4	492.5	1.5	167.5	0.34
5	485	3	191.5	0.38

3.2 Effect of Nano-Silicon on the Performance of Pervious Concrete

From the perspective of cement slurry to reach the standard consistency, there should be different water-binder ratio under different nano-silicon content. The strength and porosity of pervious concrete with different nano-silicon contents (0%, 0.5%, 1%, 1.5%, 3%) were tested under the condition that the cement slurry reached the standard consistency. Considering that the aggregate has certain absorbability, after the preliminary test, the water-binder ratio under different nano-silicon content was slightly larger than the value in Table 1 and the specific water-binder ratio was (0.28, 0.30, 0.32, 0.36, 0.4). The influence of water-binder ratio on compressive strength and voidage is shown in Figs. 1 and 2.

That the compressive strength of both 7d and 28d increased firstly and then decreased with the increasing of nano-silicon content. The compressive strength of 0.5% nano-silicon concrete at 7d and 28d was the highest, which increased by 12.66% and 7.14% compared with that of ordinary concrete, respectively. As can be seen from Fig. 3, the variation law of nano-silicon pervious concrete was higher than that with ordinary concrete. For example, the maximum total voidage occurred when the nano-silicon content was 1.5%, which increased by 24%. The maximum effective void fraction occurred when the nano-silicon content was 0.5%, which increased by 25.8%. Analysis the reasons: when the content of nano-silicon is appropriate, the surface effect plays a dominant role, which promotes the hydration reaction and improves the distribution state of cement slurry in the vertical direction of pervious concrete. With the increase of the content, the flocculation effect plays a dominant role, and the hydration reaction begins to be hindered. In conclusion, from the perspective of strength and voidage, nano-silicon content of 0.5% is optimal.

Fig. 1 Influence of nano-silicon content on compressive strength

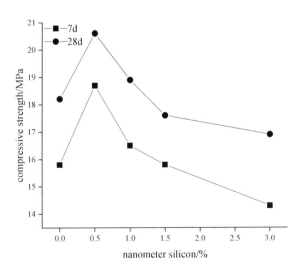

Fig. 2 Effect of nano silicon dules on void ratio

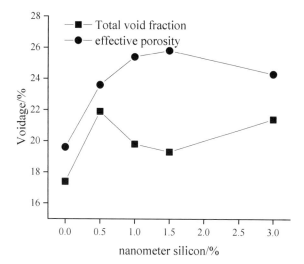

Fig. 3 Effect of glue ratio on compressive strength

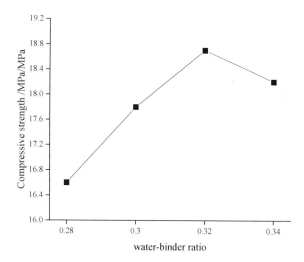

Based on the nano-silicon content of 0.5%, the influences of different water-binder ratios (0.28, 0.30, 0.32, 0.34) on the compressive strength and porosity of concrete were studied. The results were shown in Figs. 3 and 4.

It can be seen that the 7d compressive strength first increases and then decreases with the increase of water-binder ratio from the test results. When the water-binder ratio is 0.28, the strength was 16.6 MPa the lowest, which was the lowest. This is because the water-binder ratio is too small, resulting in insufficient water required for hydration reaction and low adhesion between aggregate particles. With the increasing of water-binder ratio, the overall compressive strength of pervious concrete gradually increased. When the water-binder ratio was 0.32, it reached the extreme value of

Fig. 4 Effect of glossy rate

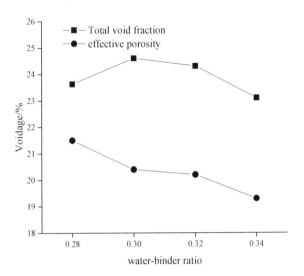

18.7 MPa, which increased by 12.7% compared with the compressive strength of concrete with the water-binder ratio of 0.28. When the water-cement ratio exceeded 0.0.32, the overall compressive strength of concrete began to decrease gradually due to excessive moisture, which was not conducive to the uniform coating of aggregate by slurry. At this time, the total void fraction and effective void fraction of pervious concrete had an obvious downward trend with the increasing of water-cement ratio.

In conclusion, from the perspective of improving the comprehensive performance of nano-silicon pervious cement concrete, the best water-binder ratio is 0.3~0.32.

4   Pervious Concrete Modified by Nano-Silicon + Polypropylene Fiber Composite

In this section, polypropylene fiber with length of 18 mm was mixed on the basis of 0.5% nano-silicon content concrete, to explore the effect of nano-silicon + fiber composite modification on the performance of concrete. The fiber content ranges from 0 to 2 kg/m^3. Considering that the amount of water used by adding fiber increases further, the water-cement ratio of 0.32 was adopted for the experimental study.

4.1   Physical and Mechanical Properties

The variation law of compressive strength and voidage of nano-silicon pervious concrete with polypropylene fiber is shown in Fig. 5 and 6. The voidage first decreases and then increases with the increase of fiber content, while the compressive strength is on the contrary. For example, when the dosage was $1.0\,kg/m^3$, the decrease of total void fraction and effective void fraction was 2% and 1.8%, which was the lowest respectively, and the compressive strength was the highest, which was increased by 29.9% and 42.2% respectively. When the content exceeds $1.0\,kg/m^3$, the void fraction began to increase and the compressive strength began to decrease. This is because the appropriate fiber content and cement slurry formed a fiber network, and the surface effect of nano silicon can enhance the strength of the network structure and reduce the porosity. However, with the addition of too many polypropylene fibers, the fiber is too dense and easy to cluster, thus weakening the compressive strength, and the porosity began to rise.

Fig. 5 The effect of fiber doping on the gap rate

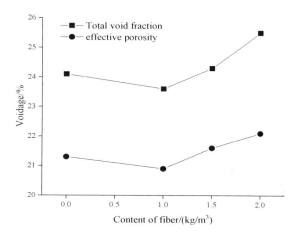

Fig. 6 Effect of fiber dosage on compressive strength

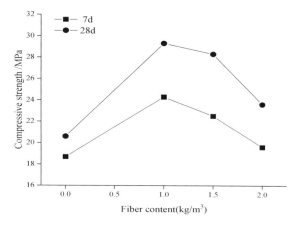

Fig. 7 The impact of compressive intensity of cyclic frozen fusion

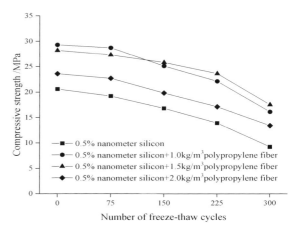

4.2 Frost Resistance

It can be seen from the results of Figs. 7 and 8 that the incorporation of polypropylene fiber greatly improves the frost resistance of pervious concrete. After 75 freeze–thaw cycles, the compressive strength of pervious concrete in each group decreased less. For example, the compressive strength of nano-silicon pervious concrete mixed with 1.0 kg/m^3 polypropylene fiber decreased by 0.6 MPa, only 2%. After 150 freeze–thaw cycles, the decline in compressive strength began to accelerate. After 225 freeze–thaw cycles, the compressive strength of nano-silicon pervious concrete mixed with 1.5 kg/m^3 polypropylene fiber was larger than that of the nano-silicon pervious concrete mixed with 1.0 kg/m^3 polypropylene fiber pervious concrete, and the compressive residual rate was 84% and 75%, respectively. After 300 freeze–thaw cycles, the residual rate of compressive strength of pervious concrete doped with nano-silicon stabilized at about 40%. The nano-silica permeable concrete mixed with 1.5 kg/m^3 polypropylene fiber had the highest compressive strength and the compressive residual rate was 62%.

5   Conclusion

(1) Nano-silicon content directly affects the water consumption of cement slurry at standard consistency, and different nano-silicon content corresponds to different water-binder ratio.

(2) By comparing the compressive strength and porosity of pervious concrete, the optimal content of nano-silicon is determined to be 0.5%. On this basis, the influence of water-binder ratio was further analyzed, and the optimal water-binder ratio range was determined to be 0.30–32.

Fig. 8 The impact of the residual rate of the reconciliation intensity of the cycle frozen fusion

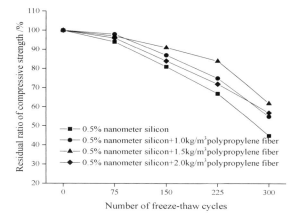

(3) When the nano-silicon content is 0.5% and a certain amount of 18 mm polypropylene fiber is mixed, the porosity of pervious concrete changes little, and the compressive strength is significantly improved, especially the late compressive strength, compared with the nano-silicon concrete only mixed 0.5%, the strength is increased by 42.2%.

(4) After mixing 18 mm polypropylene fiber on the basis of 0.5% nano-silicon content, the freezing resistance of concrete can be effectively improved. The compressive residual rate of the composite fiber concrete is stable at about 50~60%, which is higher than the single nano-silicon concrete compressive residual rate of 45%.

References

1. Singh SB, Murugan M (2021) Performance of carbon fibre-reinforced pervious concrete (CFRPC) subjected to static, cyclic and impact loads. Int J Pavement Eng 4:1–16
2. Alimohammadi V, Maghfouri M, Nourmohammadi D et al (2021) Stormwater Runoff Treatment Using Pervious Concrete Modified With Various Nanomaterials: A Comprehensive Review. Sustainability 13(15):8552
3. Mohammed BS, Liew MS, Alaloul WS et al (2018) Properties of nano-silica modified pervious concrete. Case Stud Constr Mater 409–422 (2018)
4. Pan F, Chang HL, Xin LA et al (2020) The significance of dispersion of nano-SiO_2 on early age hydration of cement pastes. Mater Des 186:108320
5. Tarangini D, Sravana P, Rao PS (2022) Effect of nano silica on frost resistance of pervious concrete. In: International conference on advances in materials science. Materials Today, pp 2185–2189
6. Ali T, Peyman M et al (2020) Evaluating the use of recycled concrete aggregate and pozzolanic additives in fiber-reinforced pervious concrete with industrial and recycled fibers. Constr Build Mater 252:118997
7. Tran TNH, Puttiwongrak A et al (2021) Microparticle filtration ability of pervious concrete mixed with recycled synthetic fibers. Constr Build Mater 270:121807

8. Akand L, Yang M, Wang X (2018) Effectiveness of chemical treatment on polypropylene fibers as reinforcement in pervious concrete. Constr Building Mater 163(FEB.28):32–39
9. Al-Hadithi AI, Noaman AT, Mosleh WK (2019) Mechanical properties and impact behavior of pet fiber reinforced self-compacting concrete (SCC). Compos Struct 224(SEP):111021
10. Banthia N, Gupta R (2006) Influence of polypropylene fiber geometry on plastic shrinkage cracking in concrete. Cem Concr Res 36(7):1263–1267
11. Ali T, Peyman M, Mahdi S et al (2020) Evaluating the use of recycled concrete aggregate and pozzolanic additives in fiber-reinforced pervious concrete with industrial and recycled fibers. Constr Build Mater 252:118997

Experimental Study on Mechanical Properties of Seawater Sea-Sand Recycled Concrete Under Sulfate Attack

Xiangsheng Tian, Yijie Huang, Jingxue Zhang, and Li Dong

Abstract The preparation of seawater sea-sand recycled concrete (SSRAC) by combining seawater, sea-sand and recycled coarse aggregate is of great significance for the utilization of marine resources and environmental protection in China. The sulfate corrosion test in this paper uses dry wet cycle to simulate the alternating dry wet environment, and compares the ordinary concrete (OC) and freshwater river sand recycled concrete (RAC) to study the mechanical property deterioration characteristics of SSRAC in dry–wet cycle (30d, 60d, 90d, 120d). The results show that with the increase of the dry–wet cycle, the apparent damage of SSRAC gradually extends from the diagonal to the periphery, and finally the cracks spread all over the whole. The mass, strength and strength corrosion resistance coefficient of SSRAC show the same law as OC and RAC, which increase first and then decrease. The resistance of SSRAC to sulfate attack is lower than OC and slightly higher than RAC, and the strength corrosion resistance coefficient is lower than 75% at 120 times of dry–wet cycle.

Keywords Seawater sea-sand recycled concrete · Sulfate attack · Wet-dry cycles

1 Introduction

With the rapid development of the construction industry, a large amount of construction waste occupies land resources and causes environmental pollution. As a new type of green concrete, SSRAC has good working and mechanical properties [1–3]. In addition, seawater and sea-sand in Marine resources is used to replace traditional freshwater and river sand, which is conducive to alleviating the problem of increasingly exhausted sand resources. Recycled coarse aggregate obtained from broken waste concrete is used to replace gravel, which is a good solution to the subsequent treatment of waste concrete.

X. Tian · Y. Huang (✉) · J. Zhang · L. Dong
Shandong University of Science and Technology, Qingdao 266590, Shandong, China
e-mail: 302huangyijie@163.com

© The Author(s) 2023
S. Wang et al. (eds.), *Proceedings of the 2nd International Conference on Innovative Solutions in Hydropower Engineering and Civil Engineering*, Lecture Notes in Civil Engineering 235, https://doi.org/10.1007/978-981-99-1748-8_16

On the other hand, the soil and groundwater in the eastern coastal areas of China contain high concentrations of sulfate, and the service life of the project is reduced due to the superposition of external environment such as long-term sulfate attack and dry–wet cycle. However, so far, the research on SSRAC by domestic and foreign scholars is not perfect, and the research on its durability is also rare. Therefore, it is of great engineering significance to study the deterioration mechanism and rule of seawater sand reclaimed concrete under sulfate attack environment to improve the durability of Marine reclaimed concrete structure and prolong its service life.

2 Experiment Materials and Methods

2.1 Materials and Mix Ratio

Three different types of concrete (OC, RAC, SSRAC) are used in the test. The cement is P·O 42.5 ordinary Portland cement. River sand and sea-sand are selected as the fine aggregate. The coarse aggregate includes gravel and recycled coarse aggregate (RCA). The physical properties of coarse aggregate and fine aggregate are shown in Tables 1 and 2 respectively. The purity of anhydrous sodium sulfate shall be Grade II analytical pure (AR). The reference water cement ratio of each concrete is 0.57, the strength grade is C30, and the mix of concrete materials is shown in Table 3. The water absorption of RCA is much higher than gravel, so an appropriate amount of additional water should be added when placing RAC and SSRAC.

Table 1 Physical properties of coarse aggregates

Material	Size (mm)	Bulk density ($kg \cdot m^{-3}$)	Apparent density ($kg \cdot m^{-3}$)	Water absorption (%)	Crushing index (%)
Gravel	5.0~25	1572	2585	0.8	8.52
RCA	5.0~25	1415	2495	6.8	11.73

Table 2 Physical properties of fine aggregates

Material	Size (mm)	Apparent density ($kg \cdot m^{-3}$)	Clay content (%)	Cl- content (%)	Shell content (%)
Sea-sand	0.15~4.75	2560	0.82	0.096	1.08
River sand	0.15~4.75	2610	2.91	–	–

Table 3 Concrete mixture ratio

Specimen	Material usage (kg·m^{-3})							
	Freshwater	Seawater	Cement	River sand	Sea-sand	Gravel	RCA	Additional water
OC	188	–	329.82	727	–	1137	–	0
RAC	188	–	329.82	727	–	–	1137	22.74
SSRAC	–	188	329.82	–	727	–	1137	22.74

2.2 Test Piece Production and Methods

Dry–wet cycle was used to simulate dry wet alternate environment in sulfate attack test. The performance of SSRAC in sulfate attack (30d, 60d, 90d, 120d) under dry–wet cycle conditions was studied by comparing OC and RAC. According to the Standard for Test Methods of Long term and Durability of Ordinary Concrete (GB/T 50082–2009) and considering the sulfate content in coastal areas of China. Determine to use 5% Na_2SO_4 solution, and the dry–wet cycle system is determined, as shown in Fig. 1.

The size of test piece is 100 mm × 100 mm × 100 mm test piece. Weigh the mass of the test piece every 10 dry–wet cycles, test the compressive strength (f_{cu}) and splitting tensile strength (f_{ts}) of the test piece every 30 dry–wet cycles, And the mechanical property test shall be carried out on the specimens with standard curing at the same age. Specific test data are shown in Tables 4 and 5.

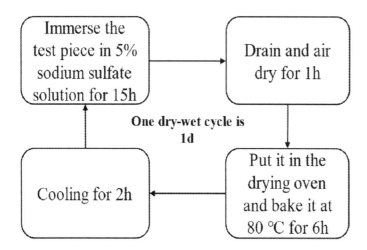

Fig. 1 The dry–wet cycle system

Table 4 Cube compressive strength (f_{cu}) at different erosion ages

Specimen	f_{cu} (MPa)								
	0d	30d		60d		90d		120d	
	W	W	SS	W	SS	W	SS	W	SS
OC	40.41	43.08	43.20	44.95	45.95	46.18	41.80	47.41	30.39
RAC	33.63	34.83	35.26	35.62	36.25	36.31	28.81	36.80	14.85
SSRAC	35.04	35.78	36.79	36.53	37.77	36.88	30.78	37.23	16.22

Table 5 Splitting tensile strength (f_{ts}) at different erosion ages

Specimen	f_{ts} (MPa)								
	0d	30d		60d		90d		120d	
	W	W	SS	W	SS	W	SS	W	SS
OC	3.48	3.67	3.81	3.75	4.11	3.96	3.01	4.14	2.46
RAC	2.82	3.10	2.97	3.24	3.10	3.33	2.07	3.34	1.29
SSRAC	2.97	3.12	3.19	3.27	3.40	3.38	2.46	3.41	1.45

Notes 0d means that the test blocks have been standard cured for 28 days but have not yet eroded. W stands for standard curing. SS stands for sulfate attack. *Definition* SSRAC-SS-60 represents 60 dry–wet cycles of seawater sea-sand recycled concrete under sulfate attack

3   Test Results and Analysis

3.1   *Apparent Damage to Concrete*

The apparent damage characteristics of concrete are shown in Fig. 2. The surface of the concrete specimen is complete and the edges and corners are clear when the dry–wet cycle is not carried out after 28 days of standard curing; After 30 dry–wet cycles, many small holes appear on the surface of the concrete specimen. When 60 dry–wet cycles are performed, some of the cement slurry on the surface of the specimen appeared spalling, and microcracks appeared at the edges and corners and gradually sanding. When the cycles were 90 times, the OC edges and corners were sanding and spalling, the surface cement slurry spalling, and the edges and corners of RAC and SSRAC were spalling. The cracks at the corners continue to extend along the corners and become more obvious. When 120 cycles are performed, OC corners are peeled off, Slender cracks appear along the corners, RAC and SSRAC cracks are all over the whole.

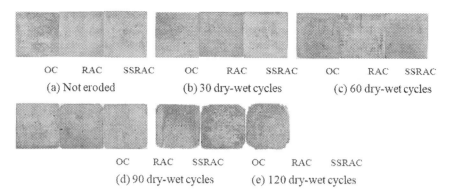

OC   RAC   SSRAC   OC   RAC   SSRAC   OC   RAC   SSRAC

(a) Not eroded   (b) 30 dry-wet cycles   (c) 60 dry-wet cycles

OC   RAC   SSRAC   OC   RAC   SSRAC

(d) 90 dry-wet cycles   (e) 120 dry-wet cycles

Fig. 2 Apparent changes

3.2 Mass

To a certain extent, the mass change can reflect the development law of performance deterioration of sulfate erosion concrete, and the change rule of specimen mass with the number of cycles is shown in Fig. 3. The mass change law of OC, RAC and SSRAC is similar, and it is mainly divided into two stages. In the first stage, the mass of concrete rises. Among them, the mass of OC, RAC and SSRAC reached a maximum value at 60 dry–wet cycles. This is due to the reaction of sulfate ions with cement hydration products to form expansive substances in the early stage of erosion, which improves the compactness of the concrete, and leads to the continuous growth of concrete mass. In the second stage, the mass of the concrete specimen deteriorates. This is due to the gradual decomposition and dissolution of hydration products such as calcium hydroxide and calcium silicate in concrete specimens, which causes surface ablation, and at the same time, as the ettringite expansion specimen appears cracks, with the increase of the number of dry–wet cycles, the cracks at the corners of the concrete continue to develop and appear peeling.

Fig. 3 The change rule of specimen mass with the number of cycles

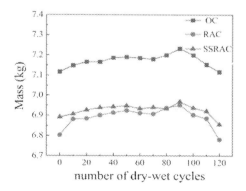

3.3 *Strength*

The change rule of cube compressive strength (f_{cu}) of different types of concrete with the number of cycles are shown in Fig. 4. With the increase of the number of dry–wet cycles, the compressive strength of OC, RAC and SSRAC showed a change law of first increasing and then decreasing, and all reached the maximum at 60 dry–wet cycles. Among them, OC-SS-60 increased by 13.28% compared with OC-W-0, RAC-SS-60 increased by 7.65% compared with RAC-W-0, and SSRAC-SS-60 increased by 7.68% compared with SSRAC-W-0 for the cubic compressive strength. This is due to the fact that sulfate ions react with cement hydration products to form expansive substances in the early stage of erosion, which fills the pores inside the concrete, resulting in the continuous growth of concrete strength [4]. With the increase of the number of dry–wet cycles, the erosion products ettringite and gypsum continue to generate, the pores of the concrete are filled, and its interior continues to expand, accelerating the formation and development of microcracks, resulting in a decrease in the strength of concrete.

The change rule of splitting tensile strength (f_{ts}) of different types of concrete with the number of cycles is shown in Fig. 5. Similar to the change law of compressive strength, the f_{ts} of OC, RAC and SSRAC first increased and then decreased with the number of dry–wet cycles, and all reached a peak at 60 times, among which OC-SS-60 increased by 17.09% compared with OC-W-0, RAC-SS-60 increased by 9.66% compared with RAC-W-0, and SSRAC-SS-60 increased by 14.05% compared with SSRAC-W-0. In the early stage of erosion, the erosion products improve the compactness of the concrete, which in turn improves the tensile strength of the splitting. The expansion stress generated by the erosion products in the later stage of erosion makes the concrete lose its tensile capacity due to the compression of the invaded layer, so that the surface of the layer not subject to sulfate erosion will generate tension, which is superimposed with the tension generated under the splitting load, resulting in a significant decrease in the tensile strength of concrete splitting in the later stage of erosion [5].

Fig. 4 The change rule of f_{cu}

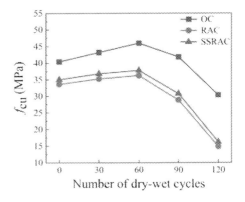

Fig. 5 The change rule of f_{ts}

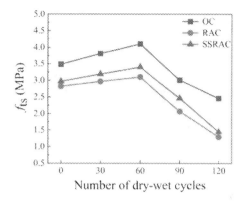

Number of dry-wet cycles

3.4 Strength Corrosion Resistance Coefficient

In order to analyze and evaluate the deterioration of concrete mechanical properties caused by sulfate attack, according to formula (1), defines the cube compressive strength corrosion resistance coefficient (K_c) and split tensile strength corrosion resistance coefficient (K_t).

$$K = \frac{f_0}{f_n} \tag{1}$$

where, K represents the strength corrosion resistance coefficient; f_0 represents the strength of concrete at the same age without sulfate attack (MPa); f_n is the concrete strength after n times of sulfate attack dry–wet cycles (MPa).

As can be seen from Fig. 6, the strength of OC, RAC and SSRAC is lower than that of not attacked concrete after 90 times of dry–wet cycles, and the corrosion resistance coefficient is lower than 75% at 120 cycles. It can be seen that after 120 dry–wet cycles of erosion, the K_c and K_t of OC are higher than those of SSRAC and RAC, and the K_c and K_t of SSRAC are slightly higher than RAC. This is because there are chloride ions in seawater sea-sand, and the transport rate of chloride ions is greater than that of sulfate ions, resulting in chloride ions entering the concrete first, which can react with the AFm to form Friedel salt [6–8], and due to the reduction of the AFm, the formation of ettringite is reduced, thereby alleviating the degree of damage of sulfate to SSRAC.

4 Conclusion

(1) With the increase of the number of dry–wet cycles, the apparent damage gradually expanded from the diagonal to the periphery. The apparent damage of OC

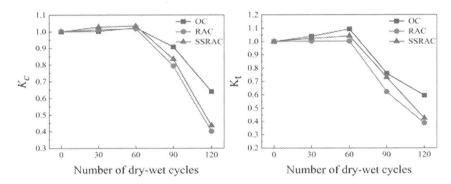

Fig. 6 The change rule of strength corrosion resistance coefficient with the number of cycles

was significantly lower than that of RAC and SSRAC, and the apparent damage of SSRAC was slightly lower than that of RAC.

(2) The mass, compressive and tensile strength of the three types of concrete increased first and then decreased rapidly with the number of dry–wet cycles. Among them, the compressive strength and splitting tensile strength of SSRAC increased by 7.8% and 14.5%, respectively, and then decreased by 57.1% and 57.4%, respectively.

(3) Under the same number of dry–wet cycles, the mass, compressive strength and splitting tensile strength of SSRAC were higher than those of RAC and lower than those of OC. Among them, the strength of OC, RAC and SSRAC at 90 dry–wet cycles is lower than that of uneroded concrete, and the corrosion resistance coefficient of each strength is lower than 75% at 120 times of dry–wet cycle.

(4) Under the action of dry–wet cycle and sulfate attack, OC has higher sulfate attack resistance than SSRAC and RAC. Due to the influence of chloride ions in seawater and sea-sand, the sulfate corrosion resistance of SSRAC is slightly better than that of RAC.

References

1. Xiao JZ, Qiang CB, Nanni A, Zhang KJ (2017) Use of sea-sand and seawater in concrete construction: current status and future opportunities. Constr Build Mater 155:1101–1111
2. Xiao JZ et al (2018) Basic mechanical properties of seawater sea-sand recycled concrete. J Archit Civil Eng 35(02):16–22
3. Huang YJ, Wang TC et al (2022) Mechanical properties of fibre reinforced seawater sea-sand recycled aggregate concrete under axial compression. Constr Build Mater 331:127338
4. Jiang L, Niu DT, Sun YZ (2014) Ultrasonic testing and microscopic analysis on concrete under sulfate attack and cyclic environment. J Central S Univ 21(12):4723–4731
5. Santhanam M, Cohen MD, Olek J (2003) Mechanism of sulfate attack: a fresh look-Part2 Proposed mechanisms. Cement Concr Res 33(3):341–346
6. Martin-Perez B, Zibara H, Hooton RD et al (2000) A study of the effect of chloride binding on service life predictions. Cem Concr Res 30(8):1215–1223

7. De Weerdt K, Orsáková D, Geiker MR (2014) The impact of sulphate and magnesium on chloride binding in Portland cement paste. Cement Concr Res 65:30–40
8. De Maes M, Belie N (2014) Resistance of concrete and mortar against combined attack of chloride and sodium sulphate. Cement Concr Compos 53:59–72

Experimental Investigation of Effectiveness of FRP Composite Repair System on Offshore Pipelines Subjected to Pitting Corrosion Under Axial Compressive Load

Chang Jia Wei, Ong Zhen Liang, and Ehsan Nikbakht Jarghouyeh

Abstract The application of FRP composite is quite well demonstrated in strengthening on RC structures but on strengthening of steel structures is still being investigated by many researchers. The aggressive marine environment can cause corrosion to the offshore pipelines which would affect the structural performance such as pitting corrosion which could be considered as the most destructive corrosion was simulated in this research. This study had attempted to carry out experimental investigation of effectiveness of FRP composite repair system on offshore pipelines subjected to pitting corrosion under axial compressive load. 3 groups of specimens have been categorized such as intact, corroded and repaired. Results such as compressive peak load, failure modes and load–displacement behavior were concerned in this study. The pitting corrosion has been investigated and proved that it has significant effect on the ultimate strength of pipelines. Number of CFRP layer such as 3 layers has been investigated which the results showed the FRP composite repair system capable of strengthening of corroded offshore pipelines. However, more detailed studies are required to be conducted in the future such as consideration of internal pressure in order to expand the understanding of this composite repair system.

Keywords FRP composite · Strengthening · Offshore pipelines · Pitting corrosion · Axial compressive · Structural behavior

1 Introduction

1.1 General

In recent decades, the development and application of fiber reinforced polymer (FRP) composite has been accelerated and widely used due to its advantages such as light weight, high tensile strength, and good durability. Fiber reinforced polymer (FRP)

C. J. Wei · O. Z. Liang · E. N. Jarghouyeh (✉)
Universiti Teknologi Petronas, Bandar Seri Iskandar, Malaysia
e-mail: ehsan.nikbakht@utp.edu.my

© The Author(s) 2023
S. Wang et al. (eds.), *Proceedings of the 2nd International Conference on Innovative Solutions in Hydropower Engineering and Civil Engineering*, Lecture Notes in Civil Engineering 235, https://doi.org/10.1007/978-981-99-1748-8_17

is a composite material that made of a polymer matrix reinforced with fiber for added strength. The fibers are usually glass, carbon, aramid, or basalt and each of the materials has its own applicability. For example, Carbon-Fiber Reinforced Polymer is applied on the strengthening of steel structures due to higher elastic modulus and tensile strength [1]. On the other hand, Glass-Fiber Reinforced Polymer is usually applied on strengthening of Reinforced Concrete (RC) structures due to higher ductility and lower cost [2].

The effectiveness of these composite on repairing of steel structure members such as risers and pipelines has been proven by several researcher. However, there is several problems regarding the behavior and performance of this composite repair systems are required further investigation to avoid any incomprehensible on the capability of this repair system. Furthermore, buckling problem might be happened on the corroded risers or pipelines under axial compressive load which induced from the internal and external pressure [3, 4]. Therefore, restoration of the ultimate strength of the members under axial compressive loads by utilizing FRPs repair system is critical to ensure the superior performance. There are various parameters such as geometry of corrosion and number of layers for this study have been adopted to assess the repair system in term of ultimate strength under axial compressive load.

From the previous research, there are several various parameters have been investigated. For example, the FRPs bonding length, level and type of defection, type of FRPs, type of loads and number of FRPs layers. A gap in knowledge has been identified based on the previous studies which is the number of FRPs layers required to restore the ultimate strength of offshore pipeline subjected to pitting corrosion which this type of corrosion is one of the most destructive corrosion which would be affecting on the effectiveness of this repair system. Therefore, further investigation is required to study the effectiveness of FRP composite repair system on offshore pipelines subjected to pitting corrosion under axial compressive load in order to understand more on the capability of this repair system.

The effectiveness of FRP composite repair system on corroded steel members is the main aim of this study to be investigated. The effect of pitting corrosion on the API 5L Grade B steel pipe under axial compressive load is investigated and the failure modes of the steel pipes by carrying out a series of axial compression tests are explored. A quantitative assessment of the structural behavior of the FRP repaired steel pipes is carried out through axial compression tests.

2 Previous Experimental Studies

2.1 Effect of Pitting Corrosion on the Buckling Strength of Thick-Wall Cylinder Under Axial Compression

Pitting corrosion is one of the most destructive corrosion due to the difficulty to be detected and predicted [5]. The effect of uniformly and randomly distributed pitting defects on the buckling strength of thick-wall cylindrical structures under axial compression have been investigated by using experiment and numerical method. The collapse pressure, F_{co} has been determined and the results within $\pm$ 3.0% error are acceptable. The collapse pressure was decreased in a range of 1.13–17.5%. The collapse of a cylinder is governed by a critical feature such as the location of pits; as the pits approach closer to the cylinder's center, the collapse pressure lowers [6]. Besides, pitting diameter also affects significantly on the collapse pressure. The buckling mode is closely related to the pitting defects in a randomly distributed corroded model. In the case of a few pits, the collapse mode and pressure are not significantly affected. As the number of pits increases, the collapse pressure falls precipitously, and the cylinder collapses in the corroded region. A cylinder with perforated pits collapses in a very different way than one with unperforated pits, and out of plane bending pressure is mostly responsible for this.

2.2 Effect of Pitting Corrosion on the Buckling Strength of Thick-Wall Cylinder Under Axial Compression

As the application of FRPs repair system focuses on integration into offshore risers, the axial load which is exerted on steel pipe due to internal and external pressure needs to be concerned in order to prevent buckling issue. According to ASME repair system, the repair laminate's design thickness must extend beyond the component's damaged region by L_{over} [5].

$$L_{over} = 2.5\sqrt{Dt/2} \qquad (1)$$

where L_{over} is the overlapping length, D is the outside diameter of the pipe (mm), t is the nominal wall thickness (mm) and L_{over} shall be at least 0.05 m. Furthermore, the total axial length of the repair, L, can be calculated as shown in Eq. (2).

$$L = 2L_{over} + L_{defect} + 2L_{tapper} \qquad (2)$$

where L_{defect} is the defect length, L_{taper} is both end taper length.

2.3 Relationship Between Axial Load and Number of FRP Layers

The relationship of axial and number of FRP layers have been investigated. The axial strength for retrofitted specimens has been improved ranges from 26 to 84% [7]. The retrofitted specimens showed greater strength than the control specimens, as well as a greater early slope. This indicates that the CFRP layers have improved their structural behavior. Furthermore, it had proved that the number of CFRP layers has a direct relationship with the increase in axial load.

3 Methodology

In this research, few parameters such as geometry of corrosion and number of CFRP layers are studied in order to find out the effect of pitting corrosion on offshore steel pipelines and the effectiveness of FRP composite repair system. The methodology of the research is explained as follow.

3.1 Specimens

API 5L grade B steel pipe with nominal yield strength of 240 MPa and tensile strength of 415 MPa according to ASTM-a106 Standard was used in this study (Outer Diameter, OD = 42.2 mm and Thickness, t = 3.56 mm). All the specimens had a length of 900 mm with a slenderness ratio of 65.6. Pitting corrosion was simulated by milling out ⌀20mm with at least 2.5 mm thickness at the middle part of the pipe. Pitting corrosion was selected as the corrosion type for this study because it is one of the most destructive corrosion due to the difficulty to be detected, predicted, and designed against. Total number of 6 pits for each corroded sample which had considered perforated and un-perforated pits and the schematic geometry of the samples with pitting corrosion were showed in Fig. 1.

A total number of 4 specimens were tested in this study. Specimens were prepared and categorized into 3 groups such as intact, corroded and repaired. Intact specimens are bare steel pipe without any corrosion and would be as control specimen in this

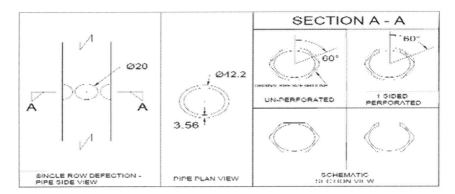

Fig. 1 A figure caption is always placed below the illustration. Short captions are centered, while long ones are justified. The macro button chooses the correct format automatically

Table 1 Summary of the specimens' details

Serial No.	Label	Specimen groups	Number of CFRP layers (nos)	Geometry of pits
1	I1	Intact	–	–
2	C1	Corroded	–	1 row unperforated
3	C2	Corroded	–	1 row with 1-sided perforated
4	R1L3	Repaired	3	1 row with 1-sided perforated

study. The specimens which the pitting corrosion had been simulated at the middle part of the pipe can be categorized as corroded specimens. Last but not least, the corroded specimens which had been repaired by using FRP composite sheet were categorized as repaired specimens. The minimum bonding length was calculated as 175 mm by using Eqs. (1) and (2). The summary of the specimens' details has shown in Table 1 where I = Intact specimen, C = Corroded Specimen, R = Repaired Specimen and L = number of CFRP layers.

3.2 Repair Process

Steel surface preparation was the first and most important step in the repair procedure. Chemical and mechanical adhesion form a bond between the FRP layer and the steel surface. Chemical adhesion was derived from the epoxy system used to wrap the fibers, while mechanical adhesion was ensured by surface preparation. This had been done by using sandpaper and then later cleaned with acetone. The FRP layers were wrapped in the second stage. A resin-impregnated glass fiber reinforced polymer (GFRP) was wrapped around the prepared steel surface in order to prevent the galvanic corrosion occur between carbon fiber and steel surface. Then the carbon fiber reinforced polymer (CFRP) was wrapped afterwards as shown in Fig. 2. Stricture bands were placed on the specimens during the curing process and removed after the soft curing period. For the curing procedure to complete, the repaired specimens are kept idle for a day.

Fig. 2 Details of wrapping of FPR layers

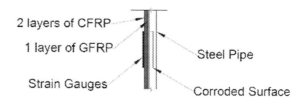

2 layers of CFRP

1 layer of GFRP

Strain Gauges

Steel Pipe

Corroded Surface

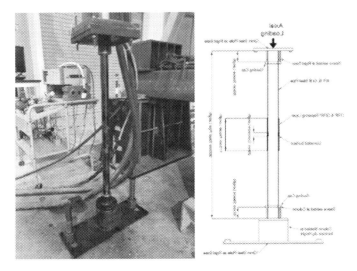

Fig. 3 Testing setup

3.3 Testing Setup

The testing setup was displayed in Fig. 3. Steel jigs had been fabricated to hold the specimens in order to prevent any accident happen during testing such as the specimens slide while they are being tested. The Amsler loading machine (500 kN) used as the testing apparatus for the study. The specimen is then inserted into the testing apparatus, where the axial force is first introduced, then raised gradually until the specimens fail altogether.

4   Results and Discussion

A series of axial compressive testing had been carried out in order to investigate the effectiveness of FRP composite repair system on offshore pipelines subjected to pitting corrosion. Results such as compressive peak load, failure modes and load–displacement behavior were concerned in this study.

4.1 Compressive Peak Load

Based on the results obtained, it can be observed that the peak load for the intact specimens is 106.71 kN. The compressive peak load of 1-sided perforated is 85.06 kN.

Table 2 Summary of the testing results

Specimens label	Specimens group	Compressive peak load (kN)
I1	Intact	106.71
C1	Corroded	98.65
C2	Corroded	85.06
R1L3	Repaired	118.79

It can be observed that the peak load has been reduced 22.16% when the sample is 1-sided perforated compared to intact sample. Furthermore, the compressive peak load for 1 row un-perforated is 98.65 kN which has showed a reduction of 7.55%. This showed that the un-perforated defects with 2.5 mm thickness and ø20 mm milled out have less significant effect on the compressive peak load of steel pipe.

According to ASME repair standard, the calculated minimum repair length is 175 mm for this study. The repaired specimens are wrapped 1 layer of GFRP at first to avoid galvanic corrosion between steel and carbon. Based on the results obtained, the repaired specimen, R1L3 with 3 layers of CFRP have restored its original compressive peak load which is 106.71 kN and even improved 8.71% and 14.19% to 118.79 kN. The summary of the testing results as shown in Table 2.

4.2 Failure Modes

Several failure modes are observed which included both end deformation, bending, global buckling and FRP cracking and crushing. Figure 4a shows both end deformation which the top and bottom of pipe has been deformed from the original outer diameter, 42.2 mm to the range of 42.5–42.9 mm. All specimens are exhibited this failure mode. Bending failure mode as shown in Fig. 4b are happened on specimens I1 and C1. Specimens C2 and R1L3 exhibited global buckling as shown in Fig. 4c. Last but not least, FRP cracking and crushing failure has been exhibited on the specimens which were repaired with 175 mm bonding length of FRP such as specimens R1L3 as shown in Fig. 4d.

4.3 Load–Displacement Behavior

The comparison of load–displacement curves has been plotted as presented Fig. 5. The displacement is measured in axial direction from top to bottom of the specimen. Based on the curve, the overall reaction of the load–displacement curve was observed that it was approached typical buckling behavior. The initial slope was the elastic phase and reached a peak ultimate buckling load then followed by load drop which the structural collapse. Besides, the curve for C2 which was the 1-sided perforated

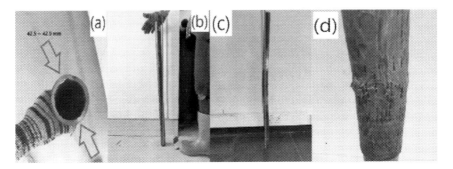

Fig. 4 **a** Both end deformation, **b** bending failure, **c** Global buckling and **d** FRP cracking failure

specimen showed a smooth peak while the rest were showed sharp peak. The failure was associated to buckling at the middle part and local deformations at top and bottom of the specimens. The ultimate load in all specimens was connected to a combination of overall buckling and member yielding. More details about the behavior of the specimens were revealed by the load displacement behavior than by the ultimate strength.

When repaired specimens reached the compressive peak load, crack sounds were heard and then the load started to drop. These crack sounds indicated that the various failures in the repair process such as FRP layers crushing and cracking. The comparison plots also demonstrated the structural recovery of the repaired specimens to their original condition. The load displacement behavior of each repaired specimen proved that, for the most part, the corroded specimens could be brought into the levels of the intact specimens. Buckling failure, which caused some areas of the structural collapse (post peak) curve to fall below the intact levels, was also seen in the restored specimens. A sudden or drastic drop in the slope of the structural collapse curve

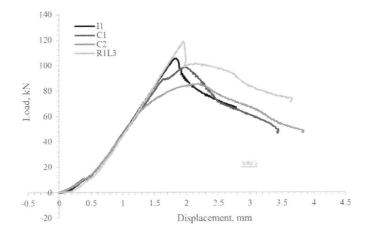

Fig. 5 Load–Displacement comparison plot

can be used to determine whether buckling failure has occurred in a specimen. The buckling failure in repaired specimens occurred at a larger displacement than it did in the similar corroded specimens, but once local buckling started, the repair could not keep up with the behavior of intact specimens.

5 Conclusion

The effectiveness of FRP composite repair system on offshore pipelines subjected to pitting corrosion was experimentally investigated by carrying out a series of axial compression tests. Previous studies had not looked into the application of FRP wraps for the repair of offshore pipelines subjected to pitting corrosion which is one of the most destructive corrosion type. However, limited technology was being used in this study. Future studies should focus on enhancing our knowledge of how these repairs work in long term in order to develop a suitable design technique for FRP-based structural retrofitting of offshore steel structures. This study can be concluded in the following:

1. Several failure modes have been observed such as top and bottom deformation, bending failure, global buckling as well as FRP cracking and crushing.
2. The corroded specimen with geometry of 1-sided perforated pits has the greatest effect on the ultimate strength of the offshore pipelines which has been decreased 22.16%. The specimens with un-perforated pits have showed less significant reduction in ultimate strength which was only 7.55%.
3. The repaired specimens with 3 layers of CFRP have been restored to intact condition. A maximum increase of 14.19% in ultimate strength was observed in these specimens. However, the repair could not keep up with the intact behavior once buckling failure had started.

Acknowledgements The author would like to acknowledge Yayasan Universiti Teknologi PETRONAS (YUTP) for supporting this study under the Fundamental Research Grant (FRG) project no 015LC0-177.

References

1. Toutanji M, Dempsey S (2001) Stress modelling of pipelines strengthened with advanced composite materials. J Thin-Walled Struct 39:153–165
2. Attari N, Amziane S, Chemrouk M (2012) Flexural strengthening of concrete beams using CFRP, GFRP and hybrid FRP sheets. Constr Build Mater 37:746–757
3. Hobbs RE (1981) Pipeline buckling caused by axial loads. J Constr Steel Res 1(2):2–10
4. Pipeline Operators Forum (2016) Specifications and requirements for in-line inspection of pipelines. www.pipelineoperators.org

5. ASME PCC-2 (2018) Repair of pressure equipment and piping. The American Society of Mechanical Engineers. Revision of ASME PCC-2-2015
6. Wang HK, Yang Y, Xu JX, Han MX (2019) Effect of pitting defects on the buckling strength of thick-wall cylinder under axial compression. Constr Build Mater 224:226–241
7. Gao XY, Balendra T, Koh CG (2013) Buckling strength of slender circular tubular steel braces strengthened by CFRP. Eng Struct 46:547–556

Research on Safety Risks and Countermeasures of Super High-Rise Steel Structure

Cuihua Ji, Yong Yin, Jianyu Yin, and Hong Zeng

Abstract Judging from the development status of most super-high-rise building industries in China, "outer frame + core tube" is basically adopted. The appearance of this core tube structure design concept puts forward higher requirements for the construction safety of the main structure. Combined with practical engineering projects, we analyze the safety risks and countermeasures of the outer frame column of the super high-rise steel structure, the safety risks and countermeasures of the cantilever steel beam, the safety risks and countermeasures of the waist truss, the safety risks and countermeasures of the waist truss, the safety risks and counter-measures of the outrigger truss, the safety risks and countermeasures of the tower crown structure are discussed one by one. It ensures the organic combination of engineering safety measures and engineering construction technology. The proposed safety measures are highly targeted and feasible, ensuring the safe construction of the project. The comprehensive application of this project confirms the feasibility of relevant countermeasures, so as to provide reference and help for subsequent similar projects.

Keywords Super high-rise · Steel structure · Safety · Risk · Measures

1 Introduction

The steel structure of this project is mainly distributed in the second phase tower, the surrounding podium and the main structure of the third phase tower. Total steel consumption is about 73340 t. The second phase tower has ninety-nine floors on the ground, with a total height of 468 m, and the height of the main structure (i.e. frame

C. Ji · Y. Yin (✉) · H. Zeng
Prefabricated Construction Applied Technology Promotion Center of Chongqing Higher
Vocational Colleges, Chongqing Jianzhu College, 857 Lihua Avenue, Nan'an District, Chongqing,
China
e-mail: 103237256@qq.com

J. Yin
University of Macau, Avenida da Universidade Taipa, Macau, China

© The Author(s) 2023 219
S. Wang et al. (eds.), *Proceedings of the 2nd International Conference on Innovative Solutions in Hydropower Engineering and Civil Engineering*, Lecture Notes in Civil Engineering 235, https://doi.org/10.1007/978-981-99-1748-8_18

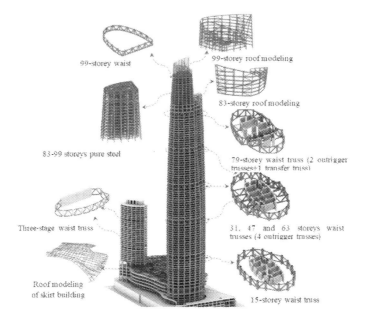

Fig. 1 Schematic diagram of steel structure engineering

structure) on the ground is 440 m, which is an outer frame with waist truss, core tube (steel support) and outrigger truss structure Among them, the number of waist trusses is 6, which are located on floors 15 to 17, 31 to 33, 47 to 49, 63 to 65, 79 to 81 and 98 to 99, respectively. The steel structures around the skirt building are mainly rigid steel reinforced concrete beams and columns, steel beams, steel columns and trusses structure. The engineering schematic diagram is shown in Fig. 1.

2   Analysis of Key and Difficult Points in Overall Safety Protection of the Project

The main safety injury forms in the steel structure construction of this project [1] are falling from a high altitude, object impact, electrical injury, fire, mechanical injury and poisoning. According to the statistical data of safety accidents in the construction industry, the accidents of falling from a high altitude and objects hitting account for 90% of the total safety accidents, which shows that the risk level of falling from a high altitude and objects hitting is extremely high.

(1)   Safety risks of falling and falling objects

    a.   When working at heights, you must wear your seat belt and safety helmet as required.

    b.   When working at heights, tools should be loaded into tool bags, which can be taken with you.

    c. When working at heights, the tool handle must wear a rope and put it on the safety belt or wrist.

    d. When working at heights, the removed connecting plates and other wastes shall be cleaned up to the ground in time, and shall not be thrown at will.

(2) Safety risks of using electrical equipment

    a. Workers use electricity illegally.

    b. Illegal use of electric welding machines or other electrical equipment.

    c. When the electric welding machine is used, it is required that the welding handle wire and the ground wire should be in place, and the welding handle wire [2] should not exceed 30 m. The length of the primary connection between the welding machine and the welding machine is not more than 5 m. If the welding wire is broken, it must be wrapped in insulating tape for three times.

(3) Safety risks of hot work

    a. Before welding and cutting operations, clean up flammable and explosive materials around and below the operations.

    b. For welding operation at high places, special personnel should be assigned to monitor the welding operation below, and fire connection or isolation measures should be taken.

    c. It is forbidden to smoke in the operation area of the construction site, especially around inflammable and explosive materials.

The hot work [3] can only be started after the hot work application is allowed. The caretaker and fire control measures should be implemented, the surrounding conditions of hot work should be checked, and inflammable and explosive articles should be cleaned up. Smoking is strictly prohibited when entering the operation area of the construction site, especially around inflammable and explosive materials. Welding and cutting operations are not allowed above flammable and explosive materials such as paint and thinner. When welding at high places, special personnel should be assigned to supervise the welding below, and fire connection or isolation measures should be taken.

(4) Safety risks of lifting operations

    a. Workers operate illegally, command illegally or violate operating rules.

    b. Safety risks of hoisting dangerous areas. In hoisting dangerous areas, there must be a special person to supervise them, and non-construction personnel are not allowed to enter the dangerous areas. Dangerous areas for hoisting should be designated as warning areas and protected by warning ropes. No one can stand under the lifting object.

    c. Safety risks of operating cranes in violation of regulations. About lifting heavy objects, the hook should be 90° with the ground, and it is forbidden to hang obliquely; Horizontal lifting is strictly prohibited.

    d. Safety risks of using hoisting equipment. At the crane station, ensure that the foundation has sufficient bearing capacity; The rotating part of the crane should have a space distance of not less than 1 m from the surrounding fixtures.

(5) Poisoning safety risks

    a. When welding the steel reinforced column in the outer frame column, set a fume extractor on the top of the column or open a fume window on the outer frame column to remove toxic gas.

    b. Welders must wear masks to enter the column for welding construction, and assign special personnel to take care of them. In case of poisoning, stop the construction immediately.

    c. When the welder is working in the column, if there is dizziness, vomiting and other phenomena, immediately stop the operation and have a physical examination.

3 Safety Risks of Columns and Countermeasures

Security risks:

Large-scale component hoisting, falling prevention from high altitude, welding poisoning and welding fire prevention.

Response measures:

a. The outer frame column component is large in volume, with the maximum weight of 53 t. Before hoisting, the hoisting lug plate must be tested for damage, and if it is damaged, the hoisting lug plate must be welded again.

b. When hoisting the hook [4], check whether the wire rope of the hook is worn or damaged, and check and calculate whether it meets the hoisting requirements according to the hoisting manual. If it does not meet the requirements, the wire rope must be replaced.

c. Before the outer frame column is hoisted, the steel column must be equipped with an installation operation platform, a ladder and safety protection measures, and the anti-falling rope must be welded, and the ladder must be tied up and dirt cleaned.

d. In the process of hoisting, the lifting of components must be smooth, and the components must not be dragged on the ground, and stay for one minute immediately after they leave the ground; When turning, you need to have a certain height. Hooking, rotating and moving are carried out alternately and slowly, and when in place, they will fall slowly to prevent the components from swinging and oscillating greatly.

e. When the outer frame is installed at high altitude, the temporary bolts must be fastened and installed, and the double-splint and self-balancing double-splint butt joint technology should be used to ensure the stable connection of steel columns. Operators must hang safety belts on the operating platform to operate, temporarily fix the components with connecting plates, and pull three cable wind ropes. Two cable wind ropes are fixed on the ear plates of the two pillars next to the next floor, and the other cable wind rope is fixed on the embedded parts of the next core tube.

f. When welding in the outer frame column, set a fume hood on the top of the column or open a fume window on the outer frame column to remove toxic gas and prevent poisoning incidents.

4 Safety Risk and Countermeasures of Cantilever Steel Beam

Individual H-beam [5] parts of cantilever steel beams are manufactured in the factory and assembled into units on site, and the units are installed on site. During installation, a hanging basket shall be installed at the node of cantilever beam. On the ring beam (main beam) hanging between the steel pipe columns of the outer cylinder, an inverted chain is set at the cantilever section of the cantilever beam to adjust its levelness. Its form is shown in Fig. 2.

Safety risks: falling prevention from high altitude and welding fire prevention.

Response measures:

a. Cantilever steel beams are hoisted with snap rings, and the snap rings must be locked when tying hooks.
b. During hoisting and installation, the installer must hang up the seat belt according to the regulations to prevent falling from the building.
c. The bottom and periphery of the hanging basket are covered with dense mesh steel wire mesh, and asbestos cloth is laid at the bottom to seal.

Fig. 2 Cantilever steel beam

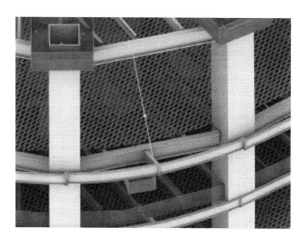

5   Safety Risk of Waist Truss and Countermeasures

Waist trusses are located in towers 15F-17F, 31F -33F, 47F -49F, 63F-65F, 79F-81F and 98F-99F. Except for floors 98F-99F, other waist trusses are of two structural floors, which are connected with outer frame columns to form an outer frame. Typical waist truss installation is carried out by sections according to parts and K-shaped units. The schematic diagram of waist truss is shown in Fig. 3.

Safety risks: hoisting of large components, falling prevention at high altitude, welding at high altitude.

Response measures:

a. Some nodes of the waist truss are assembled for K-byte point hoisting, which is large in size, high in installation precision and difficult to tie hooks. Before hoisting, the hoisting lug plate must be tested for damage, and if it is damaged, the hoisting lug plate must be welded again [6].

b. When hoisting the hook, check whether the wire rope of the hook is worn or damaged, and check and calculate whether it meets the hoisting requirements according to the hoisting manual. If it does not meet the requirements, the wire rope must be replaced.

c. In the process of hoisting, tower crane commanders must operate in accordance with relevant hoisting regulations.

d. When the waist truss diagonal web bar is installed, the upper operating surface is below the operating platform, and the operator must hang the safety belt. Use scaffolding operation platform, and use three wooden springboard as walking

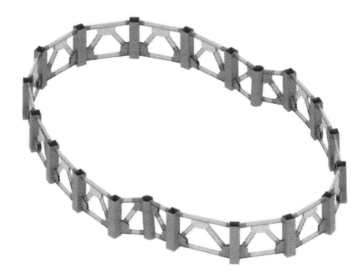

Fig. 3 Schematic diagram of waist truss

road. During welding construction, lay fire-proof asbestos cloth on the wooden springboard and cover it with wind-proof cloth.

e. During the installation of diagonal web members of waist truss, stirrups must be set up for support due to the long members and heavy weight. Stirrups are set up on the upper surface of the outer frame steel beam on the floor level, below the center point of the diagonal web member. When the diagonal web member is welded into a stable structure, the stirrup is cut off and transported to the ground for the installation of the next waist truss.

6 Safety Risks and Countermeasures of Outrigger Truss

The outrigger truss connects the outer tube column with the core tube, and the chord runs through the web wall of the core tube. The cross sections of the upper and lower chords and web members are H-shaped, box-shaped and steel plate composite cross sections. There are four types of outrigger trusses: OT1, OT2, OT3 and OT4. The outrigger trusses are distributed at 31F -33F, 47F -49F, 63F-65F, 79F-81F. The installation diagram of outrigger truss is shown in Fig. 4.

Security risks: hoisting of large components, assembling of jig and assembling of parts.

Response measures:

When the outrigger truss is a loose part, there are many hoisting times. Before hoisting, it is necessary to detect whether the hoisting lug plate is damaged. If it is damaged, it is necessary to weld the hoisting lug plate again.

a. When hoisting the hook, check whether the wire rope of the hook is worn or damaged, and check and calculate whether it meets the hoisting requirements

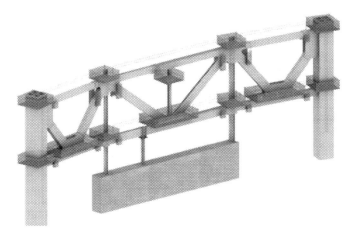

Fig. 4 Installation diagram of outrigger truss

according to the hoisting manual. If it does not meet the requirements, the wire rope must be replaced.

b. In the process of hoisting, tower crane commanders must operate in accordance with relevant hoisting regulations.

c. Because the outrigger truss is installed in pieces, the lower suspension rod of the truss must be supported by a jig.

7 Safety Risk of Tower Crown Structure and Countermeasures

The top structure of the tower is composed of steel frame and peripheral structure. The outer structure is parabola-shaped, which is composed of thirteen vertical trusses and connected steel beams between trusses. The total steel consumption of the top structure is about 300 t, and the main sections of members are H-shaped steel and steel pipes, with the largest sections of H800 × 400 × 25 × 40 and CHS300 × 15. The floor top structure has a bottom elevation of 439.750 m, a top elevation of 468.000 m and a total height of 28.250 m. The schematic diagram of steel structure at the top of the tower is shown in Fig. 5.

Safety risks: falling prevention from high altitude and welding fire prevention.

Fig. 5 Schematic diagram of steel structure at the top of the tower

Response measures:

a. The horizontal network of the floor [7]. During the construction of the steel frame inside the roof steel structure, horizontal protection is carried out by laying horizontal nets layer by layer. The steel frame inside the roof is composed of four interlayers. During the construction of each layer of steel frame, the steel beams of each part are basically installed in place and then closed with horizontal nets. Then, the installation progress is followed gradually until the whole floor is closed.

b. Outside the edge of the net. When the roof structure is hoisted, an external hook net is set at the periphery of the bottom of the roof structure. A total of two floors are set up at the outside of the safety net for 3 m, a horizontal net is laid at the bottom of the outside net, a dense mesh net is laid on the horizontal net, and a steel wire net is laid under the horizontal net to form three layers of protection. The external net support frame is made of φ 48 × 3.5 scaffold steel pipe, which is fixed on the peripheral steel column by welding and wire rope.

c. Operating platform. The peripheral structure is installed in two layers, and the channel operation platform is set up at the top of the first layer structure. The operation platform is made of steel pipes, safety nets are hung on both sides of the platform, and scaffolding boards are laid at the bottom of the platform. The platform is connected with the steel beam through a snap ring welded on the surface of the steel beam, which is mainly used for welding the upper and lower peripheral structures, and also serves as a high-altitude passage.

8 Conclusions

We studied the overall safety risks and main structural safety risks of super-high-rise steel structure construction, put forward specific safety measures, and applied them to engineering practice, and achieved good results.

Acknowledgements This paper supported by the Science and Technology Research Program of Chongqing Municipal Education Commission (Grant No.KJQN202204303).

References

1. Mo SJ (2022) Research on construction safety risk evaluation of comprehensive transportation hub engineering based on Bayesian network. Hebei University of Architecture
2. Li ZF (2019) Construction technology of starting span lifting of steel net frame. Jiangsu Build Mater S2:66–69
3. Tan X (2016) The high-pier rigid frame bridges safety production management research. Chongqing Jiaotong University
4. Guo X (2022) Preliminary study on safety supervision measures for installation and disassembly of super high-rise tower cranes. Constr Saf 37(10):35–39

5. Song N (2022) Research on construction of new flower basket cantilever scaffold in high-rise building. Dwelling 10:91–93
6. Hua JM (2019) Construction of high-rise building. Chongqing University Press
7. Lin HZ (2018) Glass curtain wall in the protection and updating and transformation of the historic buildings. South China University of Technology

Research on Axial Force Coherence of Steel Support Based on Active Control

Jianchao Sheng[ID]**, Jiuchun Sun**[ID]**, Donglai Jiang**[ID]**, Yuanjie Xiao**[ID]**,
Rundong Lv**[ID]**, and Zhe Wang**[ID]

Abstract Using servo steel support system for active control of deformation during foundation fit excavation has high superiority. To investigate the coherence of the servo support axial force, PLAXIS 3D is used to carry out numerical analysis on the coherence of the axial force applied by the construction and compare with the field test results; Field tests were conducted on the diaphragm wall joint deformation during the axial force application based on the principle of the generation of axial force coherence. The results show that the farther away from the active axial force, from which suffered get the smaller influence, and the size of the applied axial force's effect on the support in other directions in the order of horizontal, vertical, and oblique. Moreover, the higher the application position of the active axial force of the servo support, the greater the lateral axial force loss rate generated by other supports, while the opposite in vertical axial force loss rate is true. The maximum axial force loss rate is 19%. The deformation of the diaphragm wall joint in the servo steel support zone is more significant than that in the pre-stressed steel support zone. The deformation of the joint will, in turn, affect the axial force.

Keywords Servo steel support · Axial force loss · Axial force coherence · Active control · Numerical analysis · Safety of foundation pit

J. Sheng · J. Sun
Teng Da Construction Group Co., Ltd., Shanghai 610014, China

J. Sheng · D. Jiang · Z. Wang
Institute of Geotechnical Engineering, Zhejiang University of Technology, Hangzhou 310014, Zhejiang, China

Y. Xiao (✉)
National Engineering Research Center for High-Speed Railway Construction Technology, Central South University, Changsha 410004, China
e-mail: sjcpromising@foxmail.com

R. Lv
Hangzhou Geotechnical Engineering & Surveying Research Institute of China Co., Ltd., Hangzhou 410004, China

© The Author(s) 2023
S. Wang et al. (eds.), *Proceedings of the 2nd International Conference on Innovative Solutions in Hydropower Engineering and Civil Engineering*, Lecture Notes in Civil Engineering 235, https://doi.org/10.1007/978-981-99-1748-8_19

1 Instruction

Urban economic growth and population expansion have led to the vigorous development of underground transportation systems. In underground space development, safety issues occupy a more critical position. The active deformation control of the foundation pit retaining structure has become the primary goal of construction safety control [1–4]. The steel support system is widely used in deep foundation pit support systems because of its high efficiency and pre-axial force [5, 6]. Many studies have given various setting methods to determine the axial force of the support. Still, the conventional steel support will cause prestress loss due to on-site construction, environmental changes, and other factors, significantly reducing the ability to control deformation [7–9]. The servo system is applied to the steel support in the engineering field to meet the increasingly stringent safety control requirements. Through real-time monitoring and automatic compensation of axial force, remarkable results have been achieved in the lateral deformation control of retaining structures in many engineering cases [10, 11]. However, the retaining structure is often continuous, and the supporting axial force acting on the retaining structure will be transmitted through the excellent integrity of the structure itself or the rigid connection between the structures, thus affecting each other, which is the axial force coherence. If the coherence between the axial forces is ignored, it may lead to conflict disorder in the automatic regulating system and cause safety accidents. The existing research only considers the change of axial force of uniaxial steel support and does not consider the influence of transverse axial force coherence [12, 13]. In the study of transverse correlation, the fitting formula of axial force coherence under multi-point synchronous loading is given by numerical method. Still, it is not compared with the field measurement, and the deformation at the connection of retaining structure is not considered [14]. When the transverse coherence occurs, the deformation will also happen at the joint of the retaining structure. The deformation problem at the retaining structure joint in the axial force transmission cannot be ignored.

In order to achieve the refinement of servo axial force control and reduce a series of problems caused by axial force loss, it is necessary to conduct in-depth research on the law of axial force coherence. Based on the foundation pit project of the No. 2 entrance and exit of Pudong South Road Station of Shanghai Metro, this paper uses PLAXIS 3D software to analyze the axial force coherence numerically. Compared with the field test results, the axial force coherence and the deformation of the joint of the retaining structure when the servo steel support active control system is applied are discussed, which can be used for reference in similar projects.

2 Engineering Background and Model

The foundation pit of Shanghai Pudong South Road Station No. 2 entrance is located in the core area of Lujiazui, adjacent to essential facilities and structures. Micro-deformation active control is needed to ensure the safety of foundation pit and surrounding environment. The outsourcing size of the foundation pit is 60.4 m × 36.7 m, and the maximum excavation depth is 18.3 m. There are five supports in the foundation pit. The first is concrete support, and the other four are servo steel support. The thickness of retaining structure is 0.8 m. The foundation reinforcement method is high-pressure jet grouting pile sticking and strip reinforcement, and the facade is shown in Fig. 1. The servo steel support elevation layout is shown in Fig. 2. The parameters used in the calculation are given in Table 1. In the hardening model of soft soil, the dilation angle ψ is 0, the unloading–reloading Poisson's ratio v_{ur} is 0.2, the power exponent m related to the modulus stress level is 0.8, the reference stress P^{ref} is 100 kPa, the damage ratio R_f is 0.9, and R_{inter} is 0.7.

To eliminate the external influence factors, the load of the existing station building and the south district is not considered in the numerical analysis. The simulated axial force is the actual axial force applied by the field servo steel support. The 3D model is shown in Fig. 3. The excavation simulation condition is consistent with the field test condition, which is given in Table 2.

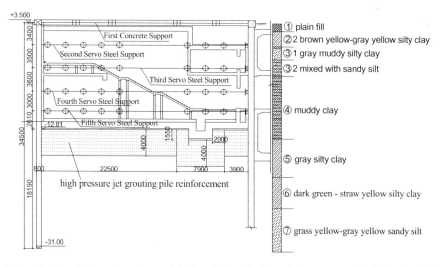

Fig. 1 Sectional schematic diagram of the foundation pit of the No. 2 entrance and exit of Shanghai Metro Pudong South Road Station

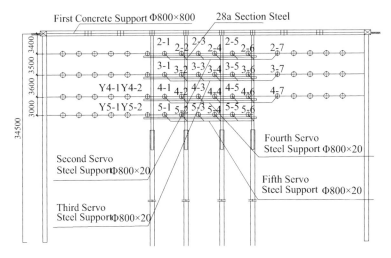

Fig. 2 Diagram of foundation pit support layout at No. 2 entrance and exit of Pudong South Road Station of Shanghai Metro

Table 1 Parameters of soil model

Geotechnical category	$\gamma/(kN\ m^{-3})$	C_{ref}/kPa	$\varphi/°$	K_0	E_s/MPa	E_{50}/MPa	E_{ur}/MPa
② brown-gray silty clay	18.2	20	17.5	0.47	9.7	9.7	48.5
③₁ gray mucky silty clay	17.5	12	19.5	0.46	9.2	9.2	46.5
③₂ mixed with sandy silt	18.6	6	29	0.45	7.5	7.5	37.5
④ muddy clay	16.7	14	12	0.58	21.6	21.6	107.8
⑤ gray silty clay	18	16	17	0.54	5.8	5.8	28.8
⑥ silty clay	19.5	46	16	0.46	8.2	8.2	41.1
⑦ sandy silt	18.5	2	31.5	0.37	13.7	13.7	68.6
⑧ silt	18.8	1	33	0.34	25.3	25.3	126.6

3 Result and Analysis

In the process of foundation pit excavation, according to the requirements of deformation control, single or multi-channel servo steel support is used to apply axial force to achieve ideal results. Since each steel support acts on the same envelope structure, the vertical coherence is bound to be generated when the axial force is applied to the single or multi-channel servo steel support, and the axial force of the steel support acting on the envelope structure will also transfer the lateral coherence through the adjacent underground continuous wall. This section will explore the

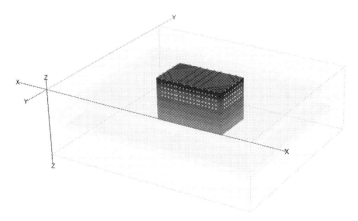

Fig. 3 Schematic diagram of numerical 3D model

Table 2 Cases adopted for numerical simulation and field test

Working condition	Content
Case 1	Influence on other support axial force after excavating to the second steel support and erecting servo steel support 3–3, 3–4
Case 2	Influence on other support axial force after excavating to the third steel support and erecting servo steel support 4–3, 4–4
Case 3	Influence on other support axial force after excavating to the fourth steel support and erecting servo steel support 5–3, 5–4

Note Supports 3–1 and 3–2 were erected in case 1; Supports 4–1 and 4–2 were erected in case 2; Supports 5–1 and 5–2 were not erected in case 3

influence of axial force application on the existing axial force according to various working conditions in numerical simulation and field test, and analyze the influence of prestressed steel support area and servo steel support area on the joint of retaining structure under different axial force application conditions of the same steel support.

3.1 Effect of Steel Support Axial Force Coherence

Figures 4 and 5 are the schematic diagrams of the influence of numerical simulation and field test on other steel supports under condition 1, respectively. Figure 6 is the comparison diagram of numerical simulation and field test under condition 1. Figures 7 and 8 are the schematic diagrams of the influence of numerical simulation and field test on other steel supports under condition 2, respectively. Figure 9 is the comparison diagram of numerical simulation and field test under condition 2. Figures 10 and 11 are the schematic diagrams of the influence of numerical simulation and field test on other steel supports under condition 3, respectively. Figure 12 is the comparison diagram of numerical simulation and field test under condition 3. From Figs. 4, 5, 6, 7, 8, 9, 10, 11 and 12, it can be seen that both the numerical simulation and the measured results show that the axial force of the second servo steel support (i.e.3–3, 3–4 support) has a significant influence on the axial force of the same support, and the vertical influence is small. The farther the support is apart, the smaller the influence on the axial force is. The field measurement results are consistent with the numerical simulation results. When the axial force of the third servo steel support (i.e.4–3, 4–4 support) and the fourth servo steel support (i.e.5–3, 5–4 support) is applied, it dramatically influences the vertical support axial force, and the horizontal support axial force has little influence. The farther the support is, the smaller the influence on the axial force is. The field measurement results are also relatively consistent with the numerical simulation results.

According to Figs. 6, 9, and 12, the axial force loss rate of servo steel support is further analyzed. The axial force loss rate obtained by the actual measurement is mostly higher than that of the numerical simulation. The axial force loss rate of horizontal support is the largest, followed by vertical support. The axial force loss rate of oblique support is the smallest. According to the analysis, the applied support axial force has the most significant influence on the horizontal support axial force, followed by the vertical steel support. The oblique steel support has the most minor influence.

The higher the application position of the active axial force of the servo support, the greater the lateral axial force loss rate generated by other supports, while the opposite in vertical axial force loss rate is true. Among them, the maximum lateral axial force loss rate generated by the second servo steel support is 19%, the third is 18%, and the fourth is 12%. This phenomenon is similar to the relationship between

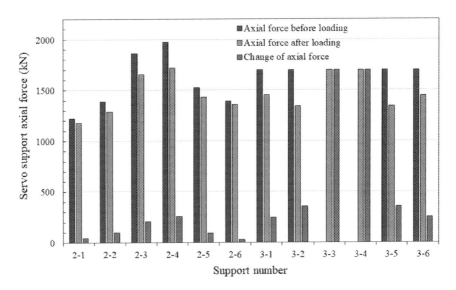

Fig. 4 Diagram of axial force coherence under numerical simulation case 1

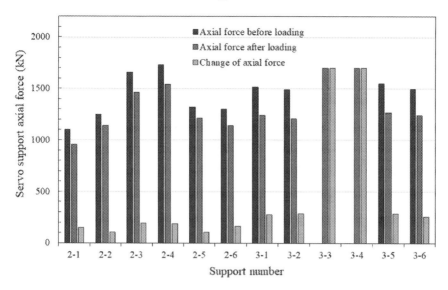

Fig. 5 Diagram of axial force coherence under field test case 1

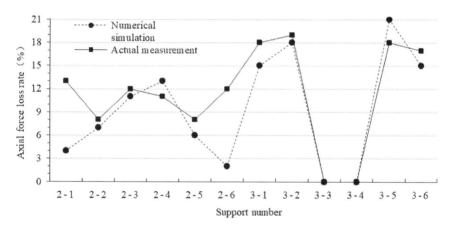

Fig. 6 Comparison between numerical simulation and field test under case 1

deformation and depth of diaphragm wall during excavation. The maximum vertical axial force loss rate generated by the second servo steel support is 12%, the third is 14%, and the fourth is 16%. It might be that the axial force of support close to the concrete support is less vulnerable.

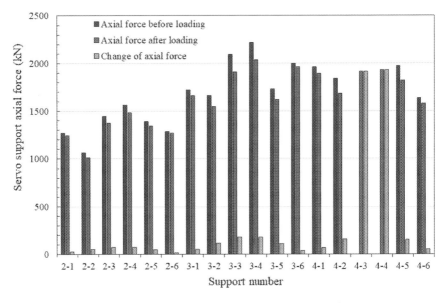

Fig. 7 Diagram of axial force coherence under numerical simulation case 2

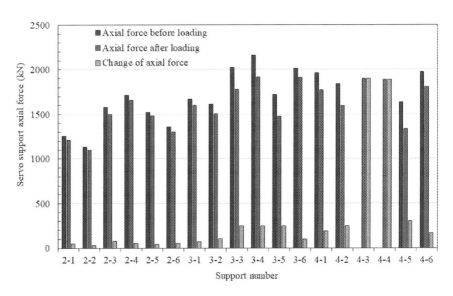

Fig. 8 Diagram of axial force coherence under field test case 2

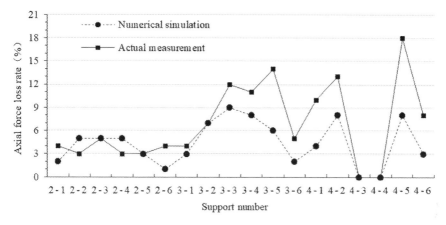

Fig. 9 Comparison between numerical simulation and field test under case 2

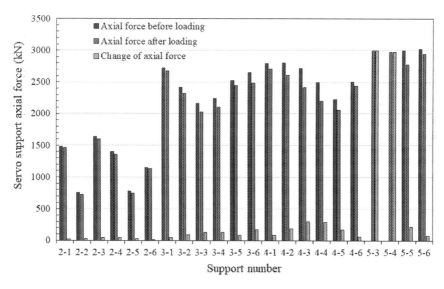

Fig. 10 Diagram of axial force coherence under numerical simulation case 3

3.2 Study on the Deformation of Diaphragm Wall Joints with Axial Force Coherence

The vertical coherence of the axial force of the servo steel support is mainly produced by the overall stiffness of the diaphragm wall to transfer the force. The integrity of a single diaphragm wall and the connection between the diaphragm walls together form the stiffness of the enclosure structure. Therefore, this section focuses on the change of the joints between the diaphragm walls when the axial force of the steel

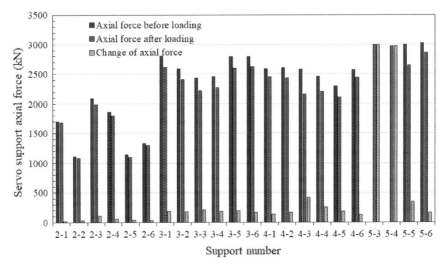

Fig. 11 Diagram of axial force coherence under field test case 3

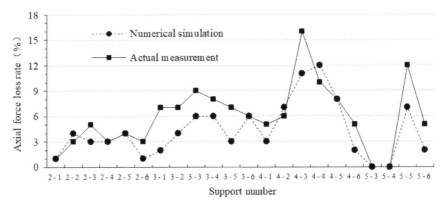

Fig. 12 Comparison between numerical simulation and field test under case 3

support is applied and studies the deformation of the joints between the diaphragm walls when the axial force of different steel supports is applied, which is helpful fine control of the axial force of the support.

In order to explore the influence of prestressed steel support and servo steel support on the displacement of underground continuous wall joints, the depth of servo steel supports joint gauge and prestressed steel support joint gauge is 11.5 m. The test conditions of the underground diaphragm wall joint meter are shown in Table 3, and the corresponding axial force is the same as that in Sect. 3.1. The horizontal deformation of the wall joint after the axial support force is applied as shown in Fig. 13.

Table 3 Cases adopted in the joint meter test of diaphragm wall

Working condition	Content
Case 1	Excavation to the third steel support, prestressed steel supports Y4-1, Y4-2 erected and applied axial force
Case 2	Excavation to the third steel support, servo steel supports 4–3, 4–4 erected and applied axial force
Case 3	Excavation to the fourth steel support, servo steel supports 5–3, 5–4 erected and applied axial force
Case 4	Excavation to the fourth steel support, prestressed steel supports Y5-1, Y5-2 erected and applied axial force

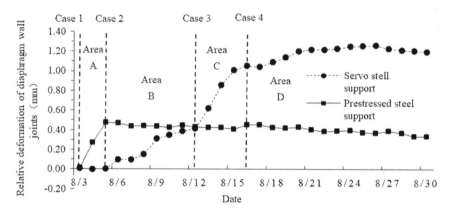

Fig. 13 Horizontal deformation of diaphragm wall joint

According to the working conditions, the above diagram is divided into four areas: A, B, C, and D. A pre-stressed steel support crack gauge in area A has been installed. Before it installs, pre-stressed steel support axial force of Y4-2 has been applied, and the other one has not been done. After the other one axial force of the pre-stressed steel support is applied, the crack deformation of the diaphragm wall in pre-stressed steel support area increases from 0 to 0.46 mm. Besides, the servo steel support area joint meter was not installed then.

The crack deformation of the diaphragm wall in area B gradually decreases, and the change is small. The servo steel support crack gauge is installed. Before it installs, the servo steel support axial force of 4-4 has been applied, and the other one has not been done. After the other one axial force is applied, the crack deformation of the diaphragm wall in servo steel support area increases from 0 to 0.41 mm. Area C shows that the deformation of pre-stressed steel support crack gauge does not change. After the 5-4 servo steel support axial force is applied, the deformation shown by servo steel support crack gauge increased from 0.41 to 1.04 mm. In area D, the axial force of the pre-stressed steel support is applied. The crack deformation of the diaphragm wall in pre-stressed steel support area is linearly decreasing. The relative deformation in the servo steel support area increases from 1.04 to 1.21 mm, but then converges.

There is a noticeable increase in the horizontal deformation of the joint in the servo steel support area. After the axial force of the servo steel support is applied, the horizontal deformation at the joint position of the diaphragm wall in the servo area continues to increase, and the axial force of the servo steel support will increase the joint crack of the diaphragm wall. The horizontal deformation of the wall joints in the pre-stressed steel support area is small. The horizontal deformation between the wall joints does basically not increase after the prestressed steel support is applied. The above results show that the horizontal deformation of the wall joint in the servo steel support zone is greater than that in the prestressed steel support zone. Applying the axial force of the servo steel support will affect the diaphragm wall joints. Therefore, it is necessary to combine the corresponding axial force application method to reduce the horizontal deformation of the diaphragm wall joints during the use of the servo steel support to prevent it from being too large and causing engineering hazards such as water leakage.

In addition, according to Figs. 6, 9, and 12 in Sect. 3.1, it can also be obtained that the axial force loss rate of the actual measurement is mostly higher than the axial force loss of the numerical simulation. We infer that the relative deformation between the diaphragm wall is not considered in the Plaxis 3D software. According to the data obtained from the actual construction site, there is an apparent horizontal deformation at the joint position in the servo zone. The deformation of the diaphragm wall leads to a change in the length of the steel support, resulting in a higher loss rate of the axial force than the axial force loss rate obtained by numerical simulation. That is, the deformation of the wall joints crack will in turn affect the axial force.

4   Conclusion

Based on the concept of axial force coherence of steel support in servo system, the influence of coherence on support in different directions and distances is analyzed according to the consistency between numerical simulation and field monitoring. Combined with the deformation of the joint of underground diaphragm wall measured on site, the influence of axial force coherence of steel support is analyzed, which provides theoretical for fine deformation control of foundation pit. The following conclusions can be drawn:

(1) The axial support force can affect other support axial forces of the retaining structure at different depths and distances. The farther away from the active axial force, the smaller the influence. The axial force of the support has the most significant influence on the axial force of the horizontal steel support, followed by the vertical steel support, and the influence on the oblique steel support is the smallest.

(2) The higher the application position of the active axial force of the servo support, the greater the lateral axial force loss rate generated by other supports, while the opposite in vertical axial force loss rate is true. The maximum lateral axial force loss rate generated by the second servo steel support is 19%. The maximum vertical axial force loss rate generated by the fourth servo steel support is 16%.

(3) The deformation of the diaphragm wall joint in the servo steel support zone is more significant than that in the pre-stressed steel support zone. The deformation of the joint will, in turn, affect the axial force, which is an interactive relationship. The relative displacement between diaphragm walls caused by a single application of servo axial force is 1.41 mm.

References

1. Wang G, Chen W, Cao L et al (2021) Retaining technology for deep foundation pit excavation adjacent to high-speed railways based on deformation control. Front Earth Sci 9:735315
2. Zhao J, Tan Z, Yu R et al (2022) Deformation responses of the foundation pit construction of the urban metro station: a case study in Xiamen. Tunn Undergr Space Technol 128:104662
3. Liu L, Cai G, Liu S et al (2021) Deformation characteristics and control for foundation pits in floodplain areas of Nanjing, China. Bull Eng Geol Env 80:5527–5538
4. Wang Q, Qian H (2018) Research on deformation characteristics of foundation pit support structure. IOP Conf Ser Mater Sci Eng 452:022101
5. Di H, Guo H, Zhou S et al (2019) Investigation of the axial force compensation and deformation control effect of servo steel struts in a deep foundation pit excavation in soft clay. Adv Civ Eng
6. Li H (2014) Research of the underground water level prediction model in deep foundation pit engineering. Appl Mech Mater 20140819:881–884
7. Cui X, Ye M, Zhuang Y (2018) Performance of a foundation pit supported by bored piles and steel struts: a case study. Soils Found 58(4):1016–1027
8. Wang X, Song Q, Gong H (2022) Research on deformation law of deep foundation pit of station in core region of saturated soft loess based on monitoring. Adv Civ Eng
9. Cao H, Sun J (2019) Study on determination method of axial force of steel support servo system in soft soil foundation pit. Chin J Build Constr 41(5):754–758
10. Jia J, Xie X, Luo F et al (2009) Support axial force servo system in deep excavation deformation control. J Shanghai Jiao Tong Univ 43(10):1589–1594
11. Gu G (2010) Application of hydraulic servo system in deep foundation pit construction. Chin J Constr Mech 31(12):49–51
12. Tanner Blackburn J, Finno RJ (2007) Three-dimensional responses observed in an internally braced excavation in soft clay. J Geotech Geoenviron Eng 133(11):1364–1373
13. Sun J, Bai T, Liao S (2021) Active control of deep foundation pit deformation based on coherence of supporting axial force. Chin J Undergr Space Eng 17(2):529–540
14. He J, Liao S, Sun J et al (2020) Study on axial force interference of soft soil deep excavation under multi-point synchronous loading. Chin Civil Eng J 53(7):99–107

Analysis of the Discriminability of High-Temperature Performance Indices of Modified Asphalt Mixtures

Mengjun Gu

Abstract Different indices have various capabilities to evaluate the high-temperature performance of modified asphalt mixtures. This study aims at investigate the discriminability of high-temperature performance indices. The values of five indices were determined from wheel tracking test, Marshall test and uniaxial penetration test, including dynamic stability (DS), comprehensive stability index (CSI), maximum rutting depth (RD), Marshall stability (MS), and uniaxial penetration strength (UPS). The discriminability of five indices was further examined by entropy weight method and CRITIC method, respectively. The results show that DS, MS and UPS are not appropriate to evaluate the high-temperature performance of modified asphalt mixtures, but CSI and RD show preferable distinguishing ability to evaluate the high-temperature performance. As a consequence, RD is recommended to be employed as secondary index to supplement the CSI in wheel tracking test. The findings of this study will contribute to the optimization of evaluation on high-temperature performance of modified asphalt mixtures.

Keywords Road engineering · Modified asphalt mixture · Discriminability · Entropy weight method · CRITIC method

1 Introduction

The high-temperature performance of asphalt mixtures is of great significance to the service of asphalt pavement, which directly determines whether it will occur serious bleeding, slippage, rutting and other distresses. To prevent adverse rutting distress, many types of additives were utilized to prepare modified asphalt mixtures, including SBS, rubber powder, polyethylene, etc. [1]. Obviously, those modified asphalt mixtures have remarkable and excellent high-temperature performances, whereas their evaluation methods remain to be enhanced because existing laboratory tests show high discreteness, poor uniformity, and low accuracy [2].

M. Gu (✉)
Wuhan University of Technology, Luoshi Road 122, Wuhan 430070, China
e-mail: 759024839@qq.com

© The Author(s) 2023
S. Wang et al. (eds.), *Proceedings of the 2nd International Conference on Innovative Solutions in Hydropower Engineering and Civil Engineering*, Lecture Notes in Civil Engineering 235, https://doi.org/10.1007/978-981-99-1748-8_20

Engineering practice has proved that the asphalt mixtures meeting the dynamic stability requirements in the specification could also occur severe rutting distress in the actual pavement [3]. The contradiction often occurs when the dynamic stability is large but the actual rutting depth is large. The possible reason for this is that the conditions of wheel tracking test cannot completely consider actual temperature, speed, and loads of the pavement. It is also believed that the wheel tracking test can only distinguish the rutting resistance of asphalt mixtures with relative small stiffness, but cannot successfully evaluate the rutting resistance of mixtures with high modulus [4].

In order to overcome the issues mentioned above, many researches attempt to upgrade the existing indices to improve their performance. Peng et al. [5] proposed a new concept of rutting coefficient WRI based on the limitation of dynamic stability, which considers cumulative deformation and maximum deformation, thus improved the evaluation efficiency of anti-rutting indices. Du and Dai [6] held the view that the dynamic stability only incorporated the growth rate of shear flow deformation in the stable consolidation period, but ignored the deformation in transition period. So comprehensive stability index (CSI) was developed to evaluate the rutting resistance of asphalt mixtures. Zhou et al. [7] found that the maximum rutting depth performed well correlation with the creep slope of asphalt mixtures, so it is recommended to employ the maximum rutting depth as supplementary evaluation index in wheel tracking test.

It is concluded that these findings aim at establishing new indices or simply validating their performance, but rarely compared their discriminative ability on high-temperature performance, in particular modified asphalt mixtures. In order to enhance the discriminability of high-temperature indices with higher preciseness, the DS, CSI, RD, MS and UPS of different modified asphalt mixtures were obtained by conducting wheel tracking test, Marshall test, and uniaxial penetration test. Further discriminability study on those five indices were compared by entropy weight method and CRITIC method. The outcomes of this study can offer recommendations on electing indices for evaluating high-temperature performance of modified asphalt mixtures.

2   Materials and Experimental Program

2.1   Test Materials

The PR PLAST.S (hereinafter referred to as PR) anti-rutting additive produced by a French company was used to modify asphalt mixtures, which presented as black solid particles with a particle size of about 4 mm. Relying on its inter-locking, reinforcement and cementation effects, the high-temperature performance of asphalt mixture can be greatly improved by directly mixing PR into asphalt mixture. Total 10 types of asphalt mixtures were designed in accordance with *Technical Specification*

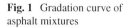
Fig. 1 Gradation curve of asphalt mixtures

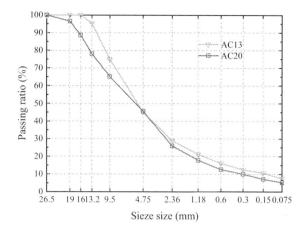

Table 1 Volume parameters of asphalt mixtures

Mixture	VFA/%	VMA/%	VV/%	VA/%	Theory maximum relative density/g cm^{-3}	Bulk relative density/g cm^{-3}
AC13	75.0	16.4	4.1	12.3	2.639	2.531
AC20	71.6	14.8	4.2	10.6	2.573	2.465

for Construction of Highway Asphalt Pavements (*JTG F40-2004*), including AC13, 0.2% PR-AC13, 0.4% PR-AC13, 0.6% PR-AC13, 0.8% PR-AC13, AC20, 0.2% PR-AC20, 0.4% PR-AC20, 0.6% PR-AC20, 0.8% PR-AC20. The gradations of AC13 and AC20 were shown in Fig. 1. SK-70# asphalt, limestone mineral powder, and aggregates were used to prepare PR-AC13 mixtures and PR-AC20 mixtures with different content of PR modifiers. The PR content refers to the mass ratio of modifiers to asphalt mixtures. The optimal asphalt contents of AC13 and AC20 were 4.7% and 4.1%, which were determined via Marshall method. The volume parameters of asphalt mixtures were shown in Table 1.

2.2 Experiments

The wheel tracking test was generally used to evaluate the high-temperature stability of the asphalt mixture for its simplicity and good relationship with real pavement. Wheel tracking test has the relatively comprehensive process of consolidation, shearing, and flow deformation, thus the above mixtures were examined by wheel tracking test under 0.7 MPa wheel-pressure and 60 °C. Finally, dynamic stability (DS) was determined.

In addition, another index comprehensive stability index (CSI) was proposed to consider the transition and stable phase of consolidation, showing good relationships with other indices [6, 8, 9]. CSI was described as:

$$CSI = \frac{(t_2 - t_1)NC_1C_2}{d_1(d_2 - d_1)} \tag{1}$$

where N is the rolling speed of wheels; C_1 and C_2 are coefficients related to machine type and specimen. d_1, d_2 are deformations corresponding to time t_1, t_2, respectively.

Relative to DS, d_1 is brought into the calculation of CIS again, thus the transition and stable stage during consolidation deformation is taken into account. On the other hand, the maximum rutting depth (RD) was also selected as an evaluation index. Marshall stability (MS) was also taken to evaluate high-temperature of asphalt mixtures.

The uniaxial penetration test was conducted by universal testing machine to evaluate the high-temperature stability of the asphalt mixtures under 60 °C. The uniaxial penetration strength was accordingly determined as below.

$$\sigma_p = P/A \tag{2}$$

$$UPS = k\sigma_p \tag{3}$$

where σ_p denotes penetration stress, MPa; UPS denotes the uniaxial penetration strength; P denotes the maximum penetration load, N; A is the cross-sectional area of indenter, mm^2; and k denotes strength coefficient.

Based on the experiments above, the indices DS, CSI, RD, MS, and UPS were selected as indices to evaluate the high-temperature stability of modified asphalt mixtures. Moreover, in order to verify the consistency of different indices that characterize the high-temperature properties of mixtures, the Pearson Correlation Coefficient was further calculated to indicate the correlation of five indices.

3   Analysis of High-Temperature Performance of Mixtures

Figure 2a–e are the results of different types of asphalt mixtures, and five indices are compared in the bar chart, including DS, CSI, RD, MS, and UPS. Relative to AC13 and AC20 mixtures, the incorporation of PR modifier causes five high-temperature performance indices of asphalt mixtures to increase with varying degrees. The high-temperature stability of 10 types of mixtures generally improves with the increase of PR content. It is notable that the high content (0.8%) of the PR modifier results in a considerable increase in both DS and CSI, as well as remarkable decrease in maximum rutting depth. Whereas the MSs of 10 types of asphalt mixtures varied slightly, even for higher content of PR modifier. Thus it implies that the PR modifier shows different synergistic effects on five indices, thus the applicability of five indices is expected to be verified regarding high-temperature stability of asphalt mixtures.

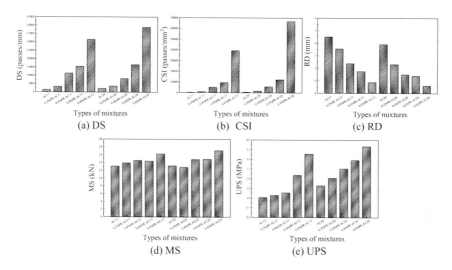

Fig. 2 Results of high-performance tests for different types of asphalt mixtures

The correlation between each index was evaluated with Pearson's Coefficient, as shown in Table 2. It is found that the DS, CIS, and MS have a strong correlation between each other, because their Pearson's Coefficients are greater than 0.9. However, this kind of result is not consistent with the above findings via bar charts. The poorest correlation occurs between CSI and RD, and the absolute value of their Pearson's Coefficient is 0.706, which means the comprehensive stability index (CSI) cannot completely reflect the deformation properties of modified asphalt mixtures under high-temperature conditions. It deduced that the introduction of d_2 adversely influenced the correlation between CSI and RD compared with DS.

Consequently, the results of the correlation analysis have several differences from the findings in the bar chart. So the discriminability of these indices remains to be examined through more advanced methods in need, thus determining the most appropriate index to evaluate high-temperature stability of modified asphalt mixtures.

Table 2 Pearson's coefficients between different indices

Pearson's coefficient	DS	CSI	RD	MS	UPS
DS	1	0.9554	− 0.8245	0.9495	0.8611
CSI	0.9554	1	− 0.706	0.9014	0.8184
RD	− 0.8245	− 0.706	1	− 0.8277	− 0.8957
MS	0.9495	0.9014	− 0.8277	1	0.8223
UPS	0.8611	0.8184	− 0.8957	0.8223	1

4   Determination of Discriminability

4.1   Calculation with Entropy Weight Method

The entropy weight method was widely used to distinguish the discriminability of indices for its simplicity and objectiveness. Entropy weight method was used to calculate the weight of each index.

An initial matrix composed of m evaluated objects and n indices $A = (a_{ij})_{m \times n}$ was constituted, where a_{ij} was the value of the jth index of the ith evaluated object. And the column formed by RD belonged to the inverse index, so the elements in the column were processed with isotropization transformation. After that, the matrix $X = (x_{ij})_{m \times n}$ was determined. Then the normalization of data was conducted to eliminate the influence of dimension, as given by:

$$\delta_{ij} = x_{ij} / \sum_{i=1}^{m} x_{ij} \tag{4}$$

Entropy value E_j and entropy weight W_j of index j were calculated as:

$$E_j = - \sum_{i=1}^{m} \delta_{ij} \ln \delta_{ij} / \ln m \tag{5}$$

$$W_j = 1 - E_j / \sum_{j=1}^{n} 1 - E_j \tag{6}$$

Finally, discriminability of index j is calculated in accordance with Eq. (8).

$$D_j = W_j \left(\max_{i=1}^{m} \delta_{ij} - \min_{i=1}^{m} \delta_{ij} \right) \tag{7}$$

4.2   Calculation with CRITIC Method

Entropy weight method simply considers the discrete degree of data without accounting their conflicts. Criteria importance though inter-criteria correlation (CRITIC) method is an objective method to determine the weight of indices based on the contrast intensity and conflicts between indices, which is applicable for the evaluation of multiple-index. Standard deviation was used to indicate the contrast intensity, as given by:

$$\sigma_j = \sqrt{\sum_{i=1}^{m} \left(\delta'_{ij} - \overline{\delta'}_j \right)/(m-1)} \qquad (8)$$

The conflicts between indices were characterized based on their correlation, as given by:

$$f_j = \sum_{i=1}^{m} \left(1 - r_{ij} \right) \qquad (9)$$

where r_{ij} was the correlation coefficient of index i and j, and Pearson's Coefficient was adopted.

Furthermore, the relative significance of index j was described by information-carrying capacity C_j. The higher C_j means larger amount of information, which implied that the corresponding index had the higher significance, as described by:

$$C_j = \sigma_j \sum_{i=1}^{m} (1 - r_{ij}) = \sigma_j f_j \qquad (10)$$

Thus, the objective weight of index j was determined by:

$$W_j = C_j / \sum_{i=1}^{m} C_j \qquad (11)$$

4.3 Analysis of Discriminability

Tables 3 and 4 present the key results from the entropy weight method and CRITIC method, respectively. It clearly shows that the discriminability (D_j) determined by entropy weight method and CRITIC method is particularly different. For entropy weight method, the D_j of CSI is the greatest value among four indices, which is 0.2561. It implies the index CSI has a satisfactory ability to distinguish the high-temperature performance of modified asphalt mixtures. However, the D_j of CSI is close to 0, which means the good and poor performance of modified asphalt mixtures cannot be well separated. In addition, the sequence of discriminability of five indices in descending order is CSI > DS > RD > UPS > MS.

With regard to CRITIC method, the discriminability of five indices differs from that of entropy weight method. The sequence of discriminability of five indices in descending order is RD > CSI > DS > UPS > MS. In particular, the RD has the biggest D_j among five indices, even up to 0.6679. Consistent with the results determined by entropy weight, the MS still has poor ability to distinguish the high-temperature

Table 3 Results determined from entropy weight method

Items	DS	CSI	RD	MS	UPS
Entropy (E_j)	0.8257	0.6401	0.9146	0.9983	0.9645
Entropy weight (W_j)	0.2654	0.5479	0.1301	0.0026	0.0540
Discriminability (D_j)	0.0748	0.2561	0.0290	0	0.0065

Table 4 Results determined from CRITIC method

Items	DS	CSI	RD	MS	UPS
Information-carrying capacity (C_j)	0.0263	0.0542	0.0972	0.0037	0.0248
Entropy weight (W_j)	0.1275	0.2628	0.4713	0.018	0.1203
Discriminability (D_j)	0.0359	0.1228	0.6679	0.0005	0.0146

performance of modified asphalt mixtures due to extremely low D_j (approaching to 0). CSI has relative acceptable discriminability compared with DS, MS, and UPS, which reaches as high as 0.1228.

It was concluded that the D_j of RD remarkably varied when calculating by entropy weight method and CRITIC method. The major reason is attributed to the correlation between RD and other indices (especially for CSI) is rather poor, as examined in previous sections. The poor correlation causes large conflicts between indices, resulting in the increase of RD's significance on high-temperature performance. Nonetheless, RD is still an important index to reflect the deformation properties of modified asphalt mixtures under high-temperature.

It is notable that the CSI was always an acceptable index to evaluate the high temperature of modified asphalt mixtures, which is consistent with the findings obtained by Fang et al. [9]. The DS always shows weak performance on evaluating the high-temperature stability of modified asphalt mixtures even though it has been widely used for design and construction of asphalt pavement. The MS and UPS are not acceptable indices to evaluate the high-temperature stability due to their low discriminability. However, the RD can be employed as secondary index to supplement the CSI in wheel tracking test due to their good discriminability.

5 Conclusions

(1) High modifier content lead to a considerable increase in both DS and CSI of modified asphalt mixtures, but the MS varied slightly for 10 types of asphalt mixtures. Five indices have different variation trend for different modified asphalt mixtures.

(2) Five indices generally have relative strong correlation between each other except for CSI and RD, thus CSI cannot comply with the deformation properties of modified asphalt mixtures.

(3) The sequence of discriminability of five indices in descending order determined by CRITIC method is RD > CSI > DS > UPS > MS. And the sequence of discriminability of five indices in descending order obtained from entropy weight method is CSI > DS > RD > UPS > MS.

(4) DS, MS and UPS are poor in evaluating the high-temperature performance of modified asphalt mixtures. CSI is recommended to evaluate the high-temperature performance of modified asphalt mixtures. And RD is expected to be employed as secondary index to supplement the CSI in wheel tracking test.

References

1. Zhou ZG, Chen GH, Zhang HB, Ling YY (2021) Study on the preparation and properties of modified asphalt by rubber powder/SBS and high viscosity modifier composite. Mater Rep 35(06):6093–6099
2. Zhang HZ, Wang D, Yang YH (2021) High temperature performance evaluation indices of asphalt mixtures. J Build Mater 24(06):1248–1254
3. Guan HX, Zhang QS, Liu J (2011) Rutting test improving methods of asphalt mixture. J Traffic Transp Eng 11(03):16–21
4. Chen K (2008) Study on domestic rutting test and Hamburg Wheel Tracking Test. M.S. thesis, Chang'an University, Xi'an, China
5. Peng B, Yuan WF, Chen ZD (2005) Evaluation of anti-wheel-rutting performance of asphalt mixture by using wheel-rutting index. J South Chin Univ Technol (Nat Sci Ed) 33(12):84–86
6. Du SC, Dai JL (2006) Permanent deformation evaluation index of asphalt mixture. Chin J Highw Transp 19(5):18–22
7. Zhou D, Zhou ZG, Liu JY (2022) Comparative studies on laboratory rutting test and evaluating indicator of asphalt concrete. J Railway Sci Eng 19(08):2287–2294
8. Wang W, Chen ZD, Meng QY, Zhou WF (2011) Test of thermal stability of PE-Y modified asphalt mixture. J Highw Transp Res Dev 28(12):21–26
9. Fang H, Barugahare J, Mo LT (2018) Rutting evaluation index analysis of asphalt mixture based on rutting test. J Wuhan Univ Technol (Transp Sci Eng) 42(01):17–20

Curvature Ductility of Confined HSC Columns

Haytham Bouzid⦿, Benferhat Rabia, and Tahar Hassaine Daouadji

Abstract To avoid brittle collapse of reinforced concrete (RC) structures, RC elements such as beams, columns, and shear walls are invited to ensure a minimum level of ductility. In this paper, an analytical method for predicting the curvature ductility factor of confined RC columns is developed. The stress–strain model of confined concrete provided by Eurocode 2 is adopted, and the effective lateral confining pressure is calculated according to the Eurocode 8. The curvature ductility factor is defined by the ratio of ultimate to yield curvature. In this context, a new hypothesis is adopted to calculate the yield curvature while the ultimate curvature is calculated based on axial load and the mechanical ratios of tension, web, and compression reinforcement. The results showed that the developed method has an excellent performance compared to the experimental results collected from previous researches, where the mean value and the standard deviation of the ratios predicted to experimental factors are equal to 1.02 and 0.17, respectively. Moreover, the calculated coefficient of determination is very close to 1.

Keywords Ductility · Columns · Reinforced concrete · Confinement · High strength concrete

1 Introduction

Ductility is a vital parameter in seismic design. The ductility of reinforced concrete (RC) structure depends on the ductility of its elements such as columns. In addition to their great potential in restoring the load-bearing capacity, high ductile RC columns

H. Bouzid (✉)
Tissemsilt University, 38000 Tissemsilt, Algeria
e-mail: haytham.bouzid@univ-tiaret.dz

B. Rabia · T. H. Daouadji
Tiaret University, 14000 Tiaret, Algeria

H. Bouzid · B. Rabia · T. H. Daouadji
Laboratory of Geomatics and Sustainable Development, Tiaret University, Tiaret, Algeria

© The Author(s) 2023 253
S. Wang et al. (eds.), *Proceedings of the 2nd International Conference on Innovative Solutions in Hydropower Engineering and Civil Engineering*, Lecture Notes in Civil Engineering 235, https://doi.org/10.1007/978-981-99-1748-8_21

provide adequate ductility to the overall structure [1]. From here, it appears the great importance of RC columns ductility. In this field, the study of RC columns ductility requires a deep analysis at the limit states, where it is essential to know the behavior of materials such as concrete and steel [2]. Practically, the RC column cross section is subjected to lateral pressure resulted from the transverse reinforcement, named confinement [3, 4]. Under confinement, significant enhancements are occurred on concrete behavior, where higher strength and higher critical strains are achieved [5, 6]. These changes on concrete characteristics affect the overall behavior of RC elements, so it is necessary to take into account these new characteristics in structural elements studying [7].

Most of the research carried out on the flexural behavior of RC columns are experimental works, where the curvature ductility factor was obtained from the moment–curvature (M–φ) diagram. The yield curvature was measured when the tension reinforcement yielded in tension while the ultimate curvature was measured when the concrete strain reaches its ultimate value. In this field, experimental researches proved that the increase of lateral pressure enhances the curvature ductility of RC columns [8, 9].

On the other hand, a few theoretical researches were conducted on the curvature ductility of RC columns. The lack of theoretical researches in this field pushed the researchers to undertake the current study. A new hypothesis is introduced in order to calculate the yield curvature while the ultimate curvature is calculated according to the proposition of Fardis [10]. In this context, the model of confined concrete given by Eurocode 2 [5] is adopted for the present study. Finally, experimental data are collected in order to show the reliability of the developed model in predicting the curvature ductility factor of confined RC columns made with normal strength (NSC) and high strength concrete (HSC).

2 Stress–Strain Relationship for Confined Concrete Under Compression

Similar to unconfined concrete, Eurocode 2 [5] provides a parabola-rectangle diagram for confined concrete under compression as shown in Fig. 1. Compared to unconfined concrete, an increase in stress (σ_c) and strain (ε_c) can be noticed. This amelioration is due to the effective lateral confining pressure (σ_2).

The compressive strength of confined concrete (f_{cc}) is given by the following expression:

$$f_{cc} = \begin{cases} f'_c\left(1.125 + 2.50\frac{\sigma_2}{f'_c}\right) & \text{for } \sigma_2 > 0.05f'_c \\ f'_c\left(1.0 + 5.0\frac{\sigma_2}{f'_c}\right) & \text{for } \sigma_2 \le 0.05f'_c \end{cases} \tag{1}$$

where f'_c = compressive strength of unconfined concrete.

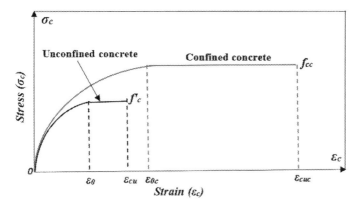

Fig. 1 Stress–strain (σ_c–ε_c) model for confined concrete [5]

The strains at reaching the maximum strength (ε_{0c}) and at ultimate (ε_{cuc}) are given as follows:

$$\varepsilon_{0c} = \varepsilon_0 \left(\frac{f_{cc}}{f'_c} \right)^2 \tag{2}$$

$$\varepsilon_{cuc} = \varepsilon_{cu} + 0.2 \frac{\sigma_2}{f'_c} \tag{3}$$

where ε_0 and ε_{cu} = strains at reaching the maximum strength and at ultimate of unconfined concrete, respectively.

On the other hand, the effective lateral confining pressure (σ_2) is calculated as follows:

$$\sigma_2 = 0.5 \, \alpha \, \omega_{wd} f'_c \tag{4}$$

where ω_{wd} = mechanical volumetric ratio of confining hoops within the critical region and α = confinement effectiveness factor. These two parameters can be calculated according to Sect. 5.4.3.2.2 of Eurocode 8 [11].

3   Curvature Ductility Factor

The curvature ductility factor (μ_φ) is calculated by dividing the ultimate curvature (φ_u) by the curvature at yielding (φ_y):

$$\mu_\varphi = \frac{\varphi_u}{\varphi_y} \tag{5}$$

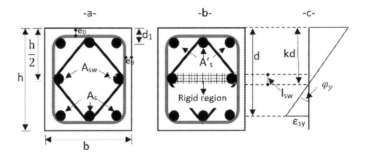

Fig. 2 Column cross section at yielding

3.1 Curvature at Yielding

The yield curvature (φ_y) is calculated when the tension reinforcement reaches their yield strength. As shown in Fig. 2b, the web reinforcement (A_{sw}) creates a rigid region in the middle of the column cross section, where they prevent the increase of the tension region depth. In this context, the neutral axis depth (kd) is assumed to be greater than (h/2), where it is taken as follows:

$$kd = \left(\frac{h}{2} + \frac{l_{sw}}{2} \right) \tag{6}$$

Hence, the yield curvature is written as follows:

$$\varphi_y = \frac{\varepsilon_{sy}}{d - \left(\frac{h}{2} + \frac{l_{sw}}{2} \right)} \tag{7}$$

where h = height of column cross section; l_{sv} = length of the rigid region created by web reinforcement; $\varepsilon_{sy} = f_y/E_s$ = strain corresponding to yield strength of steel reinforcement; d = distance measured from the extreme compression fiber to the centroid of the tension reinforcement; f_y and E_s = yield strength and modulus of elasticity of tension steel reinforcement; b = width of column cross section; d_1 = distance measured from the extreme compression fiber to the centroid of the compression reinforcement; A_s, A'_s and A_{sw} = areas of tension, compression, and web reinforcement.

3.2 Curvature at Ultimate

The ultimate curvature is calculated when the strain in the extreme compression fiber of the confined zone reaches the ultimate value (ε_{cuc}). Figure 3 shows the strain distribution at this stage. Based on this figure, the ultimate curvature is written as

Fig. 3 Column cross section at ultimate

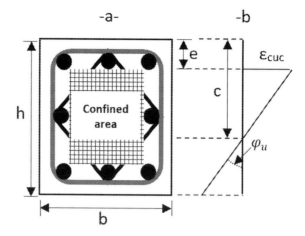

follows:

$$\varphi_u = \frac{\varepsilon_{cuc}}{c - e} \tag{8}$$

where c = neutral axis depth at ultimate and e = distance measured from extreme compression fiber of concrete to confined zone extremity.

In this case, the neutral axis depth (c) was given by Fardis [10] as follows:

$$\frac{c}{d} = \frac{(1 - \delta_1)(v - \omega_s + \omega'_s) + (1 + \delta_1)\omega_{sw}}{(1 - \delta_1)\left(1 - \frac{\varepsilon_{0c}}{3\varepsilon_{cuc}}\right) + 2\omega_{sw}} \tag{9}$$

where

- $\delta_1 = d_1/d$
- $v = N/b_0\,d_0\,f_{cc}$: ratio of the axial load (N)
- $\omega_s = A_sf_y/b_0\,d_0\,f_{cc}$, $\omega'_s = A'_sf_y/b_0\,d_0\,f_{cc}$, $\omega_{sw} = A_{sw}f_y/b_0\,d_0\,f_{cc}$: mechanical reinforcement ratios of tension, compression, and web reinforcement, respectively
- $h_0 = h - 2(e_0 + d_{bh}/2)$, $b_0 = b - 2(e_0 + d_{bh}/2)$, $d_0 = d - 2(e_0 + d_{bh}/2)$
- d_{bh}: diameter of transverse reinforcement
- e_0: concrete cover.

4 Method Verification

In this section, the performance of the presented method is tested against large experimental results collected from the literature. In this field, 30 experimentally tested columns are selected from previous researches. The details and properties of the selected specimens are shown in Table 1.

The predicted curvature ductility factors ($\mu_{\varphi,pre}$) of the selected specimens are presented in Table 2 with those obtained during experiments ($\mu_{\varphi,exp}$). The comparison shows a good agreement between the predicted and the experimental values, where the mean value (MV) and the standard deviation (SD) of the errors ($\mu_{\varphi,pre}/\mu_{\varphi,exp}$) are equal to 1.02 and 0.17, respectively. In the same context, Fig. 4 shows the predicted and the experimental results head-on. The dispersion of the points affirms the good agreement between the predicted and the experimental results. Moreover, the value of the coefficient of determination ($R^2 = 0.90$) confirms this deduction, where it is very close to 1. Consequently, these statistical results indicate the good performance of the proposed method in predicting the curvature ductility factor of NSC and HSC columns.

5 Conclusion

The current research presented a detailed method in order to predict the curvature ductility factor of confined RC columns made with NSC and HSC. Eurocode 2 [5] and Eurocode 8 [11] were used to calculate the strength and strains of concrete under confinement, i.e., under lateral pressure due to transverse reinforcement. In this field, the results obtained by using the proposed method were compared with those extracted from experimental researches. The statistical comparison showed that the proposed method has a good performance and accuracy in predicting the curvature ductility factor of confined RC columns with concrete strength up to 100 MPa.

On the one hand, the present study provided a simple method that can be used with high accuracy. On the other hand, it can be a starting point for studying the ductility of RC columns confined with other types of materials or under active confinement. Moreover, it motivates researchers to undertake other studies on different confined RC elements such as beams.

Table 1 Properties and details of the selected beams

Research	Columns	f'_c (MPa)	-b -h (mm)	ρ_l (%)	f_y (MPa)	$N/f'_c A_c$	d_{bh}/S_h (mm/mm)	f_{yh} (MPa)
Sheikh and Yeh [12]	E2	31.4	305 305	2.44	414	0.61	3/114	490
	F4	32.2				0.60	10/95	
	F6	27.2				0.75	13/173	
	E10	26.3				0.77	10/64	
	E13	27.2				0.74	13/114	
	D5	31.2		2.58		0.46	10/114	
	D15	26.9				0.75	10/114	
Sheikh and Khoury [13]	FS-9	32.4	305 305	2.44	414	0.76	10/95	507
	ES-13	32.5				0.76	13/114	464
	AS-17	31.3				0.77	10/108	507
	AS-18	32.8				0.77	13/108	464
	AS-19	32.3				0.47	10/108	507
Sheikh et al. [14]	AS-18H	54.7	305 305	2.44	414	0.64	13/108	464
	AS-20H	53.7				0.64	13/76	464
	AS-17H	59.1				0.65	10/108	507
Ho and Pam [15]	R6	72.6	325 325	0.86	250	0.11	6/100	460
	R8	74.6					8/175	
	R10	72.4					10/220	
	R12	77.8					12/85	
Pam and Ho [8]	B1	50	325 325	6.1	460	0.61	12/70	531
	B2	83.3		2.4		0.33	12/70	531
	B3	77.8		0.9		0.12	12/85	339
	B4	80.6		2.4		0.31	12/105	531
	B5	56.1		6.1		0.59	12/110	531
	B6	96.4		2.4		0.34	12/90	531
	B7	94.7		6.1		0.35	12/100	531
	B8	85		6.1		0.63	12/120	572
Karen et al. [9]	N1	64,1	150 260	1.74	500	0.36	8/100	500
	N2	71,7				0.34	8/50	
	N12	33.57				0.38	8/100	

where ρ_l = ratio of longitudinal reinforcement; A_c = concrete area (b × h); S_h = center-to-center spacing of transverse reinforcement; f_{yh} = yield strength of transverse steel reinforcement

Table 2 Comparison between predicted and experimental results

Columns	$\mu_{\varphi,exp}$	$\mu_{\varphi,pre}$	$\mu_{\varphi,pre}/\mu_{\varphi,exp}$
E2	5.4	5.64	1.05
F4	6.8	6.24	0.92
F6	4.8	3.92	0.82
E10	3.9	5.84	1.50
E13	5.5	5.42	0.99
D5	10.6	9.32	0.88
D15	5.5	6.96	1.27
FS-9	7.5	7.54	1.01
ES-13	6	6.20	1.03
AS-17	10.5	8.00	0.76
AS-18	14.5	12.24	0.84
AS-19	10	8.94	0.89
AS-18H	11	9.16	0.83
AS-20H	13.5	14.13	1.05
AS-17H	5	4.49	0.90
R6	7.3	9.25	1.27
R8	7.2	8.07	1.12
R10	7.7	8.84	1.15
R12	12.8	14.03	1.10
B1	2.37	2.36	0.99
B2	2.94	4.08	1.39
B3	3.72	3.66	0.98
B4	8.1	7.79	0.96
B5	10.1	9.60	0.95
B6	13	13.04	1.00
B7	9.1	9.25	1.02
B8	10.1	9.84	0.97
N1	9.8	10.88	1.11
N2	10.6	8.90	0.84
N12	7.1	6.53	0.92
MV			**1.02**
SD			**0.17**

Fig. 4 Comparison between predicted and experimental results

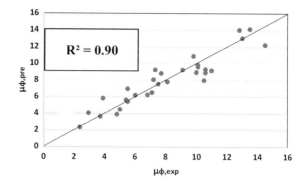

References

1. Polienko W, Holschemacher K, Henning VD (2022) Column confinement with textile-reinforced concrete (TRC). In: Proceedings of the fourth European and Mediterranean structural engineering and construction conference. ISEC Press, Leipzig, Germany
2. Bouzid H, Kassoul A (2018) Curvature ductility prediction of high strength concrete beams. Struct Eng Mech 66(2):195–201
3. Biskinis DE (2007) Resistance and deformation capacity of concrete members with or without retrofitting. Doctoral thesis, Civil Engineering Department, University of Patras, Patras, GR
4. Guadagnuolo M, Alfonso D, Anna T, Giuseppe F (2020) Experimental behavior of concrete columns confined by transverse reinforcement with different details. Open Constr Build Technol J 14:250–265
5. Eurocode 2, EN 1992-1-1 (2003) Design of concrete structures. Part 1-1: General rules and rules for buildings. European Committee for Standardization. Brussels, Belgium
6. Halima A, Kassoul A, Bouzid H (2019) New model for confinement of reinforced concrete columns with an ultra-high strength close to 200 MPa. Eng Struct 199:109594
7. Breccolotti M, Annibale LM, Bruno RB (2019) Curvature ductility of biaxially loaded reinforced concrete short columns. Eng Struct 200:109669
8. Pam HJ, Ho JCM (2009) Length of critical region for confinement steel in limited ductility high-strength reinforced concrete columns. Eng Struct 31:2896–2908
9. Karen ECM, Bonet JL, Juan NG, Pedro SR (2013) An experimental study of steel fiber-reinforced high-strength concrete slender columns under cyclic loading. Eng Struct 57:565–577
10. Fardis MN (2009) Seismic design, assessment and retrofitting of concrete buildings: based on EN-Eurocode8. Springer Science + Business Media B.V., New York
11. Eurocode 8, EN 1998-1 (2003) Design of structures for earthquake resistance. Part 1: General rules, seismic actions and rules for buildings, Brussels, Belgium
12. Sheikh SA, Yeh CC (1990) Tied concrete columns under axial load and flexure. J Struct Eng 116(10):2780–2800
13. Sheikh SA, Khoury SS (1993) Confined concrete columns with stubs. ACI Struct J 90(4):90-S44
14. Sheikh SA, Shah DV, Khoury SS (1994) Confinement of high strength columns. ACI Struct J 91(1):91-S11
15. Ho JCM, Pam HJ (2003) Inelastic design of low-axially loaded high-strength reinforced concrete columns. Eng Struct 25:1083–1096

Buckling Behaviour of Locally Dented GFRP Cylindrical Shells Under External Pressure—A Numerical Study

Neda Fazlalipour[ID]**, Robab Naseri Ghalghachi, and Saeed Eyvazinejad Firouzsalari**[ID]

Abstract Structural applications of composite materials are used in various structures of the oil and gas industry, water supply and sewage systems and a wide range of industries, such as marine, aerospace, and military industries. This paper aims to numerically investigate the influence of local dent caused by an indenter on the buckling behaviour of glass fabric-reinforced polymer cylindrical shells when subjected to external pressure. For this purpose, 24 finite element numerical models with five layers and a stacking sequence [30/-30/30/-30/30] were simulated in ABAQUS. The effect of dent depth (2, 4, 6 and 8 mm) and orientation (0 and 90 degrees) that was created at the mid-height, the 1/3rd and the 2/3rd of the shell height on the buckling behaviour of the composite cylindrical shells were evaluated. The results underscored that whilst the location of the local dent and the depth affected the shells' buckling capacity, the dent's orientation had minimal effect on the buckling capacity of the cylindrical shells.

Keywords Composite structures · Buckling · Local geometric imperfection · Dent · GFRP cylindrical shells

1 Introduction

Thin-walled cylindrical shells are extensively used as various industrial structures such as oil tanks, offshore platforms and silos, with these structures being susceptible to buckling when subjected to the hoop or axial compressive stresses. Fibre-reinforced polymer (FRP) composites are increasingly used in various structural applications, with the main reason for this being the high strength-to-weight and

N. Fazlalipour · R. N. Ghalghachi
Department of Civil Engineering, Urmia University, Urmia, Iran

S. E. Firouzsalari (✉)
Department of Civil and Environmental Engineering, The University of Auckland, Auckland, New Zealand
e-mail: seyv943@aucklanduni.ac.nz

© The Author(s) 2023
S. Wang et al. (eds.), *Proceedings of the 2nd International Conference on Innovative Solutions in Hydropower Engineering and Civil Engineering*, Lecture Notes in Civil Engineering 235, https://doi.org/10.1007/978-981-99-1748-8_22

stiffness-to-weight ratios of FRP composites when compared to conventional structural materials. Several studies have been performed on the modelling and analysis of composite laminated cylindrical shells. The theories used in these analyses mainly extend the theories developed for isotropic shell models. Fully anisotropic laminated cylindrical shells have recently attracted the attention of researchers to achieve an optimum design of composite laminated shell structures [1–7].

In cylindrical shells, mechanical damages may occur due to the physical con-tact of shells with various vehicles travelling close to shell locations. Additionally, the overturning of adjacent shells can cause mechanical damage to the shells. Among various types of mechanical damage, the dent is regarded as the most crucial damage [8], which leads to local stress/strain concentration.

The cylindrical shells are subjected to external pressure when the shell is discharged; consequently, a vacuum is developed inside the shell [9]. Various researchers have investigated the buckling response of shell structures under uniform external pressure. Tsouvalis et al. [10] reported that initial imperfections had less effect on the buckling capacity of composite cylindrical shells subjected to external pressure when compared to counterpart cylindrical shells manufactured from isotropic materials. Considering the effects of nonlinear pre-buckling deformations and initial imperfections, Hui-Shen [11] proposed a boundary layer theory to predict the buckling and post-buckling response of anisotropic laminated thin-walled cylindrical shells under external pressure. Seong-Hwa Hur et al. [12] experimentally and numerically investigated the buckling and post-buckling behaviour of composite laminated cylindrical shells subjected to external hydrostatic pressure and showed the difference between numerical models and the associated experiments was limited to 15.5% if the initial imperfections were not considered. Lopatin [13, 14] proposed an analytical solution to predict the buckling capacity of composite cylindrical shells with various boundary conditions. Shen and Pan [15] reported considerable load-bearing capacity of geometrically imperfect composite cylindrical shells beyond the buckling when subjected to uniform external pressure.

There is a paucity of research concerning the effect of local geometric imperfections on the buckling behaviour of glass fibre-reinforced polymer (GFRP) cylindrical shells subjected to external pressure. This paper aims to numerically investigate the effect of local dents caused by external interferences on the buck-ling behaviour of GFRP cylindrical shells when subjected to external pressure. For this purpose, 24 finite element numerical models were developed in ABAQUS [16] then the effect of dent depth made at the 1/3rd, half, and 2/3rd of the shell on the buckling response of the cylindrical shells was investigated.

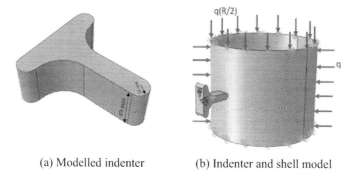

(a) Modelled indenter   (b) Indenter and shell model

Fig. 1 Details of cylindrical shell and indenter models

2 Finite Element Analysis

2.1 Finite Element Modeling

Four-node quadrilateral element (S4R) was used to model five layers of the composite cylindrical shells [14]. The layers of the composite shell were modelled considering the stacking sequence ([30/-30/30/-30/30]) of the composite material relative to the x-axis. The indenter was modelled as a discrete rigid surface, and a frictionless contact algorithm was used to simulate the contact between the cylindrical shell and the indenter. The dimensions of the indenter are shown in Fig. 1a [17].

Convergence and mesh refinement studies were performed to ensure high accuracy in analysis, and a uniform mesh of 10 mm elements was used to model the shells. The bottom edge of the shell was constrained against vertical and radial displacements. The top edge of the shell was only restricted to radial displacements (see Fig. 1b). The shell edge was subjected to a uniformly distributed transverse load proportional to the external pressure [18]. The linear buckling analysis (Eigenvalue) and nonlinear Riks static analysis were carried out to analyse the stability of cylindrical specimens. In nonlinear buckling analysis, the load applied to the structure began from zero and gradually increased with a specific increment. Additionally, the Arc Length method was used to determine the length of the increments. Two steps are defined in the FE model to simulate the effect of dent and external pressure stages, namely:

(i) Linear general static analysis: Locally geometrical imperfection in the form of linear displacement of the indenter was applied in specific locations [19].

(ii) Riks analysis: External pressure was applied to the shell models in the form of a linear distributed load on the top edge of the cylindrical shell and uniform radial external pressure on the circumference of the shell model.

The mechanical properties of the composite material, including elastic modulus, Poisson's ratio, and shear modulus in both longitudinal and transverse directions, were defined. To investigate the performance of composite materials, the linear elastic behaviour of undamaged materials, using Hashin theory which includes four failure

Table 1 Mechanical properties of cylindrical shell material [21]

Elastic modulus E_{11} (GPa)	41.498
Elastic modulus E_{22} (GPa)	13.730
Shear modulus G_{12} (GPa)	4.85
Poisson's ratio v_{12}	0.267
Tensile strength (MPa)	1167.87
Compression strength (MPa)	583.93
Shear strength (MPa)	65

modes, i.e., fibre tension and compression, matrix tension and compression, were modelled [20]. For the case of Hashin damage modelling, the beginning of shell damage is associated with the reduction in stiffness matrice of the shell.

Specifications of the Models

Twenty-four cylindrical shells were modelled, with all the shell models same geometrical specifications, i.e., a diameter of 400 mm, a height of 300 mm, and a wall thickness of 1 mm, which was composed of a five-layer composite laminate with a stacking sequence of [30/-30/30/-30/30]. The cylindrical shells were subjected to local denting load at 1/3rd, half, and 2/3rd of the shell height, where the dent depth was equal to 2, 4, 6, or 8 mm. The mechanical properties of GFRP composite are listed in Table 1. Each cylindrical shell model was assigned with a code "S-x-y-z", where S means the cylindrical shell, x is the height of the indenter relative to the bottom edge of the shell (100, 150, or 200), y is the value of the indenter displacement (2, 4, 6, or 8 mm) and z is the orientation of the indenter relative to the axial axis of the shell (0° or 90°). For example, the S-100-2-0 refers to a shell model containing a 2 mm depth of dent at the height of 100 mm, where the dent was caused by an indenter that had no orientation relative to the shell axial axis.

3 Numerical Analysis and Discussions

In Fig. 2, the failure mode of S-100-4-90 and S-100-6-90 shell models are shown, where for both models the indenter location and the orientation was the same but the value of dent depth was different. Both models failed due to fibre compression mode [22], and the area of damage increased with increasing the dent depth. Additionally, for the case of S-100-6-90 model and beneath the indenter, the fibres' damage value was approximately 80%. For a dent depth of 6 mm, placement of 6 mm for the indenter in the first step before applying of the external pressure, the fibers have been damaged by about 80% under the pressure caused by the indenter (see Fig. 3).

In Table 2, the buckling capacity of the cylindrical shells with various dent depths and denting orientations are compared with the buckling capacity of counterpart cylindrical shells with no denting. The highest reduction in the buckling capacity of the shells was for a shell with a dent depth of 2 mm at the height of 100 mm (1/3rd

Fig. 2 Failure of S-100-4-90 and S-100-6-90 shell models

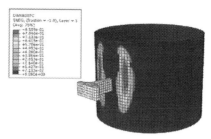

(a) Dent depth = 4 mm

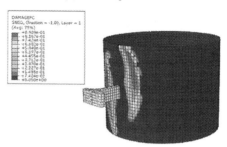

(b) Dent depth = 6 mm

Fig. 3 Fiber damage before applying external pressure

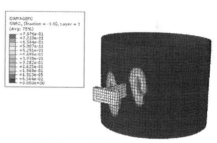

(a) Indenter with 90° orientation

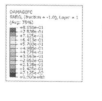

(b) Indenter with 0° orientation

of the shell height). Beyond 2 mm dent depth, the reduction in the buckling capacity of the shells was approximately constant. It can be deduced that a local dent in the proximity of the support had the largest effect on the buckling capacity of the shells (about 25%). Still, the depth of the local dent had a negligible impact in decreasing the buckling capacity of the shells.

In Fig. 4, buckling capacities versus dent depth and indenter orientation at the mid-height of the shells (150 mm from the shell bottom) and at the 2/3rd of the shell height (200 mm from the shell bottom) are shown. The results indicated that the buckling capacity decreased by increasing the dent depth. The reduction for the case of shell models subjected to an indenter with an orientation of 90° was slightly more than the counterpart shell models with an indenter orientation of 0° (about 1%).

Table 2 Comparison of buckling capacity of cylindrical shell models

Model	Displacement of indenter (mm)	Orientation of indenter (°)	Buckling capacity (kPa)	Difference with perfect shell (%)
S-100-2-0	2	0	24.55	25.38
S-100-2-90	2	90	24.55	25.38
S-150-2-0	2	0	30.84	6.26
S-150-2-90	2	90	30.99	5.80
S-200-2-0	2	0	30.94	5.95
S-200-2-90	2	90	31.26	4.98
S-100-4-0	4	0	24.47	25.62
S-100-4-90	4	90	24.49	25.56
S-150-4-0	4	0	24.75	24.77
S-150-4-90	4	90	24.96	24.13
S-200-4-0	4	0	25.02	23.95
S-200-4-90	4	90	24.73	24.83
S-100-6-0	6	0	24.36	25.95
S-100-6-90	6	90	24.44	25.71
S-150-6-0	6	0	19.61	40.39
S-150-6-90	6	90	19.95	39.36
S-200-6-0	6	0	19.93	39.42
S-200-6-90	6	90	20.77	36.86
S-100-8-0	8	0	24.31	26.10
S-100-8-90	8	90	24.39	25.86
S-150-8-0	8	0	19.50	40.72
S-150-8-90	8	90	19.58	40.48
S-200-8-0	8	0	19.37	41.12
S-200-8-90	8	90	19.55	40.57

Fig. 4 Buckling capacity
versus dent depth and
indenter orientation

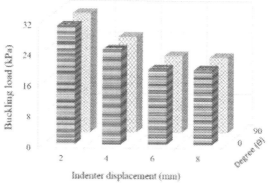

(a) Local dent at shell mid-height

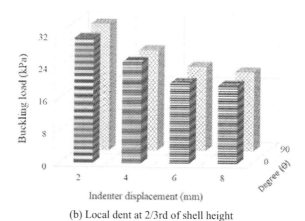

(b) Local dent at 2/3rd of shell height

4 Conclusions

This research aimed to numerically evaluate the effect of local depth caused by an indenter on the buckling behaviour of glass fabric-reinforced polymer (GFRP) cylindrical shells when subjected to external pressure. The key remarks of this research are listed below:

- The failure mode of both models is fibre compression type, and the area of damage increased with increasing the value of dent depth.
- For a dent depth of 8 mm and before applying the external pressure, the fibres were damaged by approximately 80% due to local stresses beneath the indenter.
- The effect of indenter orientation on the buckling capacity of the cylindrical shell models was negligible.

- Whilst the location of the dent depth significantly affected the buckling capacity of the cylindrical shells, the dent depth had a negligible influence on the shells' buckling capacity.
- The effect of dent depth on the reduction in the buckling capacity of the cylindrical shells increased from shell mid-height towards the 2/3rd of shell height.
- By placing the indenter in the middle and near the upper edge, the buckling capacity decreases with the increasing of the amplitude imperfection.

References

1. Weaver PM, Driesen JR, Roberts P (2002) Anisotropic effects in the compression buckling of laminated composite cylindrical shells. Compos Sci Technol 62:91–105
2. Weaver PM, Dickenson R (2003) Interactive local/Euler buckling of composite cylindrical shells. Comput Struct 81:2767–2773
3. Wong KFW, Weaver PM (2005) Approximate solution for the compression buckling of fully anisotropic cylindrical shells. AIAA J 43:2639–2645
4. Semenyuk NP, Trach VM (2007) Stability and initial postbuckling behavior of anisotropic cylindrical shells under external pressure. Int Appl Mech 43:34–328
5. Semenyuk NP, Trach VM, Zhukova NB (2008) Stability and initial postbuckling behavior of anisotropic cylindrical shells subject to torsion. Int Appl Mech 44:41–60
6. Takano A (2008) Improvement of Flügge's equations for buckling of moderately thick anisotropic cylindrical shells. AIAA J 46:903–911
7. Takano A (2011) Buckling of thin and moderately thick anisotropic cylinders under combined torsion and axial compression. Thin-Walled Struct 49:304–316
8. Firouzsalari SE, Showkati H (2013) Thorough investigation of continuously supported pipelines under combined pre-compression and denting loads. Int J Press Vessels Pip 104:83–95
9. Taraghi P, Showkati H, Firouzsalari SE (2019) The performance of steel conical shells reinforced with CFRP laminates subjected to uniform external pressure. Constr Build Mater 214:484–496
10. Tsouvalis NG, Zafeiratou AA, Papazoglou VJ (2003) The effect of geometric imperfections on the buckling behaviour of composite laminated cylinders under external hydrostatic pressure. Compos B Eng 34:217–226
11. Shen HS (2008) Boundary layer theory for the buckling and postbuckling of an anisotropic laminated cylindrical shell. Part II: Prediction under external pressure. Compos Struct 82:362–370
12. Hur SH, Son HJ, Kweon JH, Choi JH (2008) Postbuckling of composite cylinders under external hydrostatic pressure. Compos Struct 86:114–124
13. Lopatin AV, Morozov EV (2015) Buckling of the composite sandwich cylindrical shell with clamped ends under uniform external pressure. Compos Struct 122:209–216
14. Lopatin AV, Morozov EV (2017) Buckling of composite cylindrical shells with rigid end disks under hydrostatic pressure. Compos Struct 173:136–143
15. Shen KC, Pan G (2021) Buckling and strain response of filament winding composite cylindrical shell subjected to hydrostatic pressure: numerical solution and experiment. Compos Struct 276:114534
16. ABAQUS, Version 6.14-4 (2014) ABAQUS/standard user's manual. ABAQUS Inc., USA
17. Ghanbari Ghazijahani T, Jiao H, Holloway D (2014) Experiments on dented cylindrical shells under peripheral pressure. Thin-Walled Struct 84:50–58
18. Fazlalipour N, Showkati H, Ghanbari-Ghazijahani T (2022) Experiments on welded shells with section alteration under axial and peripheral pressure. J Constr Steel Res 193:107277

19. Fazlalipour N, Showkati H, Eyvazinejad Fioruzsalari S (2021) Buckling behaviour of cylindrical shells with stepwise wall thickness subjected to combined axial compression and external pressure. Thin-Walled Struct 167:108195
20. Naseri Ghalghachi R, Showkati H, Eyvazinejad Firouzsalari S (2020) Buckling behaviour of GFRP cylindrical shells subjected to axial compression load. Compos Struct 113269. https://doi.org/10.1016/j.compstruct.2020.113269
21. Rafiee R, Habibagahi MR (2018) Evaluating mechanical performance of GFRP pipes subjected to transverse loading. Thin-Walled Struct 131:347–359
22. Naseri Ghalghachi R, Showkati H (2021) Experimental and numerical investigation of fracture and buckling behavior of chopped GFRP cylindrical shells subjected to axial compression load. Sharif J Civ Eng 37:125–134

Research and Application of Ultra High Performance Concrete in Engineering Projects in Japan

Jian Zhou

Abstract Japan is located in the international seismic zone, and is also a resource intensive country. It has unique features in the research and application of ultra-high performance concrete materials. The paper analyzes and summarizes the engineering application cases of concrete strength grade above 150 N/mm^2, including structural system, mix design, production process, strength grade, etc. Silica fume composite cement, with strict calculation of sand and stone gradation, improves the compactness beyond the conventional concrete; The super high performance water reducing agent can greatly reduce the water cement ratio and improve the working performance, especially the expansion degree; Organic fiber and steel fiber are especially important in fire resistance, explosion resistance and ductility. Ultra-high performance concrete could improve the seismic performance of building structures and the utilization rate of building area, and new materials could provide more choices for design and engineering application.

Keywords UHPC · Mix proportion · Curing process · Compressive strength · Engineering application

1 Introduction

In Japan, the research of high-strength concrete began around 1970 years ago. At that time, admixtures with better water reducing performance than general water reducing agents were successively developed in Japan. The concrete in Japan is developing towards high strength concrete with low water consumption and low water cement ratio. The strength of concrete is greatly promoted by the use of new water reducing agents and the application of silica fume. Because silica fume has strong pozzolanic

J. Zhou (✉)
Harbin Institute of Technology, Shenzhen 518055, China
e-mail: superzcn@126.com

China Construction Science & Technology Group Co., Ltd., Beijing 100195, China

China Construction Science & Technology Group North China Co., Ltd., Beijing 100176, China

© The Author(s) 2023
S. Wang et al. (eds.), *Proceedings of the 2nd International Conference on Innovative Solutions in Hydropower Engineering and Civil Engineering*, Lecture Notes in Civil Engineering 235, https://doi.org/10.1007/978-981-99-1748-8_23

properties and is a mixture of ultra-fine particles, it makes the concrete structure compact, and it is easy to produce high-strength concrete above 50 N/mm^2, so it is widely used in actual structures and prefabricated components (originally PC piles). Ultra-high strength concrete, which exceeds 100 N/mm^2, has developed since 1990. With the further development of research, the research and application from 100–150 to 150–200 MPa has gradually developed, creating another miracle in the field of materials and structures. Under a series of studies and engineering applications, a series of mature systems and measures have been formed for mix proportion design, structural design, production management and quality management of ultra-high performance concrete.

Through reading and sorting out the literature, this article has sorted out six important application cases of ultra-high performance concrete, including project overview, mix proportion design, strength, providing learning and reference for domestic and foreign scholars, and contributing to the development and application of ultra-high performance concrete.

2   Engineering Project Cases

The perfect combination of ultra-high performance concrete and high-strength reinforcement (yield load of 440 MPa) has made high-strength and high-performance materials an important step forward. They have been used in many building structures in Japan, and have been widely used in 100–150 MPa [1–4]. Even 150–200 MPa [5–11] ultra-high strength concrete has been used in actual engineering projects. See Table 1 for the engineering projects applying 150–200 MPa UHPC.

Table 1 shows the basic information of six 150–200 MPa ultra-high performance concrete projects, which are mainly applied to the structural columns at the bottom core of the frame structure system of super high-rise buildings to improve the utilization rate of building area and the seismic performance of buildings. The buildings involved are widely used, but the common point is that the building structure is more regular.

3   Concrete Raw Materials and Mix Proportion

3.1   Concrete Raw Materials

According to statistics, the requirements for raw materials of ultra-high performance concrete above 150 MPa are obtained, and the raw materials are shown in Table 2.

Table 1 The engineering projects applying 150–200 MPa UHPC

Project	Name	Location	Structure	Height	Time	Purpose
N0.1 [5]	Abe Yeqiao Building	Abeno, Osaka	[> 0] steel structure frame system [< 0] steel tube concrete frame system	B5(− 30.47 m) F60(300.00 m)	2010.1–2014.3	Department store, office, hotel, art gallery
N0.2 [6]	Real Estate Development Project	Tokyo	A steel tube concrete frame system	B2 F55 + F2(180.00 m)	2013.5–2013.7	Residence
N0.3 [7]	Actual Buildings	Otsuka, Inzai City	A reinforced concrete frame system	B1,F3 Column: $\varphi = 250$ mm, $L = 4500$ mm	2010.2	Takenaka Research and Development Institute
N0.4 [8, 9]	New Project in the Second Urban Area of Daqiao District	Tokyo	A reinforced concrete frame system	B2(− 17.30 m) F42 + F1(155.27 m)	2011	Residential buildings, shops, offices, public facilities
N0.5 [10]	Research Achievements of Dacheng Construction Technology Center	Chengtian City, Chiba County	Experimental study	–	2007.6–2007.9	For performance evaluation and quality management
N0.6 [11]	The Kosugi Tower	Zhongwanzi, Zhongyuan District, Kawasaki City	A reinforced concrete frame system	B2 F49 + F2(160 m)	2008.6	Concentrated residence

Table 2 Main parameters of concrete raw materials

Introduction	Cement	Fine aggregate	Coarse aggregate	Admixture	Fibre
N0.1 [5]	Silica fume composite cement (SFCS)	Crushed sand: 70%, FM2.77 Mountain sand: 30%, FM2.01 (ρ = 2.60 g/cm^3)	Real product rate: 60.4%; 2015:1505 = 1:1 (volume ratio) (ρ = 2.62 g/cm^3)	Polycarboxylic acid high performance water reducer	–
N0.2 [6]	Composite low heat silica fume cement (ρ = 3.08 g/cm^3) (silica fume 10%)	Hard sandstone crushed sand (ρ = 2.59 g/cm^3)	Hard sandstone gravels (ρ = 2.64 g/cm^3)	① Water reducer (SSP-104 modified 3) ② Sra	Steel fiber: φ = 0.16 mm L = 13 mm
N0.3 [7]	Silica fume composite cement (SFCS) (ρ = 3.01 g/cm^3)	Crushed sand: FM2.78 (ρ = 2.63 g/cm^3)	Crushed stone (ρ = 2.62 g/cm^3) (maximum size 20 mm)	Polycarboxylic acid high performance water reducer (SP1, SP2)	PP: 48 μm, 10 mm Steel fiber: φ = 0.38 mm L = 30 mm
N0.4 [8, 9]	① Composite medium heat silica fume cement (ρ = 3.21 g/cm^3) ② High strength admixture (ρ = 2.51 g/cm^3)	Crushed sand (ρ = 2.63 g/cm^3)	Crushed stone (ρ = 2.64 g/cm^3)	① Reduced polycarboxylic acid high performance water reducer ② Lime expansion agent	–
N0.5 [10]	① Ordinary Portland cement ② Mineral powder ③ Silica fume	Crushed sand	Crushed stone	Low shrinkage polycarboxylic acid high performance water reducer	–
N0.6 [11]	① Ordinary Portland cement ② Mineral powder ③ Silica fume (ρ = 2.99 g/cm^3)	Crushed sand (ρ = 2.63 g/cm^3)	Crushed stone (ρ = 2.62 g/cm^3)	① Polycarboxylic acid high performance water reducer ② Low alcohol shrinkage reducer	Ethylene ethanol copolymer fiber

Table 2 shows that, Silica fume composite cement is widely used in Japan, which has better hydration reaction and grading density, while ordinary Portland cement is mainly used in China; The compactness of concrete is well guaranteed by strictly calculating the proportion of fine aggregate and coarse aggregate. Low water cement ratio is one of the technical requirements of ultra-high performance concrete. The development and application of ultra-high performance water reducing agent greatly improve the working performance of low water cement ratio concrete, which is the key factor. Fiber is an important factor in the application of ultra-high performance concrete to building structures. Organic fiber could improve the fire and explosion resistance of concrete, and steel fiber could improve the ductility and ultimate strength of concrete. The strict requirements for various materials have realized the application of ultra-high performance concrete.

3.2 Mix Proportion

Mix proportion is the key link to realize the application of ultra-high performance concrete materials. The mix proportion parameters are listed in Table 3.

Table 3 shows that, the aggregate grading of ultra-high performance concrete is relatively strict. The proportion of silica fume composite cement to coarse and fine aggregates makes it have a high compactness, so as to achieve high strength; High performance water reducing agent makes concrete with low water cement ratio still have high workability. Raw materials and mix proportion are the key to achieve ultra-high performance, and have important guiding significance.

4 Concrete Performance

Different curing processes are selected for different production methods, the curing process is the key to ensure the strength and working performance of concrete. Table 4 shows different manufacturing, curing process, strength, etc.

Table 4 shows that, whether cast-in-situ or factory prefabricated, heating curing is an optimal and even necessary curing process, which can guarantee the optimal performance in the fastest and best way. Heating curing includes warm water tank curing (40 °C), heating plate curing (90 °C) and steam curing (90 °C). The warm water tank curing and steam curing are applicable to the production process of the supporting prefabrication plant, the heating plate curing and steam curing are applicable to the supporting on-site casting production process, and the steam curing (90 °C + 120 h) is the most selected curing process. The test strength meets the requirements of design strength, and reasonable production process and curing process are effective guarantees for compressive strength.

Table 3 The concrete mix ratio parameters

Introduction	W/C (%)	Expansion (cm)	Sand ratio (%)	Quality (kg/m³)					Additive	
				W	C	Sand (crushed)	Sand (mountain)	Stone (crushed)	Admixture	Fibre (vol.%)
N0.1 [5]	0.170	65	31.2	165	971	299	130	857	–	–
N0.2 [6]	0.125	–	32.4	155	1240	350		744	10	Steel 0.5%
N0.3 [7]	0.117	65	30.2	150	1282	320		739	SP2	PP 0.11% Steel 0.5%
N0.4 [8, 9]	0.140	70	31.2	150	729	629 sand	312 Admixture	620	2.8%	–
N0.5 [10]	0.150	75	–	–	–	–	–	–	–	–
N0.6 [11]	0.160	70	40.2	150	938	550	817	–	PP	–

Table 4 The performance analysis of concrete obtained under different production processes for different mix ratios

Introduction	Construction measures	Curing conditions	Design strength (N/mm²)	Test strength (N/mm²)
N0.1 [5]	Pumping construction	Steam curing (90 °C)	164.0/91d	165.0/91d
N0.2 [6]	① Prefabricated ② Cast in situ	① Steam curing (90 °C) ② Panel heater (90 °C)	200.0/28d	206.5/28d
N0.3 [7]	Cast in situ	Steam curing (90 °C)	200.0/91d	229.0/91d
N0.4 [8, 9]	Prefabricated	Steam curing (90 ± 5 °C + 120 h)	200.0/91d	209.0/91d
N0.5 [10]	Prefabricated	① Warm water at 40 °C ② Adiabatic curing ③ Standard curing	① 150.0/28d ② 150.0/56d ③ 150.0/91d	① 171.0/28d ② 165.0/56d ③ 164.0/91d
N0.6 [11]	Cast in situ	① Standard curing ② Simple thermal insulation maintenance	① 150.0/7d ① 150.0/91d ② 150.0/91d	① 154.0/7d ① 164.0/91d ② 160.0/91d

5 Conclusion

With the development of high-performance materials, ultra-high performance concrete has been applied in Japanese building structures; At the same time, it lays a foundation for the development of new materials and new structural systems.

Silica fume composite cement material and solid grading analysis theory are the key technical supports for the development of ultra-high performance concrete in Japan; The development of super high performance water reducing agent is the core technology for the development of super high performance concrete, which solves the problems of high viscosity and low flow performance; Organic fiber and steel fiber solve their fire resistance, explosion resistance and ductility.

The curing process is the key to its ultra-high compressive strength. Through comparative analysis, it is found that steam curing (90 °C + 120 h) is more reasonable and suitable for large-scale and standardized production and application.

The research and application of ultra-high performance concrete in building structures in Japan have formed perfect technical standards and management measures, which provide guidance for our research and application.

Acknowledgements Thanks to the post doctoral innovation practice base jointly established by China Construction Science & Technology Group Co., Ltd. and Harbin University of Technology (Shenzhen), which provides me with resources and platforms for further learning and growth. I will certainly apply what I have learned to practice and make contributions.

References

1. Xiujie HY, Bangyan HH, Zhiming OC, Arayama M (2004) Construction of high-rise apartment building using high strength concrete of Fc 130 N/mm^2. Fortification Rec 42(10):44–49
2. Ito H, Chayama K, Hirai K, Shogo K (2004) Structural design and construction of high-rise office building with hybrid structure using F = 100 N/mm^2 high strength concrete Singapore. Spec Issue Present Status Future Dev Diversified Hybrid Struct 52(1):95–101
3. Tatsuno T, Yamamoto Y, Kawai K, Kawamoto S (2012) Design and construction of three dimensional urban space constructed by ultra high strength concrete. Fortification Rec 50(7):621–627
4. Mutsu Y, Otsuka K, Ichinomiya T, Sakurada M (2009) Summary of prestressed concrete technology association 47(2):7–13
5. Aoki Y, Iwayama T, Yamada Y, Koichi N (2012) Construction of high strength concrete CFT column of 150 N/mm^2. Fortification Rec 50(8):683–688
6. Munehiro U (2018) Construction of super high strength concrete with 200 N/mm^2 design strength to super high rise apartment. Constr Mach Constr 70(4):21–25
7. Mitsui T, Kojima M, Yonezawa T, Suda M, Haizo M (2010) Design reference strength: 150–200 N/mm^2 development of ultra high strength fiber reinforced concrete and its application to actual building. Tech Rep Collect Jpn Archit Soc 16(32):21–26
8. Yamamoto Y, Nakajima T, Watanabe S, Shimizu Y (2011) Application of super high strength precast concrete design reference strength 200 N/mm^2 in high-rise reinforced concrete buildings. Fortification Rec 49(8):37–42
9. Hattori A. Design and construction of precast concrete column based on design standard strength FC = 200 N/mm^2. Constr Plan Build 11(5):19–23
10. Jinouchi H, Terauchi R, Obama T, Hattori A (2008) Performance evaluation and quality control of high strength precast concrete based on design strength 150 N/mm^2. Tech Rep 46(7):24–29
11. Jinai H, Kuroiwa S, Watanabe S (2007) Production and construction of low shrinkage type ultra high strength concrete of 150 N/mm^2. Dacheng Constr Technol Rep 40:131–138

BIM Electromechanical Pipe Synthesis Experience Summary in Residential Basement

Ting Lei, Ming Jiang, and Jian Yu

Abstract BIM technology has been widely used in practical engineering construction projects, especially in the application scenario of mechanical and electrical pipeline synthesis. In residential projects, the electromechanical pipeline in the basement part is the most complex and concentrated in the whole project. Therefore, this paper studies and summarizes the comprehensive application of electromechanical pipeline of BIM technology in residential basement project. This paper mainly studies the implementation process of BIM electromechanical pipeline comprehensive application of four actual residential basement construction projects, analyzes the problems found and solved in the comprehensive application of electromechanical pipeline in residential basement based on BIM, and summarizes the key points that the project construction party needs to grasp in order to complete such projects with high efficiency, high quality and high standard. In addition, according to the implementation experience of actual engineering projects, this paper also puts forward some suggestions on the implementation process of BIM-based residential basement electromechanical pipeline comprehensive project.

Keywords BIM · Pipeline synthesis · Residential basement

1 Instruction

Pipeline synthesis is to find and check the loopholes in the design in advance by adjusting the spatial location relationship of electromechanical pipelines, so as to coordinate resources, improve space utilization, and thus facilitate construction and save materials and manpower. Compared with other construction methods, the application of BIM technology in basement pipelines can significantly reduce the problem

T. Lei (✉) · M. Jiang
CITIC General Institute of Architectural Design and Research Co., Ltd., Hubei 430014, Wuhan, China
e-mail: 601864043@qq.com

J. Yu
Beijing Glory PKPM Technology Co., Ltd., Beijing 100013, China

© The Author(s) 2023
S. Wang et al. (eds.), *Proceedings of the 2nd International Conference on Innovative Solutions in Hydropower Engineering and Civil Engineering*, Lecture Notes in Civil Engineering 235, https://doi.org/10.1007/978-981-99-1748-8_24

of intersection circles and errors, reduce the amount of design changes, and improve the net height of the finished surface of the basement and the overall appearance [1], while Huang Ping, Zeng Zhusheng, Zheng Yang, and Wu Wei, all based on actual projects, have studied the application of BIM technology in basement pipeline predetermined holes, gravity drainage design, pipe well deepening design, basement pipeline synthesis, outdoor pipeline synthesis, construction process BIM specific application and other scenarios [2–5], many aspects repeatedly verified that compared with the traditional two-dimensional CAD pipeline synthesis, the application of BIM technology in the basement mechanical and electrical pipeline synthesis can indeed improve construction efficiency, project quality. However, the actual project application landing process is not smooth, there are still many problems, the construction unit as the main BIM pipeline synthesis, due to the lack of qualification can not change the original design too much, design modification after re-flip mold workload is huge, and frequent changes in the design drawings, resulting in the original BIM work falls short, BIM repeated design reduces efficiency and increases the cost of use, and the design as the main BIM pipeline synthesis. Designers understand the site environment with differences and not timely, also brings resistance to the application of BIM pipeline comprehensive landing [6], which needs to optimize the implementation process to solve. In addition, the current pipeline synthesis application research summarizes the common pipeline layout principles and avoidance principles, while the actual project pipeline synthesis layout is mostly to avoid collisions, the lack of overall layout plan, resulting in the BIM model does not meet the construction requirements, and ultimately difficult to land.

In order to effectively promote the application of BIM electromechanical pipeline synthesis in actual projects, this paper takes several actual residential basement pipeline synthesis projects as the basis, analyzes the common problems of BIM electromechanical pipeline synthesis in residential basement, summarizes the key points to be grasped in the actual pipeline synthesis projects, summarizes the common residential basement pipeline synthesis layout scheme, and inductively proposes the implementation process for the smooth and efficient implementation of BIM electromechanical pipeline synthesis results, and gives targeted specific suggestions.

2   BIM Mechanical and Electrical Pipeline Comprehensive Common Problem Analysis

The main value point of BIM MEP pipeline synthesis in residential basement is to take the clear height of functional area as the premise, meet the actual construction requirements, combine the cost elements such as materials and measure costs, and integrate the problems that will be encountered during the construction process using design and construction drawings in the traditional mode by adjusting the spatial location relationship of MEP pipelines, so as to improve space utilization, coordinate resources and achieve high-quality construction. This paper summarizes

Table 1 The actual project electromechanical pipe synthesis problem situation

Project name	Area (m²)	Total number of problems (pc)	Percentage of civil engineering drawing problems (%)	Percentage of mechanical and electrical drawing problems (%)	Percentage of comprehensive pipeline problems (%)
Project I	About 70,000	115	66	17	17
Project II	About 90,000	117	78	10	12
Project III	About 90,000	132	38	34	28
Project IV	About 60,000	130	19	33	48

several common problems in residential basement by studying the problem reports of several residential basement pipeline synthesis projects and analyzing the types and quantity distribution of problems that are front-loaded in the process of MEP pipeline synthesis.

The scale of the actual residential basement project studied in this paper is around 60,000–90,000 m², and the types of problems antecedent to its electromechanical pipeline synthesis process are mainly civil drawing problems, electromechanical drawing problems and pipeline synthesis problems, the number and proportion of which are detailed in Table 1.

The problems of drawings include various kinds of labeling problems, at the same time, there are many inconsistencies in plan, elevation, section and system diagrams, in addition, there are also problems of mismatching drawings of each floor, each single unit and human defense, inconsistency between civil construction and mechanical and electrical openings, etc.; the comprehensive problems of pipelines include the problems of insufficient net height, modification of pipeline routing, modification of original design dimensions and modification of original design openings. Among them, the problem of insufficient net height in the integrated pipeline problem mostly occurs in the second floor and below, and the corresponding upper floors are mostly large equipment rooms; the problem of pipeline routing modification mostly occurs in the location where pipelines are concentrated and complicated, and the routing of electromechanical pipelines is optimized under the premise of meeting the professional design function; the problem of original design size modification mostly occurs in the structural beam size and air duct size modification; the problem of original design hole modification mostly occurs in the structural The original design holes are mostly at the holes of structural shear wall, man-proof wall and outdoor structure wall, and the original design holes are inconsistent for the electromechanical pipelines after the general pipeline synthesis, so it is necessary to reissue the hole retention diagram, otherwise it is necessary to increase more electromechanical pipeline bends and may sacrifice the local net height.

The total number of problems in the actual residential basement project studied in this paper is about 120, except for the projects with high quality drawings and simple mechanical and electrical pipelines. The larger the number of problems, the higher the overall quality of the drawings and the more complex the project, the greater the value of the project and the more cost-effective the application of BIM for E&M pipeline synthesis.

3 BIM Mechanical and Electrical Pipeline Comprehensive Implementation Key Points

This paper studies the implementation process of BIM electromechanical deepening of several well-known domestic real estate residential basement projects, and summarizes the key points that need to be grasped to make a good pipeline synthesis project, mainly three aspects of civil model analysis, building functional partition analysis, and electromechanical pipeline synthesis program determination.

3.1 Civil Model Analysis

The civil model analysis mainly includes several aspects such as floor height, structure form and floor slab form. The floor height analysis is mainly used to determine whether there is a partial mezzanine, and the floor height of each floor, in order to predict whether the civil construction conditions are sufficient to meet the basic requirements of the net height of the basement, and for the first floor of the basement, we need to focus on the height of the top slab from the outdoor ground, and check whether the cladding layer can meet the space required for the drainage pipes to discharge the outdoor; the structural form analysis should first determine whether the structure is a frame beam system or a beamless floor cover system. If it is a frame beam system, the height of the main beam and its height from the building surface of this floor should be analyzed; if it is a beamless floor cover system, the height of the column cap of the large surface and the height of the floor slab bottom from the building surface of the current floor should be analyzed; the analysis of the floor slab form mainly includes the analysis of whether the slope of the floor slab, the main thickness of each floor slab and the height of the floor slab bottom, as well as whether there is a large surface of the floor slab drop area and other issues. The following two typical cases are used for detailed explanation.

Case one, the project underground total construction area of 90,905 m², underground a total of two layers, the top of the first floor from the cladding layer to complete the surface height of 2000 mm, enough to meet the drainage pipe out of the outdoor, structural form are frame beam structure system, to the second floor underground, for example, the floor height of 3700 mm, the main beam, the bottom of the secondary beam from the floor building surface height of 2950, 3050 mm, floor slab without slope. Building slab thickness of 100 mm, structural slab thickness of 120 mm, the bottom of the slab from the bottom of the main beam has 530 mm of space, enough for the sprinkler nozzle arrangement and the upper open smoke vent beam nest arrangement, there is a defense area, and the human defense area structural floor slab thickening.

Case two, the project underground total construction area of 90,740 m^2, underground a total of two layers, the top of the first floor from the cladding layer to complete the surface height of 1800 mm, enough to meet the drainage pipe out of the outdoor, structural form of the first floor for the frame beam structure, the second floor of the ground floor for the beamless floor cover, to the second floor of the ground, for example, the floor height of 3600 mm, most of the area for the column cap structure, only the building single area with structural beam, structure conventional The bottom of the column cap is 3130 mm from the building surface height of this floor, and the bottom of a few local unconventional column caps, the bottom of which varies from 2970 to 3110 mm from the building surface height of this floor, the floor slab has no slope, but is mainly divided into two plates, the building slab thickness is 70 mm, the thickness of the structural plate has two thicknesses of 160 and 200 mm, of which the 200 mm thick floor slab area is the human defense area, and the floor slab The bottom from the bottom of the conventional column cap has 240 mm of space, only enough for the sprinkler nozzle arrangement, in addition to the local structure floor slab drop plate area, basically a single area and upstairs projection surface for the engine room area.

3.2 Building Functional Area Analysis

Considering the different net height requirements of different functional areas in basement, it is necessary to classify the common building space according to different functional areas, so as to prepare for the subsequent combination of mechanical and electrical pipeline specific situation to propose a suitable program. The general functional areas of residential basement are garage area, driveway area, Chai Fa maintenance channel, truck operation channel, single entry channel, etc. The net height requirements of these functional areas of the project are analyzed and summarized in advance to lay the foundation for the subsequent comprehensive program of electromechanical pipelines. In addition, there are two other points need to focus on research, one is the fire protection partition and smoke protection partition, understand clearly in advance the project fire protection partition and smoke protection partition, it is very critical to modify the original design routing in the integrated pipeline arrangement, especially the smoke exhaust system; second is the human defense area, understand whether the project has a human defense area, if there is a human defense area, determine which electromechanical pipelines and equipment in the area are installed in advance, and which are installed again in wartime, in order to reserve space. Based on the actual project, this paper analyzes and summarizes the headroom requirements of different building functional partitions in residential basements, as shown in Table 2.

Table 2 Building functional zoning headroom requirements

Building functional area	Net air demand	Note
Garage area	2200 mm	–
Driveway area	2400 mm	–
Large equipment transport channel	3000 mm	Generally on the first floor
Light truck operating channel	3500 mm	Generally on the first floor
Entrance lobby and entrance walkway	2600–2800 mm	General first floor to take the upper limit, the rest of the layer to take the lower limit

3.3 Comprehensive Electromechanical Pipeline Program Determination

Determine the comprehensive scheme of mechanical and electrical pipelines in residential basement. To carry out mechanical and electrical pipeline synthesis, first analyze the situation of mechanical and electrical pipelines, then determine the scheme of comprehensive arrangement of mechanical and electrical pipelines based on civil model and building functional area analysis, combined with special requirements of Party A.

First of all, we need to analyze the mechanical and electrical pipeline situation in the residential basement, the general residential basement mechanical and electrical pipeline system types are shown in Table 3. Throughout the above system, the electromechanical pipeline can be generally divided into three categories: air duct, water pipe and bridge, the highest height of air duct of general residential basement exhaust and smoke system, mostly 500 mm (including 50 mm insulation layer); the diameter of water pipe is mostly DN150, generally not more than DN200, the height of bridge is mostly 100 mm, generally not more than 200 mm. open, and avoid unnecessary cross, no beam building cover structure form try to consider the electromechanical pipeline to avoid the column cap, pipeline layout, as far as possible the same professional and the same system pipeline layout together, at the same time for large equipment transport channel and light truck running channel, need to specifically consider can go several layers of electromechanical pipeline, and whether the duct needs to move to the parking space, to avoid the channel area; Secondly, the vertical space, the air duct system Secondly, the vertical space, air duct system, consider adding support hangers and flanges, occupy 600 mm high space, water pipes and bridges consider occupying 300 mm high space.

Then, based on the civil model and the analysis of the functional area of the building, and combined with the special requirements of Party A, to determine the comprehensive arrangement of mechanical and electrical pipeline program, pipeline inter section, basically follow the principle of large pipe let small pipe, pressure pipe let gravity pipe, water pipe let bridge, weak power bridge let strong power bridge, while minimizing the turning bend, comprehensive consideration to avoid,

Table 3 Types of electromechanical pipeline systems

Professional name	System type name	Note
HVAC	Air and smoke exhaust system	The maximum size of the duct is mostly 500 mm (including 50 mm insulation)
HVAC	Make-up air system	
HVAC	Pressurized air supply system	
HVAC	Air delivery system	
HVAC	Manned air duct system	
Drainage	Spraying system	Water pipe diameter is mostly DN150
Drainage	Fire hydrant system	
Drainage	Water supply system	
Drainage	Pressure sewage/wastewater system	
Drainage	Gravity dirt/waste water systems	
Drainage	Rainwater system	
Strong electricity	High voltage bridge	Except for the lighting bridge height of 50 mm, the rest of the bridge height is mostly 100 mm and 150 mm
Strong electricity	Fire fighting power bridge	
Strong electricity	Non-fire power bridge	
Strong electricity	Charging pile bridge	
Strong electricity	Lighting bridge	
Weak electricity	Automatic fire alarm bridge	
Weak electricity	Weak power bridge	
Weak electricity	Operator bridge	

this paper to the actual project as a case, specifically summarize the six typical residential basement pipeline comprehensive program.

Program one, as shown in Fig. 1, the structure is a frame beam structure, the main beam (i.e., large beam) bottom distance from the floor building surface height minus the net height of demand is still left 600 mm when the ducts are arranged against the bottom of the beam, the sprinkler system combined with subcontracting and intensive degree, it is recommended that a separate layer, occupying 300 mm space, the rest of the water pipe bridges go one layer, occupying 300 mm space, lighting bridges against the bottom of the ducts, go to the bottom, the water pipe bridges meet the ducts, turning bend to the beam nest to avoid.

Program two, as shown in Fig. 2, when the implementation of program one, water pipes and bridging lines are dense, can not go a layer, you can take the bridging and spraying the same layer of the main, spraying branch pipe to go inside the beam nest, the water pipe to go its lower layer, the rest of the same program one.

Program three, as shown in Fig. 3, the structure is a frame beam structure, the main beam (i.e., beam) bottom from the floor building surface height minus the net height of demand is still 800 mm and above, the sprinkler system stick to the bottom of the beam alone go one layer, air ducts go under the spray, air duct top from the bottom of the beam to leave a space of 200–250 mm, the bridge go one layer (in the

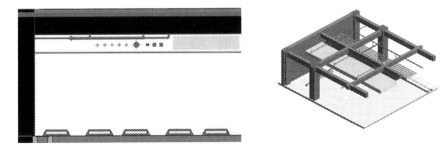

Fig. 1 Scheme 1 cross-sectional view and Scheme 1 3D diagram

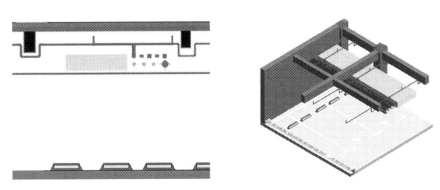

Fig. 2 Scheme 2 cross-sectional view and Scheme 2 3D diagram

spray pipe next layer), occupy 300 mm space, water pipes go its lower layer, occupy 300 mm space, lighting bridges against the bottom of the air ducts, go the bottom layer, water pipe bridges encounter air ducts, turning bend to the beam nest to avoid.

Program four, as shown in Fig. 4, similar to the case of program three, only when the water pipe and bridge longitudinal and transverse pipelines are a lot, you can also

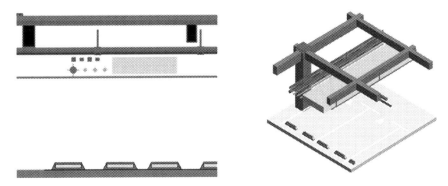

Fig. 3 Scheme 3 cross-sectional view and Scheme 3 3D diagram

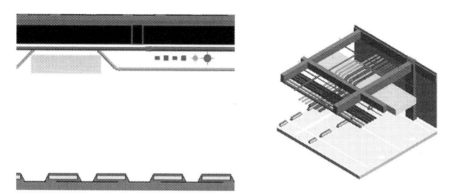

Fig. 4 Scheme 4 cross-sectional view and Scheme 4 3D diagram

consider the water pipe and bridge to take the same layer, and in the longitudinal, it takes a standard high level, in the transverse, it takes another standard high level, has reduced the number of horizontal and vertical cross when the pipeline bending, the rest with program three.

Program five, the structure is a beamless floor cover, when the bottom of the floor slab is less than 450 mm space from the bottom of the conventional column cap, combined with the bottom of the conventional column cap from the height of the building surface at this level minus the net height of the demand, the pipeline layout program is similar to the frame beam structure.

Program six, the structure is a beamless floor cover, when the bottom of the floor slab is greater than 450 mm from the bottom of the conventional column cap space, the sprinkler system can be placed in this layer of space, the rest of the pipeline as far as possible to avoid the column cap, the layout program is similar to the frame beam structure.

4   BIM Mechanical and Electrical Pipeline Comprehensive Implementation Process Proposal

In the actual implementation of the integrated electromechanical pipeline project, in order to implement the BIM integrated electromechanical pipeline results smoothly and efficiently, it is necessary to understand the demand opinions of Party A, the designer and the constructor, to integrate the opinions of all parties into the integrated pipeline model, and to obtain the unified approval of all three parties, in order to meet Party A's headroom requirements, to reduce design changes, to ensure that the constructor is willing to refer to the model construction, and to improve construction quality and progress. After the practice and summary of several residential basement projects, this paper proposes the implementation process of BIM-based residential basement E&M pipe synthesis project.

The specific implementation process and recommendations are as follows: generally a residential basement area of about 100,000 m^2, the entire BIM-based mechanical and electrical pipe synthesis requires about 50 working days, the implementation process in order for the BIM preliminary modeling, pipe synthesis of the initial row, program communication, pipe synthesis of the detailed row, detailed modeling, problem pinning, the results of 7 parts, and finally complete the implementation of the residential basement mechanical and electrical pipe synthesis project, and the time allocation of each part are recommended to be 5, 6, 2, 10, 5, 12, 10 days. BIM preliminary modeling is mainly to establish civil framework and electromechanical main model according to the construction drawings drawn by the design, during the process of recording the problems of errors and omissions in the drawings, and feedback to the design to confirm the modification of the drawings; the preliminary arrangement of pipe synthesis is mainly to analyze the civil model, building functional area analysis and electromechanical pipeline situation analysis, for The main line of electrical and mechanical, in the local scope of the layout of several options to meet the net height requirements; program communication is mainly the A party, the design party, the construction party gathered in a meeting to report the preliminary layout of the program, and combined with the views of the three parties to finalize the layout program; pipe synthesis of the detailed layout is based on the determination of the initial layout program, combined with the latest version of the stable construction drawings, the entire scope of the project pipeline comprehensive work, and record Modify the original design and problem areas (if there is no stable version of the construction drawings, only stay in the program communication, grasp the big picture net height can be, is not recommended to carry out pipe synthesis detailed arrangement); detailed modeling is to establish the branch level end model, room model, model of large sample expression, etc., in order to improve the overall model; problem pinning is mainly to solve the problems recorded in front one by one, and consensus with the A party, the design party, the construction party Solution; the results are mainly based on the completed model of mechanical and electrical pipeline synthesis, a full set of problem reports, pipeline synthesis plan, section drawing, single professional drawing, structure stay hole drawing, secondary masonry stay hole drawing, detail node 3D drawing and other results.

5   Conclusion

BIM technology is an important means of digital transformation in the construction industry, and the value of the application scenario of mechanical and electrical pipeline synthesis in the project has been widely recognized by the industry, but there is still a lot of resistance in the process of specific project implementation. In this regard, BIM technicians are required to continuously improve their professional knowledge and summarize project experience to refine the key issues that need to be grasped in the project, actively innovate to seek value points and continuously optimize the implementation management process, so as to further improve space utilization, coordinate resources, facilitate construction and save materials and manpower for construction projects.

References

1. Zhu DY (2022) Application of integrated BIM technology for basement pipelines in residential projects. Jiangxi Build Mater 04:150–152
2. Huang P, Lin GH, Pang YL (2020) Application of BIM technology in basement pipeline hole reservation. Sci-Tech Dev Enterpr 09:66–68
3. Zeng ZS, Han LL (2020) Comprehensive application of BIM for pipelines in basement and high-rise residential buildings. Installation 10:79–81
4. Zheng Y, Shi HW (2018) Integrated application of BIM for pipelines in basement and high-rise residential buildings. Manage Technol SME 04:169–170
5. Wu W, Wei JP, Cai C (2020) Application of BIM in the process of basement integrated pipeline construction. House 35:59–60
6. Chen XF (2016) Discussion on the comprehensive application of BIM in basement pipelines. Fujian Build Mater 03:34–36

Finite Element Analysis of Steel Tube Bundle Composite Shear Wall with Different Constructions

Shengwu Wan and Xueyuan Cheng

Abstract In order to study the seismic performance of the steel tube bundle composite shear wall with different constructions. Based on the failure tests of three composite shear wall specimens without steel tube bundle end, with studs and stiffener, a feasible numerical model is established by ABAQUS. The hysteresis curve and energy dissipation coefficient, skeleton curve and ductility of the shear wall were further analyses by varying the axial compression ratio and the spacing of the internal diaphragm. The results show that the stud group and the stiffener group can effectively improve the horizontal bearing capacity, energy dissipation capacity, bearing capacity and ductility of the shear wall. Under the condition of low axial compression ratio, the horizontal bearing capacity of the member with stud is increased by 6.3%; Under the condition of high axial compression ratio, the horizontal bearing capacity of members with stiffeners is increased by 4.5%; The change of axial compression ratio and the spacing between inner diaphragms of shear walls with stiffeners has little effect on their energy dissipation capacity and ductility.

Keywords Steel tube bundle composite shear wall · Seismic performance · Finite element simulation · Inner spacer spacing · Axial compression ratio

1 Instruction

Compared with ordinary reinforced concrete members, steel tube bundle composite shear wall, as a new type of shear wall member, has higher bearing capacity and energy consumption capacity, and is suitable for standardized design, factory production and assembly construction.

Wu et al. [1], Huang et al. [2] studied the double steel plate shear wall with structural measures through axial compression test and quasi-static test, and concluded that the ductility of the shear wall with structural measures under periodic load is

S. Wan · X. Cheng (✉)
School of Urban Construction, Wuhan University of Science and Technology, Wuhan, Hubei, China
e-mail: 599282382@qq.com

© The Author(s) 2023
S. Wang et al. (eds.), *Proceedings of the 2nd International Conference on Innovative Solutions in Hydropower Engineering and Civil Engineering*, Lecture Notes in Civil Engineering 235, https://doi.org/10.1007/978-981-99-1748-8_25

higher than that of the general shear wall structure. Takeuchi et al. [3], Yan [4, 5], Ozaki [6], etc. have studied the influence of steel plate spacing, thickness, etc. On the shear wall components of steel plate shear wall components with structural measures through experiments. The latest research status in China mainly focuses on the seismic performance [7, 8], shear performance [9] and compression bending performance [10] of steel tube bundle composite shear wall, and the research on structural measures [11, 12] is less.

In this paper, based on ABAQUS, the steel tube bundle composite shear wall with stud and stiffener at the end is simulate the low-reversed loading tests., and the feasibility of the model is verified by comparing the hysteresis curves of simulation and test. Further changing the axial compression ratio and the spacing between inner diaphragms of steel tube bundle composite shear walls, the influence of various parameters on the seismic performance of steel tube bundle composite shear walls with different structures is analyzed.

2   Overview of Model Design

Three groups of 9 test pieces are designed, Control group (steel tube bundle composite shear wall), shear bolts group (set the specification of two rows to A8* 40 mm@120 mm shear bolts) and stiffeners group (set two groups of 30 * 3 mm stiffeners and arrange them in full length). Among them, each group is also provided with a reference specimen with different axial compression ratio and spacing between inner diaphragms. See Table 1 for dimensions of each test piece. In the finite element simulation, the material properties, specimen size and loading method in the test in literature [12] are adopted.

3   Establishment and Verification of Finite Element Model

3.1   Material Constitutive Relation

The ideal elastic–plastic constitutive relation double line model is adopted for steel. The model assumes that the steel is isotropic and satisfies Mises yield. The elastic modulus of steel is $E_s = 2.03e5$ MPa, Poisson's ratio is 0.3. Considering that the outer steel plate of this model has a restraining effect on the internal concrete and the outer steel plate is rectangular, the constitutive relation of confined concrete is used. The stress–strain relationship of concrete under compression is taken as per [13].

The concrete damage model established in this damage model of concrete by Lubliner [14] and Lee [15], the elastic stiffness degradation caused by concrete damage is described by damage factors (dc, dt).

Table 1 Design parameters of each component

Group	Model no	Axial compression ratio	Wall section size
Blank	CSW1	0.3	103 * (100 + 180 * 3 + 100)
	CSW1-1	0.1	103 * (100 + 180 * 3 + 100)
	CSW1-2	0.3	103 * (100 + 90 * 6 + 100)
Stud	CSW2	0.3	103 * (100 + 180 * 3 + 100)
	CSW2-1	0.1	103 * (100 + 180 * 3 + 100)
	CSW2-2	0.3	103 * (100 + 90 * 6 + 100)
Stiffener	CSW3	0.3	103 * (100 + 180 * 3 + 100)
	CSW3-1	0.1	103 * (100 + 180 * 3 + 100)
	CSW3-2	0.3	103 * (100 + 90 * 6 + 100)

3.2 Unit Selection and Meshing

All grids in the model are structured, C3D8R concrete and C3D8I steel pipe bundle, it can effectively avoid shear self-locking and excessive deformation. The global size of the mesh is 120 mm. The stud is shear bolts as a non independent entity to grid it, meshing is C3D8R and the approximate global size is 1 mm.

3.3 Contact Treatment and Boundary Conditions

In the model, the steel tube bundle, loading beam and foundation slab are considered as a whole, so there is no contact between them. The normal behavior of the contact property between the steel tube bundle and the filled concrete is defined as hard contact, the tangential behavior is defined as penalty contact, and the friction coefficient is taken as the reference value of 0.6. The stiffener is set to be embedded in the concrete in a built-in coupling mode. The stud is embedded into the filled concrete of the steel tube bundle in a built-in coupling mode.

According to the test method, the boundary condition of the model is that the foundation plate is set as fully consolidated, the reference point is set at the centroid position on the right side of the loading beam, the reference point is coupled with the entire end face on the left side of the loading beam, and the low circumferential

Fig. 1 CSW1 hysteretic curve comparison between test and finite element simulation

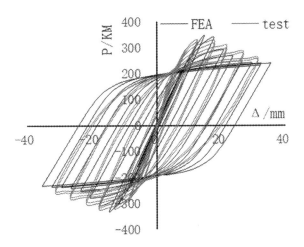

displacement load acts on this reference coupling point. A reference point is set above the loading beam and coupled with the upper end face of the loading beam. The axial pressure acts on this reference coupling point.

3.4 Finite Element Model Verification

Use the above modeling methods to model and analyze CSW1, CSW2, and CSW3, and compare the finite element simulation results of the hysteresis curve with the experimental results. The results are shown in Fig. 1. It can be seen from Fig. 1 that the shape of the hysteresis curve obtained by numerical simulation and test is similar, and the change rule of the curve is also the same, except that the curve values are different. The error of the ultimate load is within 5% and the simulation result is higher than the test result, because the finite element model is the ideal model of the test, and the bond slip between the steel tube bundle and the core concrete is not considered.

4 Finite Element Simulation Results and Analysis

4.1 Hysteresis Curve and Energy Dissipation Capacity

In order to further accurately analyze the influence of axial compression ratio and spacing of inner diaphragms on the energy dissipation capacity of steel tube bundle

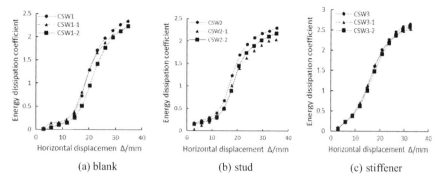

Fig. 2 Energy dissipation coefficient—displacement diagram

composite shear wall with different structural measures, energy dissipation coefficient is used to measure. See Fig. 2 for the energy dissipation coefficient displacement diagram of 9 test pieces.

It can be seen from Fig. 2 that by comparing the blank group, shear bolts and stiffeners with changed axial compression ratio, it can be seen that when the axial compression ratio decreases, the energy dissipation coefficient decreases, indicating that the energy dissipation capacity decreases. By comparing the shear wall groups with different structures in each group, it can be found that the energy change curves of the blank group and shear bolts are quite different, which indicates that changing the axial compression ratio and the spacing of inner diaphragms has a greater impact on the energy dissipation coefficients of the blank group and the stud group. The energy dissipation coefficients of the three specimens of the stiffening group are greater than 2.5. The energy dissipation capacity of the stiffening group is stronger than that of the other groups, and the axial compression ratio and the spacing of inner diaphragms have little influence on it.

4.2 Skeleton Curve and Bearing Capacity

It can be seen from Fig. 3 that when the axial compression ratio decreases and the spacing between inner diaphragms remains unchanged, the slope of the curve descent section of the blank group and shear bolts group becomes smaller, and the trend tends to be flat. The ductility of the specimen increases. However, the curve of stiffening rib group did not change significantly, and the ductility did not improve significantly. When the spacing between inner diaphragms decreases and the axial compression ratio remains unchanged, the ductility of blank group and shear bolts group increases, while that of stiffener group does not change significantly.

It can be seen from Table 2 that when the axial compression ratio decreases and the spacing between inner diaphragms remains unchanged, the ultimate load of the blank group decreases by 3.66%, the ultimate load of the shear bolts group decreases

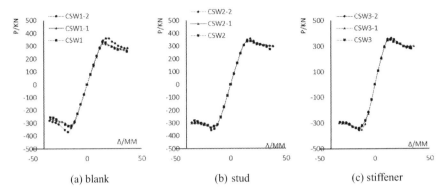

(a) blank   (b) stud   (c) stiffener

Fig. 3 Skeleton curves of each group

by 2.37%, and the ultimate load of the stiffener group decreases by 0.70%; It shows that the change of axial compression ratio has influence on the ultimate bearing capacity of blank group and stud group, and the influence on the ultimate bearing capacity of shear wall with stiffening at the edge can be ignored. When the spacing between the inner diaphragms decreases and the axial compression ratio remains unchanged, the ultimate bearing capacity of the blank group increases by 12.06%, the ultimate bearing capacity of the shear bolts group increases by 6.39%, and the ultimate bearing capacity of the stiffener group increases by 4.50%; It shows that the increase of embedded steel plates will improve the ultimate bearing capacity of shear walls in varying degrees, and the blank group has the greatest impact.

Table 2 Mechanical properties parameters

Model No	P_y/kN	Δ_y/mm	P_u/kN	Δ_u/mm	μ
CSW1	283.53	10.31	315.07	27.02	2.62
CSW1-1	272.49	10.24	303.96	31.45	3.07
CSW1-2	306.20	10.43	340.62	33.49	3.21
CSW2	293.25	9.28	320.29	33.76	3.15
CSW2-1	286.42	9.15	312.59	34.29	3.70
CSW2-2	304.32	9.96	332.57	37.94	3.80
CSW3	299.26	8.65	323.42	29.66	3.40
CSW3-1	296.53	8.37	321.16	29.80	3.56
CSW3-2	305.58	9.74	335.91	29.86	3.46

4.3 Ductility

In this paper, displacement ductility coefficient is used μ to represent the ductility of shear walls with different structures at the edges, which is expressed as $\mu = \Delta_u/\Delta_y$. See Table 2 for displacement ductility coefficient of each specimen.

It can be seen from the displacement ductility coefficient in Table 2 that when the axial compression ratio decreases and the spacing between inner diaphragms remains unchanged, The ultimate bearing capacity of CSW1-1 is less than that of CSW1, but its displacement ductility coefficient is 17.18% higher than that of CSW1; At the same time, the ultimate bearing capacity of CSW2-1 is lower than that of CSW2, but its ductility coefficient is 18.97% higher than that of CSW2; The ductility coefficient in the stiffening rib group is increased by 4.70%. Blank group and stiffening group can improve the ductility of steel tube bundle composite shear wall by reducing the axial compression ratio. When the distance between inner diaphragms decreases and the axial compression ratio remains unchanged, the displacement ductility coefficient increases. The displacement ductility coefficient of CSW1-3 is 22.52% higher than that of CSW1, and that of CSW3-2 is 1.7% higher than that of CSW3. It shows that increasing the number of internal diaphragms can significantly improve the ductility of blank group and shear bolts group.

5 Conclusion

(a) The ductility of the three groups of steel tube bundle composite shear walls in this paper can be improved to varying degrees by reducing the axial compression ratio. The ductility of the blank group and the stud group is significantly improved, and the energy dissipation coefficient is reduced, indicating that the energy dissipation capacity is reduced.

(b) The smaller the spacing between inner diaphragms, the more the ductility increases; The spacing of inner diaphragms has a great influence on the energy dissipation coefficient of blank group and stud group, and the ultimate bearing capacity has been improved to varying degrees, among which the blank group has the largest influence.

(c) The energy dissipation capacity of steel tube bundle composite shear wall structure with stiffening rib structure mechanism is better than blank group and stud group, and the change of axial compression ratio and inner diaphragm spacing has little effect on its energy dissipation capacity.

References

1. Wu XD, Tong LW, Xue WC (2016) Experimental investigation on seismic behavior of short-leg steel-concrete-steel composite shear walls. J Tongji Univ (Nat Sci Ed) 44(09):1316–1323
2. Huang ST, Huang YS, He A (2018) Experimental study on seismic behaviors of an innovative composite shear walls. J Constr Steel Res 148:165–179
3. Takeuchi M, Narikawa M, Matsuo I (1998) Study on a concrete filled structure for nuclear power plants. Nucl Eng Des 179(2):209–223
4. Yan JB, Wang XT, Wang T (2018) Compressive behaviors of normal weight concrete confined by the steel face plates in SCS sandwich walls. Constr Build Mater 171:437–454
5. Yan JB, Wang XT, Wang T (2018) Seismic behaviors of double skin composite shear walls with overlapped headed studs. Constr Build Mater 191:590–607
6. Ozaki M, Akita SO, Suga H (2004) Study on steel plate reinforced concrete panels subjected to cyclic in-plane shear. Nucl Eng Des 228(1):225–244
7. Yu HR, Wang Y, An Q (2021) Seismic performance analysis of anchored prefabricated wall beam connection of steel tube bundle composite shear wall structure. Ind Build 51(06):84–94
8. Zhang P, Zhou XG, Miao ZH (2020) Experimental study on seismic behavior of composite shear wall with bundled steel tube and infill concrete when shear span ratio is 1.5. Build Struct 50(5):109–115
9. Miao ZH, Zhou XG, Zhang P (2018) Simulation analysis on shear resistance of steel tube-concrete composite shear wall. J Yantai Univ (Nat Sci Eng Ed) 31(1):76–82
10. Zhao YF, Zhou XG (2022) Bending performance analysis of steel tube-concrete composite shear wall. J Yantai Univ (Nat Sci Eng Ed) 35(02):235–241
11. Song WG, Zhou XG, Zhao YF (2021) Compressive and bending performance of steel tube bundle-concrete composite shear wall and stiffeners on both sides of steel tube. J Yantai Univ (Nat Sci Eng Ed) 34(3):348–354
12. Wan SW, Xu P, Xu CX (2019) Experimental seismic behavior of steel tube bundle composite shear wall with different constructions. Sichuan Build Sci Res 45(4):18–23
13. Han LH (2016) Concrete filled steel tube structure. Science Press
14. Lubliner J, Oliver J, Oiler S, Onate E (1989) Plastic-damage model for concrete. Int J Solids Struct
15. Lee J, Fenves GL (1998) Plastic-damage model for cyclic loading of concrete structures. J Eng Mech Div ASCE 124(8):892–900

Experiment and Engineering Application Study on the Compression Properties of Foundation Soil in the Process of Reinforcing Existing Buildings with Composite Piled Foundation

Qiufeng Tang, Jianxing Tong, Ning Jia, and Shengbin Zhou

Abstract When an existing building has excessive settlement and has no convergence tendency due to foundation problems, the composite piled foundation is often used to partially underpin the upper load. However, considering that the compression properties of the foundation soil have changed greatly, the compressibility index provided by the initial survey report is no longer suitable for calculating the settlement of the reinforced foundation. In this paper, a consolidation test is designed to simulate the compression state of the foundation soil in the process of building construction, reinforcement, and settlement stability. The test results show that the recompression modulus E_{rs} of the soil sample is 2.5–3.5 times the compression modulus $E_{s1\text{-}2}$. In the engineering example, the recompression modulus is used to calculate the settlement of the reinforced foundation according to the settlement calculation method of the composite foundation with settlement-reducing piles, and the results are close to the actual settlement. In actual engineering, in the absence of supplementary survey, it is recommended to obtain the recompression modulus E_{rs} through simulated test or use 2.5–3.5 times the compression modulus $E_{s1\text{-}2}$ to calculate the settlement of the reinforced foundation.

Keywords Reinforcement of existing buildings · Consolidation test · The recompression modulus · Settlement calculation · Engineering application

Q. Tang · J. Tong (✉) · N. Jia · S. Zhou
State Key Laboratory of Building Safety and Built Environment, No. 30 North Third Ring East Road, Beijing, China
e-mail: tjxcabr@sina.com

Institute of Foundation Engineering, China Academy of Building Research, No. 30 North Third Ring East Road, Beijing, China

Beijing Engineering Technology Research Center of Foundation and City Underground Space Development and Utilization, No. 30 North Third Ring East Road, Beijing, China

S. Wang et al. (eds.), *Proceedings of the 2nd International Conference on Innovative Solutions in Hydropower Engineering and Civil Engineering*, Lecture Notes in Civil Engineering 235, https://doi.org/10.1007/978-981-99-1748-8_26

1 Introduction

In actual engineering, due to problems in survey, design, construction, use, and the surrounding environment, engineering accidents caused by the settlement of existing building foundations occur from time to time [1], such as excessive overall settlement, overall inclination, and local differential settlement, which results in failure of normal use, reduction of safety reserves, reduction of service life and even destruction [2–4].

When the settlement of the existing building is too large and there is no tendency to converge, adding new piles under the foundation is a relatively safe reinforcement solution. According to the proportion of the upper load underpinned by the new piles, it can be divided into the composite piled foundation reinforcement scheme and the piled foundation reinforcement scheme. The piles and the existing foundation in the composite piled foundation share the upper load together after reinforcement. Therefore, the composite piled foundation reinforcement scheme has lower pile bearing capacity of a single underpinning pile, smaller pile diameter, and greater pile spacing (generally greater than 6 times the pile diameter), and these mean less damage to the foundation, less disturbance to the foundation soil during reinforcement, and lower reinforcement costs. All things considered it is usually the first choice.

Considering the large settlement of the foundation before reinforcement, to make the total settlement meet the allowable value of foundation deformation and normal use requirements, the reinforcement design should calculate the settlement of the reinforced foundation. Literature [5] and [6] show that during the construction process, the physical and mechanical indexes such as soil void ratio and compressive modulus have changed significantly. In addition, the stress path of foundation soil is complicated during the reinforcement process. Therefore, the compressibility index provided by the initial survey report can no longer correctly reflect the real compressive properties of the soil, which will lead to large deviations in the calculation results.

Literature [7] analyzes the stress change under the foundation when the composite piled foundation is used to reinforce the existing building: (1) The foundation settlement problem generally occurs shortly after the building is completed, and the foundation soil has not yet been consolidated; (2) Before the piles are connected to the foundation, it is considered that the load shared by the soil between the piles remains unchanged; (3) After the piles are connected to the foundation, the piles begin to share the load, resulting in a decrease in the additional stress of the soil between the piles and an increase in the additional stress of the soil below the pile tip; (4) With the further consolidation of the foundation soil, the additional stress of the soil between the piles increases slightly [8–10], and finally remains stable. Considering that the soil stress at the pile tip is difficult to measure, and the settlement calculation is very complicated, it is not easy to compare the theory with practice, so this paper focuses on the compression properties of the soil between piles.

2 Consolidation Test

Two kinds of soil samples, silty soil and silty clay, were selected as the test objects, representing high compressibility soil and medium to low compressibility soil. Two kinds of consolidation tests were carried out for each soil sample, namely the standard consolidation test and the simulated consolidation test. The test load at each level is the sum of the self-weight stress and the additional stress at the corresponding depth.

The standard consolidation test is carried out according to the *Standard for Geotechnical Testing Method* [11]. The next level of load is loaded after the settlement of the soil sample under the current level of load is stable. The loading sequence of the simulated consolidation test is loading, holding, unloading, reloading, holding until stable. The loading time of each level is proportionally scaled according to Formula 1. Considering that the compression amount in the reloading stage of the simulated consolidation test is very small, the unloading range is slightly larger than the actual situation to reduce the test error.

$$\Delta t_{sim} = \frac{\Delta t_{std} \cdot t_1}{t_2} \tag{1}$$

In the formula, Δt_{std} is the average time required for the soil sample to reach settlement stability under each load in the standard consolidation test. In the test, it takes 6–8 h for silty soil and 3–4 h for silty clay t_1 is the time required for building construction and t_2 is the time required for building settlement stability, which are both determined according to engineering experience; Δt_{sim} is the average loading time of each level of load in the simulated consolidation test. It is estimated that it takes 40 min for silty soil and 1 h for silty clay in the test.

The test load is based on the measured data of a reinforcement project in Hainan province. The size of the foundation is about 70 m × 13 m, the pressure of the soil between the piles before and after reinforcement is 130 kPa and 78 kPa respectively. The main equipment and soil samples are shown in Fig. 1 and the test results are shown in Table 1.

According to the test results, the recompression modulus E_{rs} of the simulated test is 2.5–3.5 times the compression modulus E_{s1-2} of the standard consolidation test under the test conditions.

3 Engineering Example

The engineering example is a reinforcement project in Hainan province. The project includes several buildings, each of which adopts shear wall structure, raft foundation, and composite foundation with cement-soil mixed piles. The characteristic values of bearing capacity of single pile, foundation soil and composite foundation are 300 kN, 130 kPa and 250–300 kPa respectively. The pile tip falls on the layer ④. The pile length is 15–18 m, the pile spacing is 1–1.2 m, the pile diameter is 500 mm, the pile

Fig. 1 The consolidation instrument and soil samples

Table 1 Test results

Soil sample	Sampling depth (m)	Recompression modulus E_{rs} (MPa)	Compression modulus E_{s1-2} (MPa)	E_{rs}/E_{s1-2}
Silty soil	11	5.93	2.03	2.92
	15	6.41	2.30	2.79
	20	11.71	3.28	3.57
Silty clay	6	14.75	4.17	3.54
	13	14.36	5.68	2.53
	20	14.30	5.44	2.63

body strength is not less than 5.1 MPa, and the cushion thickness is 250 mm. The physical and mechanical indexes of each soil layer are shown in Table 2.

The settlement of each building after completion is too large and there is no convergence trend. The reason for the excessive settlement is that the foundation treatment method is improperly selected, the construction quality has serious defects, and the silty soil layer is thick. The test results of borehole sampling show that the cement-soil mixed piles in the silty soil layer are uneven, broken, and low in strength, and cannot effectively transfer the upper load to the deep layer.

The reinforcement scheme is to add piles under the foundation to partially underpin the upper load, forming a composite piled foundation with the existing composite foundation. The pile is a steel pipe with 8 mm wall thickness and C30 concrete poured inside. The parameters of the underpinning piles are shown in Table 3.

Table 2 The physical and mechanical indexes of each soil layer

Layer number	Soil layer	Average thickness (m)	Natural unit weight γ (kN m^{-3})	Compression modulus s_{1-2} (MPa)	Limit shaft resistance (kPa)	Limit tip resistance (kPa)	Characteristic values of bearing capacity (kPa)
①	Fill	1.44	17.6	4.74		/	70
②	Middle sand	4.45	19.3	12.48	30	/	130
②$_1$	Fine sand	1.39	18.8	11.29	18	/	120
③	Silty soil	13.15	17.1	2.37	19	/	80
④	Gravel sand	5.39	20.7	13.92	116	2500	260
⑤	Silty clay	Unpenetrated	18.8	8.66	85	1200	280

Table 3 The parameters of the underpinning piles

Building No.	Pile diameter (mm)	Pile length (m)	Pile spacing (m)	Characteristic value of bearing capacity of single pile (kN)	Designed underpinning ratio (%)	Number of piles
A	299	18–19	1.8	300	44.2	280
B	299	18–19	1.9	300	35.9	204
C	245	14–15	1.9	250	37	195

There are two methods for settlement calculation of composite piled foundations in *Technical Code for Building Pile Foundations* [12]. One is that the settlement of the pile top, which is equal to the sum of the compression amount of the pile body and the compression amount of the soil under the pile tip, can be calculated as the settlement of the foundation. The other is that the settlement of the soil between the piles, which is equal to the settlement of the soil between the piles caused by the additional stress of the soil and the interaction between the pile and the soil, can be calculated to avoid the calculation of the plastic penetration amount of the pile tip.

The two calculation methods are mainly aimed at new buildings, but they are also applicable to the reinforcement of existing buildings with composite piled foundations in principle. In this project, the underpinning pile tip provides about 30% of the bearing capacity of a single pile, and the bearing layer is good, so the plastic penetration of the pile tip can be ignored, which means both calculation methods are applicable. However, according to the design idea of the simulated consolidation test in this paper, we can only choose the second calculation method.

The calculation parameters and results are shown in Table 4. The recompression modulus of layers ② and layer ③ is 3.5 times E_{s1-2} considering the replacement and

Table 4 The calculation parameters and results

Building No.		A	B	C
Foundation size		About 70 m × 13 m	About 60 m × 13 m	About 50 m × 13 m
Press of the soil between the piles (kPa)		78	74	88
Calculation depth (m)	Layer ②	3	3	5
	Layer ③	15	15	9.3
	Layer ④	5.4	5.4	3.9
Recompression modulus E_{rs} (MPa)	Layer ②	43.68		
	Layer ③	8.30		
	Layer ④	41.76		
Calculated settlement value of the center point (mm)		51.3	48.5	38.0
Average measured settlement value (mm)		42.9	38.5	33.0

reinforcement effects of cement-soil mixed piles, and the recompression modulus of layer ④ is 3.0 times E_{s1-2}.

The calculated settlement of the reinforced foundation is 1.15–1.26 times of the average measured settlement, which means the results are relatively accurate and safe. If we adopt the compression modulus E_{s1-2}, the results will be far greater than the actual settlement, which will cause the reinforcement scheme to be too conservative.

On the other hand, it also proves that the second method can be applied to calculate the settlement of the reinforced foundation of the existing building reinforced by composite piled foundation.

4   Conclusion

Through consolidation tests and engineering examples, this paper studies the compression properties of foundation soil in the process of reinforcing the existing building with composite piled foundation, and the following conclusions can be drawn:

(1) In the simulated consolidation test, the recompression modulus of high compressibility soil and medium to low compressibility soil at different depths is 2.5–3.5 times the compression modulus E_{s1-2} of the standard consolidation test.

(2) In the engineering example, the recompression modulus is used to calculate the settlement of the reinforced foundation according to the composite foundation with settlement-reducing piles settlement calculation method. The results

are close to the actual settlement, which prove the applicability of the test conclusion.

(3) In actual engineering, in the absence of supplementary survey, it is recommended to obtain the recompression modulus E_{rs} through simulated test or use 2.5–3.5 times the compression modulus E_{s1-2} to calculate the settlement of the reinforced foundation.

References

1. (2012) Technical code for improvement of soil and foundation of existing buildings (JGJ123-2012). China Architecture & Building Press, Beijing
2. Gu BH (2015) Review of typical cases of geotechnical engineering. China Architecture & Building Press, Beijing
3. Sun XH, Tong JX, Yang XH, Luo PF, Du SW (2017) Application of micro-piles in reinforcement of soft soil area. Chin J Geotech Eng 39(S2):91–94
4. Sun XH, Tong JX, Yang XH, Sun XZ, Zhao ZP (2016) Research on engineering application of composite foundation with settlement-reducing piles in soft soil area. Build Sci 32(S2):206–209
5. Teng YJ et al (2012) Reconstruction and reinforcement technology of existing building foundation. China Architecture & Building Press, Beijing
6. Li QR (2008) Experimental study on redesign of soil and foundation for increasing storey on existing buildings. MA thesis, China Academy of Building Research
7. Tang QF, Tong JX, Jia N, Zhou SB (2021) Experimental study on the compression properties of the foundation soil of existing buildings considering the construction and reinforcement process. Build Struct 51(S1):1723–1727
8. Yan ML, Zhang DG (2001) Technology and engineering practice of composite foundation with CFG piles. China Water & Power Press, Beijing
9. Liu DL, Zheng G, Liu JL, Li JX (2006) Experimental study on comparison of behavior between rigid pile composite foundation and composite pile foundation. J Build Struct 27(4):121–128
10. Zhou F, Zai JM, Mei GX (2005) The composite ground and the composite pile foundation. Geotech Eng Tech 19(3):141–143, 151
11. (2019) Standard for geotechnical testing method (GB/T 50123-2019). China Planning Press, Beijing
12. (2008) Technical code for building pile foundations (JGJ94-2008). China Architecture & Building Press, Beijing

The Forward Simulation on Geometric Characteristics of Adverse Buried Bodies Using Ground Penetrating Radar

Hengyi Li, Shengya He, Jianjing Zhang, Liang Ye, Haijia Wen, Congcong Li, Huchen Duan, Xie Peng, and Haitao Zhu

Abstract Adverse buried bodies near to ground surface would damage the buildings, and the detailed geometric characteristics of adverse buried bodies is essential for reducing damage and potential risk. To achieve this goal, forward simulation on geometric characteristics of three typical culverts in Chongqing were carried out in this paper. The response characteristics of (ground penetrating radar) GPR profiles caused by geometric characteristics and filling materials change were summarized, and the apexes of diffraction hyperbolas and lateral changes in the reflection pattern were used to determine the dimension and boundary of buried bodies. The comparison between GPR profiles interpretation result and the measured data in the field was used to verify the validity.

Keywords Buried body · Ground penetrating radar · Geometric characteristics · Forward simulation

1 Introduction

With the evolution of long-term complex geological environment and the development of human activities, Natural geo-bodies and man-made buried bodies appear [1]. The demand for energy leads to the continuous development of land, and the

H. Li · S. He · L. Ye
Guangzhou Metro Design and Research Institute Co., Ltd., Guangzhou 510010, China

J. Zhang
College of Civil Engineering, Southwest Jiaotong University, Chengdu 620031, China

H. Wen
College of Civil Engineering, Chongqing University, Chongqing 400045, China

C. Li · H. Duan · X. Peng (✉)
College of Civil Engineering and Architecture, Hainan University, Haikou 570228, China
e-mail: Peng_Xie@hainanu.edu.cn

H. Zhu
School of Civil Engineering, Tianjin University, Tianjin 300350, China

© The Author(s) 2023
S. Wang et al. (eds.), *Proceedings of the 2nd International Conference on Innovative Solutions in Hydropower Engineering and Civil Engineering*, Lecture Notes in Civil Engineering 235, https://doi.org/10.1007/978-981-99-1748-8_27

existence of buried bodies will adversely affect the construction and operation of infrastructure and buildings.

GPR is a non-destructive geophysical detection technology. Because of its high detection accuracy, it is widely used in many fields [2]. Underground archaeological investigation is a typical representative [3]. For hydraulic engineering, the erosion inside the dam retaining walls, structurally hidden defects and groundwater distribution characteristics are all applied [4]. In the process of geological survey, it is used to locate buried geological bodies and survey near-surface faults [5], including detection of distribution characteristics of Karst caves and sinkholes, changes in rock and soil properties and water content [6]. In civil engineering, the drawing of internal cracks in rock slopes, the evaluation of the stability of existing buildings and road foundations, and the investigation of scour pits around underwater piers have achieved good results [7].

Through the above elaboration, it can be seen that the application of GPR in practical engineering is quite extensive, and has achieved fruitful research results and economic benefits. However, through the forwarding analysis of two-dimensional GPR images, the shape of the buried object, especially the size range, is relatively less defined. Therefore, forward simulation on geometric characteristics of three typical culverts in Chongqing were carried out in this paper. The response characteristics of (ground penetrating radar) GPR profiles caused by geometric characteristics and filling materials change were summarized, and the apexes of diffraction hyperbolas and lateral changes in the reflection pattern were used to determine the dimension and boundary of buried bodies. The comparison between GPR profiles interpretation result and the measured data in the field was used to verify the validity.

2   The Basic Principle of GPR

GPR is a geophysical method propagating through high frequency electromagnetic wave. A typical GPR system consists of a laptop, a DAD control unit, and an antenna. During the whole exploration process, the transmitted electromagnetic wave propagates in the medium. Due to the different properties of the medium, the interface between different materials is reflected. The reflected signal is received and used for image storage and processing.

3   Acquisition and Interpretation of GPR Image

3.1   GPR Image Acquisition

According to the size difference of buried body, 80 and 400 MHz antennas were employed to complete the continuous survey along the measurement line parallel

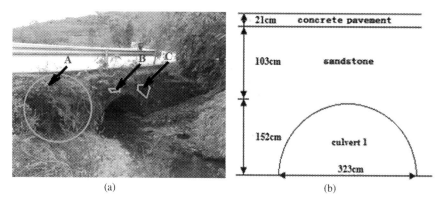

(a)   (b)

Fig. 1 Culvert 1: **a** is the picture of culvert 1 in the field and **b** is the diagram of culvert 1's dimension

to the road direction, and the supporting data was obtained using RIS software. To ensure the accuracy of the image acquisition, GresWin2 software is used to process the two-dimensional GPR image.

3.2 Interpretation of GPR Image

Figure 1a is the picture of culvert 1, and the stratum structure and engineering geological environment characteristics are shown in Fig. 1b.

Figure 2 is a two-dimensional GPR image of culvert section 1. There is a discontinuity in the two-dimensional image at a depth of 0.2 m, so that the existence of the interface L_1 can be determined. The coaxial dislocation in the marked B and C regions in the figure indicates the damage on the actual structure (regions B and C in Fig. 1a). The diffraction hyperbola is formed at the top of the culvert (red curve in Fig. 2). Due to the presence of air between the culvert and the water flow, a diffraction independent of the top of the culvert is formed (green curve in Fig. 2). It can be seen from the diffraction curve vertex marking data that the buried depth is 1.24 m below the ground. Due to the high relative permittivity of water, the electromagnetic wave has a high degree of attenuation during propagation (region D in Fig. 2). In addition, due to the relatively developed vegetation on the surface of the bedrock, and the GPR image manifests the coaxial dislocation of the local area (Region A in Fig. 2).

The bedrock of culvert 2 is dense sandstone, and the shape of culvert is irregular quadrilateral. Through the actual investigation, the horizontal distance, buried depth and specific size from the starting point of the test point are shown in Fig. 3b. From the measured two-dimensional GPR image, the abnormal region E can be delineated (Fig. 4a). Through the processing of the image, the diffraction hyperbola will be formed at the inflection point of the irregular area, and the vertex of the hyperbola will be marked (Fig. 4b): F (1.8, 1.04), G (2.15, 1.04), H (1.75, 1.6), I (2.23, 1.6). By connecting these four points, the range of abnormal section can be determined.

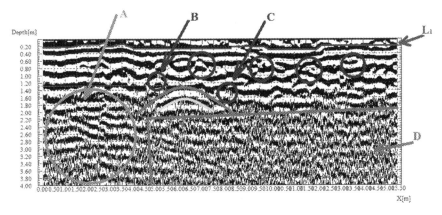

Fig. 2 The labeled GPR profile of culvert 1 after processing

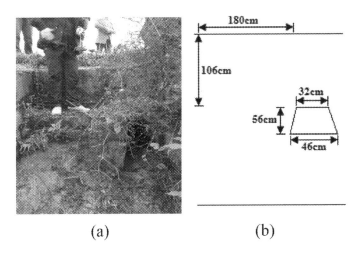

Fig. 3 Culvert 2: **a** is the picture of culvert 2 in the field and **b** is the diagram of culvert 2's dimension

Culvert 3 is a 0.35 m diameter concrete pipe buried 0.5 m below the soil level, surrounded by vegetation (Fig. 5a, b). Due to the difference in permittivity between air, concrete and rock and soil, the diffraction image appears black-white-black hyperbolic alternation. Ignoring the polarization and diffusion effects of the medium, the distance between the hyperbolic vertices in pairs represents the diameter of the concrete tube. According to the coordinates of the peak points L (1.2, 0.5) and M (1.2, 0.85), the buried body is buried 0.5 m deep, the diameter is 0.35 m, and the distance from the starting point of the measurement is 1.2 m. The position and size obtained by image analysis are consistent with the spatial position of the cross-section distri-

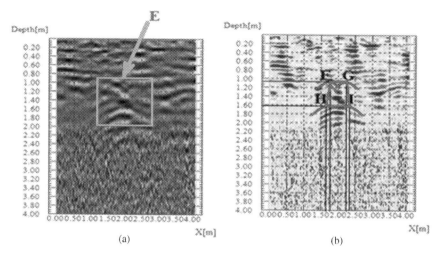

(a)   (b)

Fig. 4 GPR profile of culvert 2: **a** is the image after processing and **b** is the labeled image after processing

bution. In addition, Fig. 6 shows an anomalous region K at a depth of 0.95 m, and the cross-sectional shape of the buried body is circular and the cross-sectional diameter is 0.24 m.

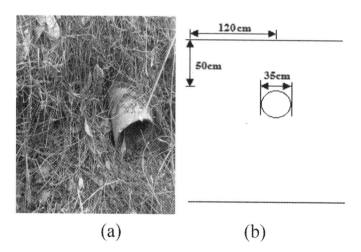

(a)   (b)

Fig. 5 Culvert 3: **a** is the picture in the field and **b** is the diagram of actual dimension

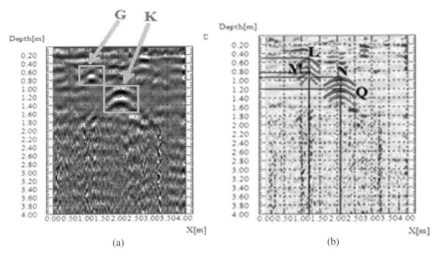

Fig. 6 GPR profile of culvert 3: **a** is the radargram after processing and **b** is the labeled image after processing

4   Conclusion

The shape of sections were corresponding to different diffraction images. An arcuate section appear a single diffraction hyperbolic. For an circle section, a pair of diffraction hyperbolas are formed at the top and bottom of the section. When the cross-section is an irregular quadrilateral, a diffraction single hyperbolic will be formed at the turning point of the irregular pattern.

Based on the vertex coordinates of the image diffraction curve and the change of the horizontal reflection image, geometric characteristics for the cross-section of buried bodies can be determined. The results indicate that the consistency of the regular graphic results is high, and there is a certain error for irregular quadrilateral, but the error value is only 2 mm.

Combing the measured data in the field with orthogonal two-dimensional GPR images is a future development direction from the perspective of engineering practice.

Funding Statement The research on which this article is based has been supported by grants: The Independent innovation fund project (Tianjin University and Hainan University) (Grant No: TDHD-2206); Special Research Project of Guangzhou Metro Design and Research Institute Co; Ltd (Grant No: KY-2021-058); Hainan Provincial Natural Science Foundation of China (Grant No: 520QN229, 422RC599).

References

1. Zhang LF, Zeng XS, Yao Y, Liao WL (2007) Summary of karst collapse research in China. Chin J Geol Hazard Control 18(3):126–130
2. Deng SK, Qi MS (1993) Application of GPR in landslide and karst area detection. Earth Sci J Chin Univ Geosci 18(3):329–338
3. Orlando L (2013) GPR to constrain ERT data inversion in cavity searching: theoretical and practical applications in archeology. J Appl Geophys 89:35–47
4. Xu XX, Zeng QS, Li D, Wu J, Wu XG, Shen JY (2010) GPR detection of several common subsurface voids inside dikes and dams. Eng Geol 111:31–42
5. Pin WT, Chih CC, Tao TL et al (2015) Applying FDEM, ERT and GPR at a site with soil contamination. A case study. J Appl Geophys 121:21–30
6. Hossain D (2013) Evaluation of ground penetrating radar and resistivity profilings for 26 characterizing lithology and moisture content changes: a case study of the high-conductivity United Kingdom Triassic sandstones. J Geophys Eng 10:065003
7. Gutiérrez F, Galve JP, Lucha P, Bonachea J, Jordá L, Jordá R (2009) Investigation of a large collapse sinkhole affecting a multi-storey building by means of geophysics and the trenching technique (Zaragoza city, NE Spain). Environ Geol 58:1107–1122

An Experimental Study on Intelligent Bearing System for Bridge Health Monitoring

Gaofei Teng, Xinning Cao, Weiming Gai, and Ruijuan Jiang

Abstract Bridge Bearings are important devices for transferring loads between the upper and lower structures of bridges. They can ensure the safety of bridges and regulate deformation. They can also prevent bridge displacement caused by temperature changes, seismic forces, and other factors in various structural types. The status of bridge bearings can effectively reflect the healthy status of the bridge. Monitoring of stresses on bridge bearings can be used to evaluate the healthy status of the bridge structure. However, bridge bearings are installed in invisible locations, and traditional manual inspection methods cannot accurately determine their damage and degree of destruction. In this paper, an intelligent bearing pair is designed to resolve this problem. The designed intelligent bearing status assessment system with multi-level threshold values can trigger a warning when the threshold is exceeded. Experiments have been conducted to verify the accuracy of its signal processing.

Keywords Bridge Intelligent Bearing · Bridge structure · Bridge health monitoring

1 Introduction

Bridges, as an important part of highways, are key hubs of transportation systems. Bridge construction has made great progress in recent years around the world. The wide application of bridges makes logistics and transportation more convenient, and traffic volume increases day by day. However, due to design defects, erosion, and aging of the stressed members, the bridges may be damaged with different degrees. As a result, bridge collapse may lead to different degrees of economic losses and human casualties. To reduce the occurrence of similar accidents, the traditional practice is

G. Teng · X. Cao
School of Mechanical, Electrical and Information Engineering, Shandong University, Weihai 264209, China

W. Gai (✉) · R. Jiang
Shenzhen Municipal Design and Research Institute, Co., Ltd., Shenzhen 518029, China
e-mail: gaiweiming@163.com

© The Author(s) 2023
S. Wang et al. (eds.), *Proceedings of the 2nd International Conference on Innovative Solutions in Hydropower Engineering and Civil Engineering*, Lecture Notes in Civil Engineering 235, https://doi.org/10.1007/978-981-99-1748-8_28

to conduct periodic inspections of bridge structures. The reliability and safety of the structure are assessed based on the inspection results. However, this approach is costly, non-real-time, and poor in integrity and real-time. Thus, it cannot effectively ensure the safety and durability of bridges during usage. With the development of electronic information technologies, intelligent health monitoring systems for bridge structures have come into being [1–4]. Unlike the static observation method of structural inspection, intelligent monitoring of bridge structures is real-time dynamic monitoring. By analyzing the real-time stress signal of the bridge, the safety status of the bridge can be obtained. The system mainly includes components such as advanced sensor subsystems [5, 6], stable data acquisition and data transmission systems [7, 8], and reliable monitoring and safety assessment platforms [9, 10]. The modules of the system operate in different hardware or software environments and work in concert to accomplish the intelligent health monitoring and safety assessment functions of bridge structures.

2 The Bearing Capacity Experiment of Bridge Intelligent Bearing

2.1 Equipment and Purpose of the Experiment

The purpose of this experiment is to verify the accuracy of signal processing of force measuring bearings and calibrate the relationship between the calibrated measuring point stress and the intelligent bearing sensor. The experiment data is used to determine the threshold values of each level for multi-level warning of the bearing status, and to provide data support for the intelligent bearing status evaluation system.

A press with a maximum value of 300 MN was used for this experiment, which can realize the functions of equal-rate loading, equal-rate displacement test, load holding, etc. The physical diagram of the pressure test machine is shown in Fig. 1. The intelligent support adopts model GPZ (II)-2.0–10%-GD, as shown in Fig. 2.

2.2 Process of the Experiment

Before the experiment starts, the sensors will be installed in the four corresponding positions of the Bridge Intelligent Bearing. The collection equipment will be installed on the test bench and powered on for testing. The test is officially conducted after the equipment works normally. The equipment arrangement is shown in Figs. 3 and 4.

During the experiment, the vertical application pressure is loaded and unloaded at 11 levels, which are 0, 0.55, 0.868, 1, 1.468, and 1.71 MN. When the pressure machine is loaded to a certain level, it is held under that load for three minutes. When

Fig. 1 Pressure machine

Fig. 2 The appearance of Bridge Intelligent Bearing

Fig. 3 The arrangement of acquisition equipment

Fig. 4 The arrangement of
Bridge Intelligent Bearing

the level of 1.71 MN grade is reached, unloading is performed in a graded manner.
The experiment test has been repeated three times.

2.3 The Results of the Experiment

Firstly, the vertical applied pressure F (MN) of the press is converted into the
compressive stress P (MPa) of the rubber sheet, as shown in Eq. (1).

$$P = \frac{F}{\pi \times \left(\frac{d}{2}\right)^2} \tag{1}$$

where P is the value of compressive stress applied vertically in MPa; F is the vertical
applied pressure in MN; d is the diameter of the rubber sheet 0.33 m.

The calculated values were used as theoretical values, and the data were compared
with the compressive stress values of the four measurement points obtained from the
acquisition equipment to obtain the relationship curves, as shown in Fig. 5.

The average error percentage of each measurement point is obtained by summing
and averaging according to the relative error, as shown in Eq. (2).

$$z = \frac{\sum_{i=0}^{n} \frac{\left(\sum_{j=0}^{m} x_i\right) - y_i}{y_i}}{n + 1} \tag{2}$$

where z, x_i, and y_i denote the average error ratio, measured value, and theoretical
value, respectively. The average error percentages of measurement points 1, 2, 3, and
4 are 8.38%, 10.60%, 8.90%, and 11.60%, respectively. The results show that the
compressive stress values from the acquisition system are highly accurate.

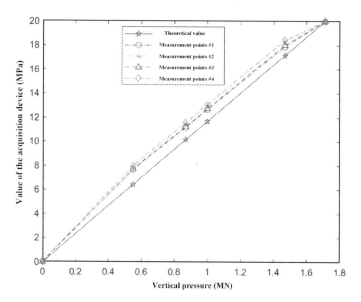

Fig. 5 The relationship curve between stress and theoretical value at each measurement point of fixed support

3 Conclusion

To realize bridge health monitoring, we designed the intelligent bearing system by mearing the force loaded on the bearings. The signal accuracy of the force measuring was verified through experiments. The whole system errors could be eliminated by data obtained from the experiment. The experiments we conducted verified the feasibility of the intelligent bearing, which provides theoretical support for the practical application of the subsequent health monitoring system.

Acknowledgements This research was funded partially by Key Coordinative Innovation Plan of Guangdong Province, Technology Developing Project of Shenzhen, grant number Key 20180126, National Natural Science Foundation of China, grant number 41904158, the China Postdoctoral Science Foundation, grant number 2019M652385, Shandong postdoctoral innovation project, grant number Key 202002004, and the Young Scholars Program of Shandong University, Weihai (20820201005), Weihai Science and Technology Development Plan.

References

1. Philip SM, Frangopol DM (2008) Reinforced concrete bridge deck reliability model incorporating temporal and spatial variations of probabilistic corrosion rate sensor data. Reliab Eng Syst Saf 93(3):394–409

2. Liuyu Z (2011) Construction of intelligent control and safety warning system for long-span continuous rigid frame bridge. In: 2011 fourth international conference on intelligent computation technology and automation, pp 372–375
3. Zhou GD, Yi TH (2013) Recent developments on wireless sensor networks technology for bridge health monitoring. Math Prob Eng 1–33
4. Yu J, Ziehl P, Zárate B et al (2011) Prediction of fatigue crack growth in steel bridge components using acoustic emission. J Constr Steel Res 67(8):1254–1260
5. Nagayama T, Spencer BF, Rice JA (2009) Autonomous decentralized structural health monitoring using smart sensors. Struct Control Health Monit 16(7):842–859
6. Yao Y, Glisic B (2015) Sensing sheets: optimal arrangement of dense array of sensors for an improved probability of damage detection. Struct Health Monit 14(5):513–531
7. Zou Z, Bao Y, Li H, Spencer BF, Ou J (2015) Embedding compressive sensing-based data loss recovery algorithm into wireless smart sensors for structural health monitoring. IEEE Sens J 15(2):797–808
8. Zou Z, Bao Y, Deng F, Li H (2015) An approach of reliable data transmission with random redundancy for wireless sensors in structural health monitoring. IEEE Sens J 15(2):809–818
9. Sun Z, Zhang Y (2016) Failure mechanism of expansion joints in a suspension bridge. J Bridg Eng 21(10):1–13
10. Siringoringo DM, Fujino Y, Namikawa K (2014) Seismic response analyses of the Yokohama bay cable-stayed bridge in the 2011 great east Japan earthquake. J Bridg Eng 19(8):1–17

Experimental Research on Road Snow Melting Performance Based on Electric Heating Tube

Zhenhua Jiang, Yan Chen, Xintong Lu, and Jianhan Hu

Abstract Cold regions and snowing weather are widely distributed in Global, which caused a series of problems in road safety, such as icy surfaces of roads, frozen-thaw Circle damage, and traffic jams. This research aims to improve the energy-efficient of electric heating snow-melting tubes for solving the icy and snowy pavement of urban roads. Numerical simulation and in-situ testing experiments is used in the research while the temperature field distribution and snow melting performance of the electric heating pipe heating road system are explored. It is found that the increasing of electric heating pipes cause a great impact on the snow melting effects. Moreover, this study lays the theoretical and experimental foundation for practical applications.

Keywords Road snow melting · Electric heating tube · Numerical simulation · In-situ testing experiment

1 Introduction

According to relevant surveys, road traffic accidents caused by snow and ice on roads account for 30% of overall traffic accidents, and the problem of road ice is gradually becoming a major traffic problem in China. Common passive snow removal methods include the removal method and melting method, which cost high, have low snow removal efficiency, use bad effects [1] and pollute the environment [2]. Compared with the traditional snow-melting methods, the active snow-melting pavement itself has the function of melting snow and ice, which is more efficient. To solve the

Z. Jiang
Jiangsu Province Communications Planning and Design Institute Limited Branch Company, Guangzhou, Guangdong, China

Y. Chen
Jiangsu Transportation Institute, Nanjing, Jiangsu, China

X. Lu (✉) · J. Hu
Intelligent Transportation System Research Center, Southeast University, Nanjing, Jiangsu, China
e-mail: 220223119@seu.edu.cn

© The Author(s) 2023
S. Wang et al. (eds.), *Proceedings of the 2nd International Conference on Innovative Solutions in Hydropower Engineering and Civil Engineering*, Lecture Notes in Civil Engineering 235, https://doi.org/10.1007/978-981-99-1748-8_29

drawbacks of passive snow removal, in recent years, methods including solar energy [3], circulating thermal fluid [4], geothermal energy and heat pumps [5], and electric heating systems [6] have also gradually attracted more and more attention.

This research adopts an electric heating tube to establish road snow melting experimental system, analyze the snow melting effect of the road snow melting system and the heat transfer on the road surface, explore the main factors affecting the road snow melting effect, and carry out indoor simulation experiments of asphalt slabs for control based on numerical simulation, to provide effective guidance for the practical application of electric heating tube snow melting road system.

The electric heating tube usually consists of five parts, which are sealing material, electric heating wire, magnesium oxide powder, metal tube sleeve, and lead rod. The metal tube sleeve mainly plays the role of protection, has the role of heat resistance, and heat conduction. The lead rod is used to conduct electricity and plays the role of connection. The role of magnesium powder is insulation and heat insulation. At the same time, the electric heating wire should be wound into a circle when used to increase the resistance, and extend the service life of the electric heating tube.

Compared with other snow melting methods, electric heating tube gain an efficiency heating and a better mechanical performance with long service life and simple installing combination. Once the current is turned on, heat can be diffused to the surface of the heating tube through the magnesium oxide crystals, making the internal temperature of the heating tube rise and exchange to the outside world, to play the role of heating.

2   Materials and Methodology

2.1   Materials

The electric heating tube snow melting system in the electric heating tube generally adopts a linear layout, snow melting system structure as shown in Fig. 1, the road structure from bottom to top is soil, lime soil, cement stabilized gravel base, and asphalt concrete, and the electric heating tube parallel to the direction of travel buried in the road concrete. This system adopts the intermittent working way: when the power is turned on, the electric heating tube generates heat energy, the temperature is raised, and the heat is transferred to the snow and ice layer through the asphalt layer so that it absorbs heat and warms up to achieve the effect of melting snow and turning ice, and the main parameters of the materials in the model are shown in Table 1.

The electric heating tube is buried as shown in Fig. 2, D indicates the buried spacing of electric heating tubes, the electric heating tube is parallel to the direction of traffic, buried around the wheel track of the traffic belt, when the power is on, the electric heating tube will transfer heat to the asphalt layer, melting the ice and

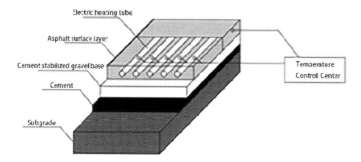

Fig. 1 Electric heated snow melting system

Table 1 Structural design parameters of each layer of road

Structural layer	Thickness (cm)	Material	Specific heat (J/kg °C)	Conductivity (J/m s °C)	Poisson ratio
Upper layer	4	Modified SMA-13	1000	1.25	0.25
Middle surface layer	6	Modified Sup-20	1000	1.05	
Lower layer	8	Heavy traffic Sup-25	1046	1.25	
Base layer	40	Cement stabilizes gravel	910	1.3	0.2
Subbase	20	Cement	940	1.4	0.3
Subgrade	250	Sand	1040	1.3	0.4
Electric heating tube	1.4	Steel G305	400	380	0.3

snow of the traffic belt, which can minimize the cost of the electric heating tube snow melting system.

The electric heating snow melting system includes a heating system experiment and control system, as shown in Fig. 1. The system adopts a controlled relay control circuit system, set temperature sensors on the electric heating tube and in the asphalt layer, and set the temperature break control threshold: lower than the rated temperature (low temperature) energized, the electric heating tube continuous power supply, higher than the rated temperature, the circuit automatically shut down, to ensure that the temperature around the electric heating tube is not too high to destroy the material allowable temperature.

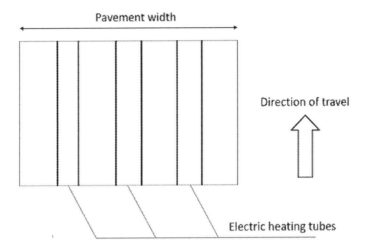

Fig. 2 Electric heating pipe layout scheme

2.2 Numerical Simulation

Assumptions of the Model. The electric heaters are laid in the asphalt concrete, and the heaters generate heat when they are energized, and the snow absorbs the heat and melts. To simplify the problem, the following assumptions are made: no consideration is given to the bending of the tube deformation due to vehicle loading at the burial depth. The temperature field of the structure is time-dependent, so the whole snow-melting heating process is a non-stationary process and ignore the heat transfer in the direction of the length of the heating tube. The ice and structural layers are assumed to be solid, and the study area is a homogeneous and isotropic material.

Initial and Boundary Conditions. The initial conditions of the model are

$$\lambda \left(\frac{\partial^2 t}{\partial x^2} + \frac{\partial^2 t}{\partial y^2} \right) = \rho c \frac{\partial t}{\partial \tau} \tag{1}$$

The initial condition is $\tau = 0, t(x, y) = t_0$ (t_0 is a known constant value). The ice melt system is an intermittent operation system.

2.3 In-Situ Testing Experiment

Considering the large temperature variation and dispersion of outdoor sites, this study proposes a large model test of large slab low-temperature indoor ice melting. By analyzing the warming situation and snow melting effect of a concrete large slab

Fig. 3 Schematic diagram of two heating pipe concrete surface temperature measuring points layout

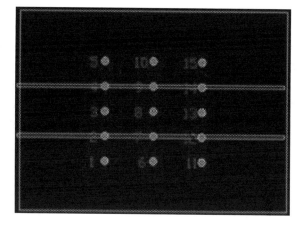

containing an electric heating tube under different input power and ambient temperature, so as get the influence of each factor on the warming effect of the concrete slab surface. The power consumption of the ice melting process and the constant temperature of the slab surface are obtained by the power-on heating and power-off cooling time within the limited temperature interval of the electric heater. The electric heaters (12 mm in diameter and 800 mm in length) were laid at a spacing of 20 cm. The experiments were conducted using a low-temperature controlled icehouse to simulate the external conditions, and the temperature was controlled from −10 to − 5 °C. To obtain the real-time data of pavement temperature change it is proposed to lay temperature measurement points on the surface of the asphalt concrete surface, including 15 measurement points for two heating tubes and 14 measurement points for three heating tubes in Figs. 3 and 4. The distance between the measurement points along the tube direction is 20 cm, and 10 cm perpendicular to the tube direction (the measurement points are arranged alternately above and between the tubes), and the measurement points are generally distributed in the middle of the concrete panel (concrete panel length 110 cm, width 80 cm).

3 Results and Discussion

3.1 Analysis of Numerical Simulation

It is concluded that as the heating time increases, the maximum and minimum surface temperatures both increase approximately according to an approximately linear law, due to the existence of heat transfer, thermal convection and thermal radiation, the temperature does not change exactly according to linear, where the average temperature difference tends to decrease with the increase in electrical heating time. The surface center temperature initially rises faster, and with the thermal conduction

Fig. 4 Schematic diagram
of three heating pipe
concrete surface temperature
measuring points layout

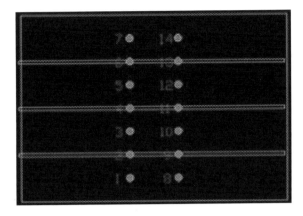

process, the internal heat storage in the structural layer, the temperature rise rate slows down, and the final curve tends to level off, and the temperature is nearly kept constant.

In this calculation model, when the electric heating tube is buried at a depth of 4 cm, the surface temperature can meet the requirements of snow melting temperature and temperature equilibrium when the initial ambient temperature is set at −10 °C and the electric heating tube is at 500 W line power with a burial depth spacing of 20 cm. And the spacing is 20 cm for further calculation. Set the spacing to 20 cm, buried depth of 8 cm, simulation, and simulation results in the buried depth of 4 cm structure for comparison, the results are shown in Fig. 5.

According to the above theory, the operational effect of the heating electric heating tube road ice system can be improved by reducing the laying depth of the heating

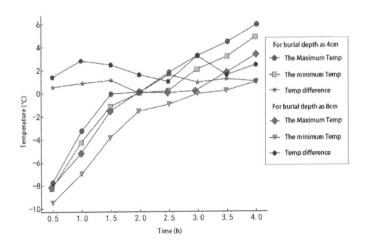

Fig. 5 Temperature curves of different buried depths

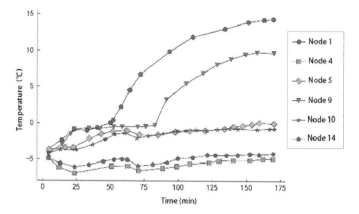

Fig. 6 Temperature change diagram of buried two electric heating tubes

electric heating tube in practical application. It is recommended to use an 8 cm burial depth and tube spacing of 20 cm.

3.2 Analysis of In-Situ Testing Experiment

According to the temperature change curves of the measured points of asphalt concrete specimens from the deployment of two electric heating tubes and three electric heating tubes in Figs. 6 and 7, it can be concluded that the heating efficiency of three tubes is higher compared with two tubes. In a short period of time, the temperature of the test points at the edge of the slab does not produce a large change, and the closer the heating tubes are, the temperature of the test points rises more quickly. The heating efficiency of three electric heating tubes is significantly higher than that of two tubes. Therefore, in the actual engineering construction project, the number of buried electric heating tubes should be given priority in the case of project funds.

4 Conclusion

This study established a solution as electric heating tube road snow melting test platform and a finished in-situ test under complex situation caused by melting during snowing, such as low efficiency, narrow scope of application, and low energy utilization of traditional road snow melting methods. In this study, the numerical simulation calculation was carried out using ABAQUS finite element software. The following conclusions can be drawn:

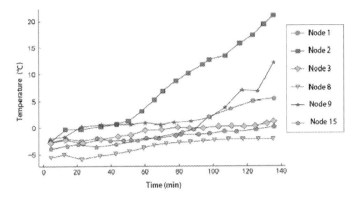

Fig. 7 Temperature change diagram of buried three electric heating tubes

(1) The heat transfer effect of the electric heating tube can effectively raise the temperature of the road and speed up the rate of snow melting on the road.

(2) The burial parameters of the electric heating tube have an important influence on the snow melting effect of the electric heating tube snow melting system. Reducing the burial depth and taking proper spacing of electric heating tube burial can improve the temperature of the road and accelerate the snow melting rate. In addition, the number of electric heating tubes also has a certain influence.

(3) In this study, the control variable method was used to optimize the burial depth and deployment spacing. Since the influence of external temperature and snow characteristics on the snow, melting performance was not considered further research and experiments are still needed.

(4) In this paper, when analyzing the influence of electric heating tube burial parameters, the control variable method is used to optimize the burial depth and deployment spacing, but from the overall consideration, the model is not well considered for snow melting performance, and the influence of external temperature and snow characteristics on snow melting performance is not considered. In addition, the phase change simulation does not consider the corresponding snow melting, which will have a certain influence on the distribution of the temperature field, which is worth further research and experiment.

References

1. Hongchao D, Wenxing M, Baode J (2005) Technology of removing snow and ice on roads and its developing trend. Constr Mach Equip 12:41–44
2. Shi, Jungwirth, Akin, et al (2014) Evaluating snow and ice control chemicals for environmentally sustainable highway maintenance operations. J Transp Eng 140(11)
3. Pan P, Wu SP, Xiao Y, Liu G (2015) A review on hydronic asphalt pavement for energy harvesting and snow melting. Renew Sustain Energ Rev 48:624–634

4. Asfour S, Bernardin F, Toussaint E, Piau JM et al (2016) Hydrothermal modeling of porous pavement for its surface de-freezing. Appl Therm Eng 82(3):493–500
5. Lai JX, Wang XL et al (2018) A state-of-the-art review of sustainable energy-based freeze proof technology for cold-region tunnels in China. Renew Sustain Energ Rev 3554–3569
6. Vo HV, Park DW, Dessouky S (2015) Simulation of snow melting pavement performance using measured thermal properties of graphite-modified asphalt mixture. Road Mater Pavement 16(3):696–706

Evaluation and Repair of Steel Structure Damaged by Fire

Hongnan Wang and Huabo Liu

Abstract In recent years, steel structures have been widely used in buildings because of their advantages in self-weight and seismic performance. However, due to the great influence of temperature on steel structure, once a fire occurs, it may lead to the destruction of steel structure and even the collapse of the building. In this paper, through the assessment of a steel structure factory after fire, the damage degree of the components and structural integrity of the building are determined, which provides a reliable technical support for the subsequent repair and retrofit work. Based on the extent of damage identified, repairs are performed on the affected members. The analysis results of the rehabilitated structure show that this method is safe and effective.

Keywords Steel structure · Assessment · Evaluation · Retrofit · Fire-damaged building

1 Introduction

Fire statistics show that in all kinds of fire accidents, construction fire is the biggest threat to human life and property safety. Among them, the possibility for factory fires is relatively high due to the special production process, lack of safety awareness and other reasons [1].

Structural steel is a kind of material widely used in factory buildings with high strength, light dead weight and superior seismic performance. However, the fire resistance is very poor, usually in the building fire within 15 min, the steel structure of the building will lose its bearing capacity, the building will soon collapse, causing huge casualties and property losses [2]. Therefore, it is necessary to evaluate the fire

H. Wang (✉)
College of Civil Engineering, Shanghai Normal University, Shanghai, China
e-mail: whn@shnu.edu.cn

H. Liu
Shanghai Key Laboratory of Engineering Structure Safety, SRIBS, Shanghai, China

© The Author(s) 2023
S. Wang et al. (eds.), *Proceedings of the 2nd International Conference on Innovative Solutions in Hydropower Engineering and Civil Engineering*, Lecture Notes in Civil Engineering 235, https://doi.org/10.1007/978-981-99-1748-8_30

damage of the building to provide reliable evidence for the subsequent repair and retrofit of the building [3]. In this study, the evaluation of fire damaged structural members is carried out, so as to determine the damage class of fire and put forward the corresponding retrofit methods.

2 Application Example

2.1 Description of the Building

The factory building is a single-storey light steel structure with portal frame. The eaves height is 8.5 m. The building is constructed in two phases. The first phase (1–21 axes) was built in 2004, with a second phase (21–24 axes) following in 2006. The building is basically rectangular in plan, as shown in Fig. 1. The building dimensions are 96 m in width × 180 m in length × 10 m in height. The top elevation of the steel beam at the eave is 8.5 m, forming a north–south 3% drainage slope. The column spacing is mostly 8 m (except for the two spaces on the west side which are 6 m).

The primary structural frame includes columns, purlin, and roof steel H-beams. The portal frame columns are all welded H-bar with constant section. The side columns are H550 × 150 × 6 × 10, H550 × 225 × 6 × 10, H550 × 250 × 6 × 10, and the middle columns are H450 × 350 × 6 × 12, H350 × 250 × 6 × 10, H350 × 250 × 5 × 10. The wind columns are H320 × 250 × 6 × 10 and H320 × 175

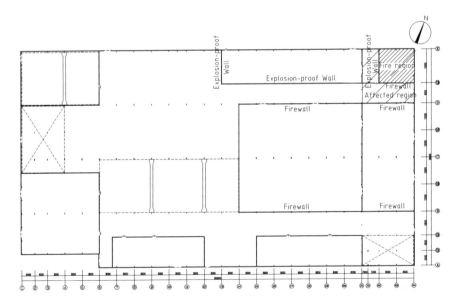

Fig. 1 Plan view of the factory building

× 6 × 10. The beams are welded H-bar with variable section, H(750–850) × 200 × 6 × 12, H(600–850) × 200 × 6 × 12, and H(600–700) × 200 × 6 × 12. Roof purlin is cold-formed thin wall Z-shaped steel (Z200 × 70 × 20 × 2.5). Steel beams and columns are generally connected with 6M20 or 8M20 high-strength bolts. The columns are connected to the foundations with a combination of base plates and anchor bolts.

2.2 Information About the Fire

On May 7, 2019, an electrical fire broke out between 22 and 24/(1/D)-E axes in the northeast corner of the second phase of the factory building (see Fig. 1). Before the fire, plastic products (motor fans) stacked in the northeast corner of the building belong to flammable substances. Because it is mainly used for storage and occupies a limited area, there is no active sprinkler system in the building. The plastic was flammable, causing the fire to spread quickly. Luckily, the fire was quickly brought under control. The fire was observed to last approximately 10 min. The fire covered about 240 m². Combustibles stored in the area were completely destroyed.

3 Damage of Steel Structural Members After Fire

A systematic investigation of the structure was carried out after the fire. It was found that the direct fire zone was limited to the 23–24/(1/D)-E axes. The reason was that the surrounding walls of the fire area were block or brick masonry firewalls or explosion-proof walls. All these walls with a 2-h fire rating effectively prevented the fire from spreading. The main damage was in the northeast corner. No significant fire-induced damage was observed in other areas. Plaster falling from the enclosure walls was observed in areas outside the fire due to high temperature.

3.1 Assessment of Damage to Columns

Though the overall structure did not collapse, many steel structural components were severely damaged in the fire region (Fig. 2). Some steel columns exposed to fire suffered reduction in performance and instability that result in localized buckling of the flanges or overall buckling. Some of the columns experienced minor damage, such as falling of fireproof coating due to elevated temperature.

Fig. 2 Damage views of building following the fire accident: **a** overall view of fire damaged area; **b** buckling and large deflection of roof steel beams and purlins; **c** flexural deformation of roof panels; and **d** falling of fireproof coating on steel columns (Images by Huabo Liu.)

3.2 Assessment of Damage to Horizontal Framing Members

The heat from the fire visibly damaged steel members located directly above. The color-coated steel sheet roofing was affected and it exhibited visually apparent deformation, spalling, and discoloration. Although there was no complete collapse, several large depressed areas were observed on the roof of the building. Inspection of the interior of the building showed that purlins were exposed to extreme temperatures and failed in a buckling mode as the steel lost its strength. The horizontal structural components supporting the roof were badly damaged. The damage to roof steel beams and purlins included (a) large in-plane deflection in the steel beams and purlins; (b) buckling of the roof beams and (c) out of plane deflection due to buckling, as shown in Fig. 2.

4 Evaluation of Steel Components

According to Standard for Appraisal of Engineering Structures After Fire (T/CECS252-2019), the steel components after fire are evaluated and rated. Fire damaged steel is typically classified in four classes:

Table 1 Evaluation and rating results of steel components after fire

Component	Axis	Damage class
Steel columns	23/E	IV
	22/E, 24/E; 24/(2/D), 24/(3/D)	III
	22/D, 23/D, 24/D	II
Steel beams	22/D-E, 23/D-E, 24/D-E	IV

- Class I—Members with no damage or slight deformations that are not easily detected by visual observation. Performance and safety of the component have not been affected by fire.
- Class II_a, II_b, or III—Members with visible deformation. According to the component fireproof coating, local deformation, overall deformation, connection damage, the most serious damage class is taken as the component damage class.
- Class IV—Severely damaged components, such as excessive overall deformation, severe residual deformation, cracking or fracture, local buckling, bolt breakage, etc.

According to the standard, the evaluation and rating results of steel components after fire are shown in Table 1.

5 Retrofit Methodology

5.1 Repairing Decisions

Based on the results of the steel component assessment, taking into account the method to be used, the extend of intervention, etc., decisions are made concerning repair of the structure after fire.

For Damage class I, only redecoration is required.

For Damage class II, superficial repair of slight damage will suffice, such as non-structural or minor structural treatment (restoring fireproof coating on steel columns, etc.).

Damage class III, generally damaged steel, will be retrofit to satisfy increased strength demands. For these columns or beams, retrofit is more economical than replacing fire-damaged components.

Damage class IV, severely deformed steel, will be removed and replaced.

5.2 Structural Shortcomings in Details

The factory building was built in 2004, when the "Technical specification for steed structure of light-weight Buildings with gabled frames" CECS102:2002 was just issued [4]. The current national standard, "Technical code for steel structure of light-weight building with gabled frames" GB51022-2015 [5], has improved in all aspects compared with the original regulations. Therefore, it does not meet the requirements of current design specifications. The main deficiencies of the detail measures are:

- Smaller section size of components compared with current standard.
- Inadequate support system to form the spatial structure system effectively.
- Lack of adequate lateral support out of plane.

To sum up, the safety of the factory building does not meet the requirements of the current relevant specifications, and retrofit measures should be taken. Treatment measures include: (1) supplement the support system, (2) adding necessary roof knee braces, (3) adding rigid tie rod between side columns to reduce the calculated length out of plane, and (4) strengthening the section of the main components, etc.

5.3 Structural Reanalysis

According to the final retrofit methodology, a structural analysis was performed to assess performance of the building and identify deficiencies (see Fig. 3), using the new components and structural information. The transverse plane model is calculated as a portal frame. The column foots are taken as rigid connection, and the beam-column joints are partially rigid and partially hinged according to the actual connection mode. The length of the roof oblique beam out of plane is determined by the spacing between the knee braces.

The results of structural calculation show that the structure has good performance and meets the requirements of current specifications. Results of the analysis are subsequently used to repair and strengthening the damaged structure.

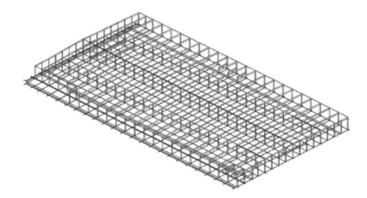

Fig. 3 Model of structure

6 Conclusion

In this study, the building was not designed in accordance with current building regulations. Structural damage due to fire further complicates retrofitting. Based on the proper assessment, an appropriate and cost-effective retrofit methodology is performed.

For the steel structure, once the fire occurs, if no fire prevention measures are taken or the fire is not put out in time, the steel structure will collapse quickly. If the fire damage is very severe, demolition is recommended on the basis of safety and life-cycle economic analysis. If the fire is not serious, the structure can be retrofitted according to the usability and durability of the structure to prevent the occurrence of secondary disasters. Post-fire evaluation is critical to the safety of a building in the future. How to make the repair and strengthening of steel structure after fire reasonable, economical and effective is a problem worth further study.

References

1. Ha T, Ko J, Lee S, Kim S, Jung J, Kim DJ (2016) A case study on the rehabilitation of a fire-damaged structure. Appl Sci 6(5):126
2. Anvari A, Mahamid M, McNallan M, Eslami M (2019) Effectiveness of damaged fire proofing in structural steel members subjected to fire. J Struc Fire Eng 10(1):24–47
3. Wiesner F, Randmael F, Wan W, Bisby L, Hadden RM (2017) Structural response of cross-laminated timber compression elements exposed to fire. Fire Saf J 9156–67
4. CCMSA (China Construction Metal Structure Association) (2002) Technical specification for steed structure of light-weight Buildings with gabled frames. Proc CECS 102. In Chinese
5. Ministry of Housing and Urban-Rural Development of People's Republic of China (2015) Technical code for steel structure of light-weight building with gabled frames. GB51022-2015. In Chinese

Research on the Pore Characteristics of Dredger Fill By Preloading

Jinfeng Tian, Huan Yan, and Guijie Zhao

Abstract This study investigated the characteristics of dredger fill after preloading at the DaLian. Granulometric composition test and X-ray diffraction were employed to determine the composition of dredger fill. Mercury intrusion porosimetry was combined to determine the related features of the pore. This paper also discussed the changes in porosity, pore diameter distribution, and morphological characteristics of the pore. Preloading influence depth was established based on pore changes. The results show that: the dredger fill contained numerous clay minerals, mainly illite and illite–smectite mixed layer; and the influence depth established based on pore changes of the area was almost 11 m. Within the influence depth, porosity decreased rapidly as the depth increased, and the main scope of pore volume fraction changes from the scope of 0.06–10 μm to the scope of 0.1–1 μm. When depth was greater than the influence depth, porosity slowly changed, pores with diameters greater than 1 μm show an increasing trend. Thus, the soil was compacted.

Keywords Dredger fill · Preloading · Mercury intrusion porosimetry · Pore characteristic

1 Foreword

With the rapid development of coastal city construction, there is a large demanding for land resources. Land reclamation has became one of the most effective ways to solve such problems. In the process of reclamation in such coastal areas, the original offshore muddy soil is often used as reclamation material [1]. The composition of soil is complex, diverse and uncertain which makes the subsequent construction difficult [2]. When the project is established, it often needs to reinforce the foundation

J. Tian
Shenyang Jianzhu University, Shenyang, China

H. Yan (✉) · G. Zhao
Changchun Institute of Technology, Changchun, China
e-mail: yanhuan@ccit.edu.cn

© The Author(s) 2023
S. Wang et al. (eds.), *Proceedings of the 2nd International Conference on Innovative Solutions in Hydropower Engineering and Civil Engineering*, Lecture Notes in Civil Engineering 235, https://doi.org/10.1007/978-981-99-1748-8_31

first. The commonly used method is preloading. The plenty engineering practices shows that the dredger fill site after reinforcement still has lots of problems, such as complex engineering geological properties and large post-construction settlement [3]. And the internal microstructure of the soil has changed completely. Therefore, it is great significance to study the dredger fill after preloading treatment from the perspective of microstructure. It has great significance to the safety of the project and the adverse effects after construction.

In view of this, the author takes the dredger fill in DaLian as the research object, and evaluates the consolidation characteristics of the dredger fill from the aspects of material composition and porosity. By comparing the basic properties and pore characteristics of dredger fill at different depths, the consolidation law is studied. And the related characteristics of porosity and pore equivalent diameter at different depths in the consolidation process are analyzed to determine the influence depth of surcharge preloading. Combined with these variation characteristics, it provides a detailed material basis for the in-depth study of the reinforcement mechanism of dredger fill.

2 Material Composition

2.1 Granularity Composition

The granularity composition of soil refers to the percentage of various sizes particles in the soil [4]. It can be seen from Fig. 1 that after adding dispersant agent that the content of clay group in the soil samples increased, while the content of silt group decreased. It shows that the dredger fill contains aggregates formed by the combination of clay and clay, clay and silt [5]. The experimental results show that the non-uniformity coefficient of soil samples is greater than 5 which show the gradation is good. The composition of different particle sizes shows that the soil sample can be pore reduction after compaction.

The four soil samples are mainly silt and clay which reached more than 90% from Table 1. The sand and clay content of the 01 soil sample are the highest. The silt

Fig. 1 Granular metric analysis curve

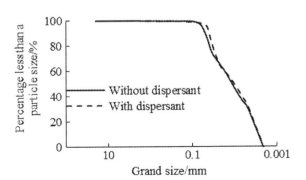

Table 1 Test results of granular metric

No.	Type	Depth (m)	Percentage grain size (%)		
			Grit 2–0.075	Silt 0.075–0.005	Clay < 0.005
01	Earth-filled	2	7.25	49.11	43.64
02	Mucky clay	6	2.48	67.55	29.97
03	Silty clay	11	1.68	82.12	16.20
04	Silt soil	17	3.15	86.64	10.21

content of soil sample has increases as the depth increasing, and the clay content showed a decreasing trend. The main reason is the different material sources.

2.2 Mineral Components

In this paper, semi-quantitative analysis and identification of some soil samples were carried out by X-ray powder diffraction [6]. From Table 2: the primary minerals of soil samples are mainly quartz, and the secondary mineral content is higher proportion, about 50%.The clay minerals is mainly illite and illite–smectite mixed layer. And contain a small amount of kaolinite and chlorite. This is also consistent with the results of particle test. The clay content of the analyzed soil sample is large. Because the clay has good hydrophilicity, it will increase the difficulty of soil drainage consolidation.

3 Microscopic Pore Characteristics

The microstructure of soil has an important influence on the macroscopic mechanical characteristics of soil. There are two aspects to study at present, the morphological

Table 2 Mineral composition test results

No.	Ineral content ω (B)/10^{-2}									
	Q	fs	Pl	Cc	Do	I/S	I	K	Ch	Am
01	21	11	9	8	3	8	28	6	4	2
02	25	4	4	6	4	13	30	8	5	1
03	22	7	12	5	4	7	31	3	5	1
04	22	6	8	6	5	8	35	3	4	3

*Q—quartz, fs—alkali feldspar, Pl—piagioclase, Cc—calcite, Do—muscovite, Am—amphibole, I/S—illite/smectite, K—kaolinite, I—ilite, Ch—chlorite

characteristics and the connection characteristics of soil. Mercury injection experiments and scanning electron microscopy are commonly used research methods [7]. Mercury intrusion test is selected in this paper.

3.1 Mercury Injection Test

During the compression and consolidation of dredger fill, the distribution and variation of pores is one of the important factors to measure the compressive deformation. The distribution and variation of pores in soil will show significant differences with the different degree of consolidation [8].

This is the No. 9500 experimental principle: suppose the pores of the soil are cylindrical holes and the radius is r, press liquid mercury into the pores by instrument. The pressure of mercury and repulsion force reached equilibrium.

$$p\pi r^2 = 2\pi r\alpha \cos\alpha \tag{1}$$

where p is pressure, r is pore radius, σ is surface tension coefficient of mercury, 0.485 N/m, α is mercury wetting angle on Materials, 130°.

3.2 Analysis of Test Results

Porosity Distribution. It can be seen from Table 3 that the dredger fill after preloading, with the depth increasing, the overall trend of porosity is decreasing. This is because the water in the dredger fill gradually was drained out of the pores under the action of the upper preloading load. Soil particles moved and some small particles was filled in the pores of the lower soil layer, which made the soil structure more and more dense. It can be seen that the porosity of soil samples decreases rapidly from surface to 11 m and it remains virtually unchanged from 11 to 17 m by comparing the rate of porosity reduction of soil samples at different depths. It can be confirmed that the preloading effect was better within a certain depth range as the depth increased. And beyond that, the effect is worse gradually.

Interval Distribution of Aperture. It can be seen from Fig. 2: the cumulative volume fraction of pores size less than 0.01 μm is small. In the range of 0.01–0.1 μm, distribution curve of 01 soil sample shows an obviously increasing trend. The curves

Table 3 Data of mercury injection

No.	01	02	03	04
Depth (m)	2	6	11	17
Porosity (%)	40.35	34.45	32.56	30.22

of 02, 03 and 04 soil samples did not change significantly. In the range of 0.1–1 μm, the curves of 02, 03 and 04 soil samples showed a significant upward trend. But the growth rate of 01 soil sample curve is slightly slow; When the aperture is larger than 1 μm, the four curves all showed an upward trend. But growth rate slowed significantly and slowly approached 100%. It can be inferred that there are mainly medium and small pores in preloading dredger fill while less macropores. It is mainly due to the high clay content of the soil.

The microstructure of dredger fill is irregularity and complexity. But its pore distribution is self-similarity. Therefore, the pores of dredger fill can be analyzed by fractal theory [9]. Assume that the pore is a sphere with radius r. V_P is the pore volume with radius less than r.

$$V_P = \int_0^r \frac{4}{3}\pi r^3 dN = Ar^{3-D} \tag{2}$$

where N represents the number of pores with radius less than r; D denotes the dimension of pore distribution; A is a constant.

In double logarithmic coordinates, draw the cumulative volume fraction content classification curve of soil sample pore size-pore. The fractal dimension D of pore distribution can be determined. If the slope of the straight line in the curve is K, then D = 3 − K [9].

It can be seen from Fig. 3, the curve changes obviously when the aperture is 0.06, 0.4, 4, and 40 μm. According to this, the curve can be roughly divided into 5 broken line segments. It shows that the distribution of pores has multifractal properties. Each line segment shows that the pore diameter has self-similar properties in the interval. Therefore, according to the self-similarity of pores and the experience of predecessors, the pore size can be divided into 5 levels: (a) Micropores: d ≤ 0.06 μm. (b) Small pores: 0.06 μm < d ≤ 0.4 μm. (c) Medium pore: 0.4 μm < d ≤ 4 μm. (d) Large pores: 4 μm < d ≤ 40 μm. (e) Extra-large pores: d > 40 μm.

Fig. 2 Cumulative pore size distribution curves

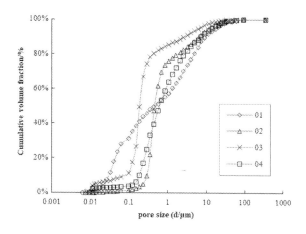

Fig. 3 Logarithmic curve of
cumulative pore size
distribution

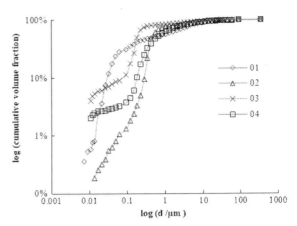

Through the cumulative distribution curve and the above pore size division, the volume fraction of the corresponding aperture interval can be obtained. The results are shown in Fig. 4. From the figure, there is a huge different in the volume fraction concentration range of the aperture interval with depth. No. 01 soil sample is taken from the surface. The concentration range of pore volume fraction is wide, roughly 0.06–10 μm. With sampling depth increasing, the concentration range of 02 and 03 is smaller than 01. Mainly concentrated in the 0.06–4 μm interval. But the concentration range of the deepest 04 soil sample has increased compared with the front two samples. The results show that: in a certain depth range, the pores larger than 4 μm in the soil sample are crushed into smaller pores with the increase of soil depth after dredger fill preloading. The pores less than 0.06 μm decreased. It may be because small pores connect into slightly larger pores with the discharge of water from the soil. As a result, the volume fraction of pores in this interval decreases. When the depth is greater than a certain value, the influence of preloading gradually decreases. The pores larger than 4 μm show an increasing trend. It has a certain relationship with the influence depth of preloading treatment.

Fig. 4 The volume fraction
of every pore diameter range

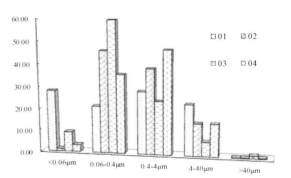

4 Conclusion

The material composition and microscopic pore characteristics of dredger fill soil at different depths after preloading treatment were analyzed. The following conclusions were reached.

There are a lot of clay minerals in the soil samples. It is mainly illite and illite-montmorillonite mixed layer minerals. However, mineral composition and content are extremely similar with depth changing which shows that the mineral content has little to do with depth.

The related parameters of pores were obtained by mercury intrusion test. The porosity of the dredger fill after preloading treatment becomes smaller with the depth increasing. The concentration range of apertures has significant differences. The influence depth of preloading in this study area is determined to be 11 m. In range of this depth, the porosity of soil sample decreases rapidly. The main concentration range of pore volume fraction changes from 0.06–10 μm to 0.1–1 μm. The change of porosity slows down beyond depth. The pores larger than 1 μm showed an increasing trend.

References

1. Liu Y, Wang Q, Xiao SF (2003) The comparative research on fundamental properties of dredger fill in different areas. Geotech Eng Tech 4:197–200
2. Han XJ (1999) Principle and application of large reclamation dredger fill foundation treatment technology. China Architecture Publishing & Media Co., Ltd., Beijing
3. Yan SW, Liu KJ et al (2010) Study of creep properties of soft clay in Tianjin Binhai New Area and no-yield-surface constitutive model. Rock Soil Mech 31(5):1431–1436
4. Song J, Wang Q et al (2011) Physical and chemical indicators of dredger fill with high clay by vacuum preloading. J Jilin Univ 41(5):1476–1480
5. Ministry of Water Resources of the People's Republic of China (1999) Standard for soil test method. China Architecture Publishing & Media Co. Ltd, Beijing
6. Song J (2011) Laboratory simulation test and PFC3D numerical analysis of high clay dredger fill in the consolidation process of step vacuum preloading. Jilin University, Chang Chun
7. Sun MQ (2013) Engineering characteristic and post-construction subsidence of soft in Tianjin Binhai New Area. Jilin University, Chang Chun
8. Niu CC, Wang Q et al (2011) Microstructure fractal feature of dredger fill during seepage flow consolidation. J Southwest Jiaotong Univ 47(1):78–83
9. Yang Y, Yao HL et al (2006) Characteristics of microcosmic structure of Guangxi expansive soil. Rock Soil Mech 27(1):155–158

Study on Comprehensive Detection and Treatment Measures of Qinlan Tunnel of Tianba Expressway Crossing Large Karst Cave

Xiaoming Wang, Yongxing Zhou, Gang Zhang, Xueqi Chen, Zhiqiang Liu, and Xiaoyong He

Abstract Aiming at the problem of comprehensive detection technology and treatment measures for tunnels crossing large karst caves, this paper systematically carried out the research on advanced detection of karst geology and treatment measures of karst caves in combination with the Qinlan Tunnel Project of Tianba Expressway in Guangxi. The comprehensive detection method of "karst geological survey + geological radar detection + advanced horizontal drilling + three-dimensional laser scanning + geological sketch" is used for advanced detection. On the basis of ascertaining the spatial location and development scale of karst caves, the scheme of tunnel excavation after advanced reinforcement and support treatment such as advanced small pipe grouting is determined. It has been successfully applied to the construction practice of Qinlan Tunnel crossing ZK89+110 large karst caves, and the treatment measures of "root pile + lining strengthening" are put forward. It effectively avoids the occurrence of disasters caused by karst caves, and has important reference significance for similar projects.

Keywords Highway tunnel · Karst cave · Advanced detection · Treatment measures

1 Introduction

Karst is a general term for the chemical effects of water on soluble rocks, supplemented by mechanical effects such as erosion, deposition and collapse, and the phenomena caused by these effects. Karst geology is widely distributed in China, mainly concentrated in Yunnan, Sichuan, Guangxi, Guizhou and other regions [1].

X. Wang · Y. Zhou
Huazhong of CCCC First Highway Engineering Group Company Limited, Wuhan 430013, People's Republic of China

G. Zhang (✉) · X. Chen · Z. Liu · X. He
School of Highway, Chang'an University, Xi'an, Shaanxi 710064, People's Republic of China
e-mail: zg15538957763@163.com

© The Author(s) 2023
S. Wang et al. (eds.), *Proceedings of the 2nd International Conference on Innovative Solutions in Hydropower Engineering and Civil Engineering*, Lecture Notes in Civil Engineering 235, https://doi.org/10.1007/978-981-99-1748-8_32

Due to the complexity of karst geology and high risk of karst caves, tunnel construction faces great risks and challenges [2]. During the construction of Maluqing Tunnel [3], Yuanliangshan Tunnel [4] and Xiaogaoshan Tunnel of Shanghai-Kunming Passenger Dedicated Railway [5], karst water and mud inrush disasters occurred, causing serious casualties and economic losses. Therefore, the use of advanced detection technology to accurately identify the spatial location and development scale of karst caves can effectively ensure the safety of karst tunnel construction.

At present, researchers have conducted many studies on karst tunnel advanced geological prediction technology and cave treatment. Zhao Shaozhong et al. studied the deformation law of surrounding rock by numerical simulation method and gave the treatment measures of karst cave in different positions, combining with the karst cave exposed in the construction of Qiyueshan tunnel [6]. Wang Shaohui et al. proposed a comprehensive treatment scheme for the combined structure of "bearing pile foundation-longitudinal and transverse frame beam structure-reinforced concrete retaining wall" in view of the extra-large karst cave exposed during the construction of Naqiu Tunnel [7]. Liu Tongjiang et al. determined the treatment scheme of "backfill ballast + upper grouting reinforcement" for the giant karst cave crossed by the alpine tunnel of Qian-Zhang-Chang railway [8]. Based on the engineering background of Shangjiawan tunnel, Li Shucai et al. proposed the cave treatment scheme of "beam plate method + backfill method" [9]. Wang Jian et al. used the geological radar advanced geological prediction method to detect the karst development of Xinjie Tunnel of Guizhou Kaili Ring Expressway, and verified the applicability of the method [10]. Combining with Yesanguan tunnel of Yiwan railway, Sun Mingbiao put forward the treatment measures such as high level drainage depressurization, curtain grouting and pipe shed pre-support, which realized the purpose of depressurization, reinforcement and seepage prevention [11]. Based on the Taiping Tunnel of Xuanhe Expressway, Liu et al. proposed a variety of technical schemes for crossing karst caves and determined the double-arch open-cut tunnel + backfill excavation scheme through multi-dimensional comparison and selection [12].

Based on Qinlan Tunnel Project of Tianba Expressway, aiming at the problem of karst disaster in the process of tunnel excavation, this paper carries out karst detection by using comprehensive means such as geological advance prediction, advance drilling and three-dimensional laser scanning. On the basis of ascertaining the spatial location and development scale of karst caves, this paper puts forward targeted treatment measures for karst caves.

2   Engineering Overview

Qinlan Tunnel is located in Hechi City, Guangxi. The overall direction of the tunnel is east–west, and the tunnel is a small clearance + separate special long tunnel. The starting and ending pile numbers of the left line of the tunnel are ZK86+164~ZK91+486, and the length is 5322.0 m. The starting and ending piles of the right line are K86+190~K91+494, and the design length is 5304.0 m. The tunnel

Fig. 1 Panorama of Qinlan tunnel

is a two-way six-lane tunnel with a maximum buried depth of about 356.67 m. The overall view of the tunnel is shown in Fig. 1.

The stratas are mainly the Fourth Series Dissolution Remnant Accumulation Layer (Q^{el}) and the Carboniferous Middle System (C_2) strata. The tunnel mainly passes through the Carboniferous Middle System (C_2) limestone, with cryptocrystalline ~ microcrystalline structure and layer thickness ~ huge thick layered structure. The rock is hard and there are karst caves developed locally. The tunnel area belongs to karst peaks and depressions, with strong mountain erosion and exposed bedrock. The ground elevation of the karst depression on the entrance side of the cave entrance is 350–370 m. And there are springs and sinkholes in the low-lying areas.

Groundwater types can be divided into loose rock pore water, clastic rock fracture water and carbonate fracture karst water. Karst in the tunnel area is relatively developed. The main karst forms are karst caves, gullies and dissolution depressions. Trenches are mostly filled with clayey soil of the fourth series residual slope layer. During the construction period, a very large cave was revealed at ZK89+110 (see Fig. 2), with a height of about 60 m. It develops from the right to the lower left, crosses diagonally with the tunnel axis, and invades the left vault range of the right cave.

Fig. 2 Situation of karst cave at left hole ZK89+110

3 Joint Detection and Analysis of Caves

A variety of detection methods were used to detect and analyze the cave at ZK89+110.

3.1 Geological Radar Method

The geological radar method works according to the principle of pulsed radar emission waves. High-frequency electromagnetic waves are sent underground in the form of broadband short pulses through the transmitting antenna. It was reflected by underground media or buried objects, which was received by the receiving antenna and recorded on the host to form a radar profile [13]. The geological radar YQZLS20112 was used to detect the tunnel of ZK89+095~ZK89+125 in the left cave, and the schematic diagram of the survey line layout was shown in Fig. 3.

The detection results show that the surrounding rock of ZK89+095~ZK89+120 section is mainly moderately weathered limestone, with medium and thick layered structure. The joints and dissolution fractures are relatively developed, mostly filled with calcite veins or argillaceous. The surrounding rock of ZK89+120~ZK89+125 section is affected by joint fractures, and the rock is loosely broken, filled with mud and mud. And dense fracture zone or karst development area is developed. The processed image of the acquired waveform is shown in Fig. 4.

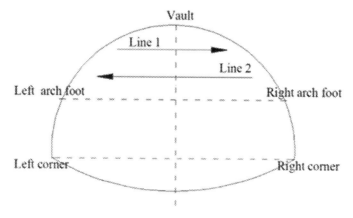

Fig. 3 Geological radar detection line layout diagram

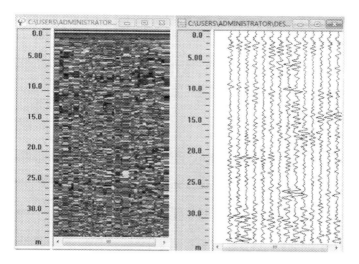

Fig. 4 Qinlan tunnel ZK89+095~ZK89+125 radar scanning map (Line 1)

3.2 Advanced Horizontal Drilling

Advanced horizontal drilling is a short-range tunnel surrounding rock geological prediction technique. A horizontal drilling rig is installed inside the tunnel and drilled to infer the geology in front of the palm according to the drilling data [14]. ZDY1250 full-hydraulic multi-functional crawler drilling rig was used to carry out geological advanced drilling work on the face ZK89+035 of the second half of the tunnel in Qinlan. Five drill holes were arranged on the face of the face. And the location of the drilling hole arrangement is shown in Fig. 5.

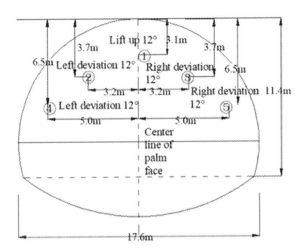

Fig. 5 Advanced drilling location diagram

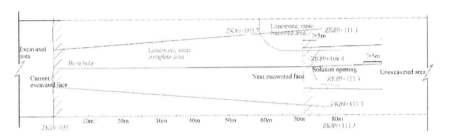

Fig. 6 Advanced driling results of ZK89+035 in left line of Qinlan tunnel

Three advanced drill holes were completed and the geotechnical characteristics revealed by each hole are shown in Fig. 6. It shows that there are clay-filled karst caves in the ZK89+108.4~ZK89+111.3 section of the tunnel. The karst cave has not been uncovered, and the total length of the karst cave section is more than 2.9 m. The rock mass around the cave is relatively complete limestone and locally broken limestone. Karst development is strong, and the stability of surrounding rock is poor. After excavation, mud gushing, block falling, cavity collapse, roof falling and other phenomena are prone to occur, and groundwater is weakly developed.

3.3 3D Laser Scanning Technology

Through the high-speed non-contact laser measurement method, the three-dimensional laser scanning technology can carry out 360° panoramic rapid scanning of complex objects in three-dimensional space. We can obtain the three-dimensional

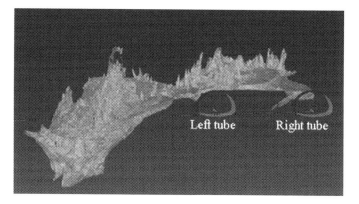

Fig. 7 Three-dimensional map of cave at ZK89+110

shape of the object to guide the construction of the project [15]. The cave was scanned by 3D laser scanning technology and its three-dimensional morphology was obtained (see Fig. 7). The results show that the large karst cave is exposed in front of the ZK89+110 face. The karst cave crosses the tunnel obliquely and intrudes into the right tunnel K89+130~K89+210 of the back section of Qinlan tunnel. The surrounding rock is mainly silty clay, which has poor stability and it is prone to collapse without support. There are many sinkholes and 2 water-dissipation caves.

4 Cave Treatment Measures

This project selects the root pile + lining strengthening scheme. The treatment facade of ZK89+110 karst cave is shown in Fig. 8. The specific design is as follows.

(1) First, remove the cave floor slag and silt layer. Then investigate the direction of water flow in karst cave, water dissipation tunnel, etc.
(2) The ZK89+100~ZK89+170 section of the karst cave is lined with S5-P. Cave internal system anchor can be canceled. The advanced support adopts the dense row of advanced small ducts. The length of the advanced small duct is 4 m, the circumferential spacing is 40 cm, and the longitudinal row spacing is 1 m. The locking feet support on both sides of the arch line is replaced by a 9 m long $\Phi 108*6$ mm locking foot steel pipe with a longitudinal spacing of 1 m. It is staggered with $\Phi 50*5$ mm locking steel tube.
(3) The foundation of the K89+130~K89+205 section of the inverted arch is reinforced by a composite foundation of tree root stumps. The root pile is 20 cm in diameter and it is plum-shaped layout. The specific design scheme is shown in Fig. 9.

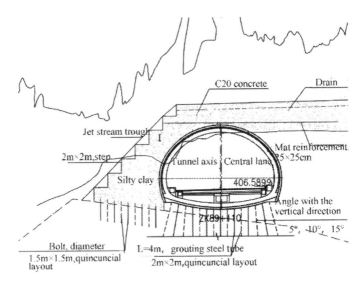

Fig. 8 ZK89+110 karst cave treatment elevation map

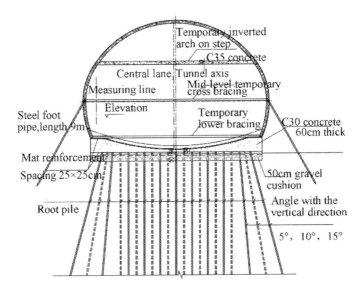

Fig. 9 Facade diagram of root pile treatment scheme

(4) The cave is treated by pumping concrete. The vault and the cavity at the waist are pumped first. Then 60 cm thick reinforced concrete was poured on the vault range. Finally, pour the concrete drainage ditch.

5 Conclusion

With the large karst cave crossed by Tianba highway Qinlan tunnel, this paper studies the comprehensive detection and treatment measures of large karst cave crossed by tunnel. The main conclusions are as follows.

(1) Aiming at the large karst caves traversed by the tunnel, the comprehensive detection method of "karst geological survey + geological radar detection + advanced horizontal drilling + three-dimensional laser scanning + geological sketching" is proposed for advanced detection.

(2) On the basis of finding out the location and development scale of karst cave, the scheme of tunnel excavation after advance reinforcement support treatment such as advance small pipe grouting is determined, which ensures that the Qinlan tunnel passes through ZK89+110 large karst cave safely.

(3) In this paper, the comprehensive treatment scheme of "comprehensive detection + advanced small pipe support + grouting reinforcement + root pile composite foundation reinforcement + lining reinforcement" is adopted. The problem of tunnel crossing karst cave treatment measures has been successfully solved, which provides an effective reference for similar projects in the future.

Acknowledgements Authors wishing to acknowledge the financial support from the Science and Technology Project of Gansu Province (21YF5FA002), and the Basic Research Program of Natural Science from Shaanxi Science and Technology Department (Grant No. 2022JM-191).

References

1. Ren R, Bai WZ, Gong LJ (2018) Study on detection and treatment technology of Karst Cave in highway tunnel. Road Constr Mach Constr Mech 10:79–83
2. Sun JT, Li ZT, Yuan JQ, Xiao DH (2021) Comprehensive advance detection and treatment technology for tunnels passing through large infilled Karst Caves. Mod Tunn Technol S1:416–425
3. Yang B (2011) Treatment techniques for Karst disaster of Maluqing Tunnel on Yichang-Wanzhou Railway. Chin J Underground Space Eng 03:581–586
4. Liu ZW, He MC, Wang SR (2006) Study on Karst waterburst mechanism and prevention countermeasures in Yuanliangshan tunnel. Rock Soil Mech (02):228–232+246
5. Zhang HG, Zhang GZ, Mao BY (2016) Mechanism analysis and water and mud breakout in the Xiaogao Mountain tunnel in Shanghai-Kunming passenger dedicated railway. J Railway Eng Soc (08):66–70+84
6. Zhao SZ, Huang X, Xu ZH, Li MH (2022) Development characteristics and treatment techniques of Karst Caves in Qiyueshan Tunnel. Tunn Constr 07:1289–1299
7. Wang SH, Chen Z, Jiang C et al (2017) Comprehensive treatment scheme and construction technology of super large karst tunnel. Tunn Constr 06:748–752
8. Liu TJ, Tang G, Wang J, Sun YF (2019) Treatment technology of Giant Karst Cave of Gaoshan Tunnel on Qianjiang-Zhangjiajie-Changde Railway. Tunn Constr 06:972–982
9. Li SC, Zhou ZQ, Ye ZH et al (2015) Comprehensive geophysical prediction and treatment measures of karst caves in deep buried tunnel. J Appl Geophys 116:247–257

10. Wang J, Wang LJ, Ma JX, He X (2020) Karst investigation and the countermeasures in Xinjie Tunnel of the Kaili Ring Highway, Guizhou Province. Geotech Eng Tech (05):276–281+285
11. Sun MB (2010) 602 Karst Cave Treatment in Yesanguan Tunnel on Yichang-Wanzhou Railway. Mod Tunn Technol 01:91–98
12. Liu XB, Wang LC, Wang YN (2022) Construction and treatment technologies of highway tunnel crossing super-large caves: a case study of Taiping tunnel project. Mod Tunn Technol S1:892–902
13. Cheng QJ (2020) Research on advanced detection and treatment technology of karst cave in tunnel. Highway 05:357–362
14. Nong J (2016) Study on Highway tunnel advanced level forecast drilling parameters of power ratio and wavelet analysis. Chang'an Univ 02:1–2
15. Hou GP (2021) Research on data processing method of tunnel deformation point cloud based on 3D laser scanning. Southwest Jiaotong Univ 05:6–8

Analysis on the Modeling of Rockbusrt Prediction in Deep Tunnels Based on Machine Learning

Yong Zhang, Haijun Liu, Yueyuan Ma, Xueqi Chen, Weinan Li, and Zheng Huang

Abstract Deep tunnels will face complex mechanical behavior problems of rock mass during construction and operation, among which the rockburst disaster is particularly prominent. Therefore, how to scientifically predict the rockburst activities of deep tunnels has become an urgent problem to be solved. According to the three aspects of energy conditions, lithology conditions and stress intensity that rockburst must meet, the strain energy storage coefficient (Wet), stress concentration factor (SCF) and brittleness index of rock mass ($\sigma c/\sigma t$) are selected as the discrimination indexes. On this basis, combining with the method of cluster analysis, some abnormal data are removed and a case database is constructed. Aiming at the defects of traditional prediction methods, this paper proposes a support vector machine method to establish the rockburst prediction model. According to the prediction model proposed in this paper to predict the rockburst situation of several deep tunnel engineering in China. The results show that the tunnel rockburst prediction model based on support vector machine is well consistent with the actual situation, indicating that the model has strong feasibility in practical application.

Keywords Deep tunnel · Rockburst · Risk prediction · Support vector machine

1 Introduction

With the rapid development of China's economy and the accelerating process of urbanization, the contradiction between the shortage of urban land resources is becoming increasingly prominent. More and more people turn their attention to the direction of underground space utilization. The rapid development of underground space construction has also brought many engineering problems. Among

Y. Zhang · H. Liu · Y. Ma · W. Li
CCCC First Harbor Engineering Company Limted, Tianjin 300461, China

X. Chen (✉) · Z. Huang
School of Highway, Chang'an University, Xi'an, Shaanxi 710064, China
e-mail: 1170979347@qq.com

© The Author(s) 2023
S. Wang et al. (eds.), *Proceedings of the 2nd International Conference on Innovative Solutions in Hydropower Engineering and Civil Engineering*, Lecture Notes in Civil Engineering 235, https://doi.org/10.1007/978-981-99-1748-8_33

them, the prediction of rockburst has always been a difficult problem in the field of rock engineering.

Rockburst is a kind of elastic deformation energy accumulated in rock mass, which is suddenly and violently released under certain conditions, resulting in rockburst and ejection phenomenon [1]. As one of the main disasters of deep tunnel, rockburst will threaten the safety of construction workers and machine facilities and affect the construction progress. Therefore, the prediction of rockburst is particularly important. The disaster-causing factors of rockburst are complex, domestic and many scholars have made in-depth research. There is no uniform standard for rockburst prediction methods, but the traditional rockburst prediction methods usually only consider a single index, and have strong subjectivity and low efficiency [2]. According to the characteristics that the damage, deformation and failure of rock is a nonlinear process, this paper attempts to establish a rockburst prediction model based on support vector machine. Compared with the traditional learning methods, its advantage is that it can use the kernel function to high-dimensional space mapping, which can better deal with nonlinear, small sample problems [3]. Feng et al. [4, 5] first applied the classification method of support vector machine to rockburst prediction. The results show that the prediction accuracy based on this machine learning method is considerable and practical.

Based on a large number of data samples collected, this paper attempts to establish a rockburst prediction model by means of support vector machine, which is used for rockburst prediction and engineering verification.

2 Establishment of Rockburst Database

2.1 Selection of Rockburst Prediction Factors

There are many factors that lead to the rockburst, and reasonable rockburst prediction factors should be chosen to establis the rockburst database [6]. The selection principles include: strong correlation with rockburst cases, high accuracy, and simple acquisition method. The factors that lead to rockburst can be divided into two aspects of internal and external factors. The internal factor refers to the rock mass itself, only the hard rock will store sufficient elastic deformation energy, so that the broken rock obtains the power of ejection, forming rockburst phenomenon. The evaluation indexes include energy storage and consumption index, elastic strain energy index, rock brittleness index, tangential stress size, rock brittleness coefficient and so on. The external factor refers to the engineering characteristics, the evaluation indexes include the buried depth of cavern, the size of ground stress, the excavation method and the excavation speed, etc.

The occurrence of rockburst must first satisfy the conditions of strength, energy, brittleness and so on, so this paper does not consider the influence of external factors on rockburst. In summary, combined with the selection principle, this paper selects the stress concentration factor SCF (σ_θ/σ_c), the strain energy storage index (Wet) and the rock brittleness index (σ_c/σ_t) as the prediction factors of rockburst. The stress concentration factor reflects the strength of the surrounding rock, and the index of Wet reflects the energy characteristics of the surrounding rock and the index of σ_c/σ_t reflects the brittle failure mechanism [7].

2.2 Collection of the Rockburst Cases

This paper refer to the collection including Erlang Mountain tunnel [8], Xuefeng Mountain tunnel [9], Gaoligongshan mountain tunnel [10], jinping II Hydropower Station [11], Qinling tunnel [12] 246 rock data, case, contains dozens of different types of rocks, rockburst grade is divided into no rockburst, minor rockburst, medium rockburst, strong rockburst level, and record each case of three predictors. These data were first recorded in Excel and then analyzed and processed with spss software. Some of the rockburst data are shown in Table 1.

The relationship between the strain energy storage coefficient (Wet) and the rockburst intensity is shown in Fig. 1. From the above scatter diagram we can intuitively see that the group of data has a good aggregation, different levels of rockburst index numerical boundaries more obvious. The strain energy storage index Wet decreases with the decrease of rockburst intensity, indicating that the lower the rockburst intensity, the smaller the energy stored in the rock mass, which is also consistent with our understanding of rockburst. However, there are still a few data that are far away from the data group. In order to ensure the accuracy of the rockburst prediction model, it is necessary to eliminate such outliers by mathematical methods.

Table 1 Data summarye domestic and foreign rockburst cases

Number	SCF	Wet	σ_c/σ_t	Rockburst intensity grade
1	0.77	5	17.5	Strong rockburst
...				
48	0.45	5.08	17.53	Minor rockburst
49	0.28	3.67	28.9	Minor rockburst
...				
129	0.32	5	21.69	Medium rockburst
130	0.38	5	21.67	Medium rockburst
...				
207	0.2	2.29	36.04	No rockburst
208	0.19	1.87	47.93	No rockburst

Fig. 1 Scatter plot of
rockburst index WET

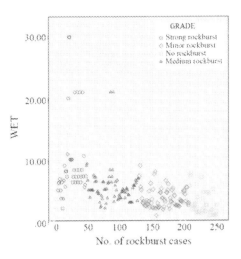

2.3   Cluster Analysis and Data Processing

This paper uses the idea of cluster analysis to analyze data. Cluster analysis is an
important research content in the field of data mining and artificial intelligence.
It is limited by the problems of dimension disaster and data scale [13]. By using
the SPSS software to realize the system clustering analysis, the pedigree diagram
between the samples under each rockburst intensity is drawn. The pedigree diagram
corresponding to the strong rockburst is shown in Fig. 2.

It can be seen from the above pedigree diagram that in the case of strong rockburst,
the No. 17 rockburst case was merged with the data group at the end, indicating that
the case belongs to the outlier data and should be eliminated. After eliminating the
discrete values and abnormal values of all the cases of intensity grade rockburst
cases, the most representative 185 data are finally obtained, including 35 data of
strong rockburst, 57 data of medium rockburst, 57 data of minor rockburst and 36
data of no rockburst.

3   Rockburst Prediction Model Based on Support Vector Machine

3.1   The Basic Idea of Support Vector Machine

The basic idea of support vector machines is to focus on the points close to the
hyperplane and divide the training number correctly in the set case, the interval from
the nearest point to the hyperplane is the largest, that is, the ultimate goal is to find an
optimal classification surface based on the training data. The problem of solving the

Fig. 2 Pedigree diagram corresponding to case of intense rockburst

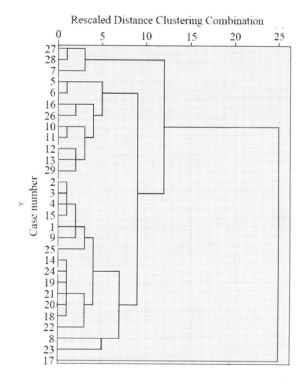

optimal classification surface of support vector machine can be transformed into the solution of the quadratic function of maximizing the classification interval of data samples [14]. The key is to obtain the target solution of the maximum classification interval. Taking two types of linear separable data as an example, one type of data is represented by circle, and the other type of data is represented by square. The optimal classification line is shown in Fig. 3.

3.2 Support Vector Machine Prediction Model

In this paper, the Libsvm software package is used to construct the rockburst prediction model of support vector machine, and the corresponding program is compiled in MATLAB. The detailed steps are as follows:

(1) Generate test set and training set of samples

After setting up the test environment, we import the Excel files of the rockburst case database into Matlab, randomly generate the training set and the test set, a total of 185 data, of which 80% (148) is used for learning and training, and the remaining 20% (37) data is used for model prediction.

Fig. 3 Sample diagram of
optimal classification line

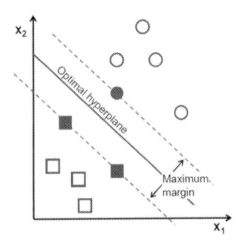

(2) Normalize the input sample data

When using the constructed support vector machine model to predict, the sample data
should be normalized to eliminate the interference of dimension on the comparability
of the original data. The data are normalized by $[-1,1]$ interval here.

(3) Select the kernel function

Support vector machine kernel function has RBF kernel function, package linear
kernel function, polynomial kernel function, etc. The RBF kernel function requires
fewer parameters and the parameter constraints are simple. Based on the above
advantages, this paper selects the RBF kernel function to apply.

(4) Parameter optimization

In the regression modeling of support vector machine, there are many important
parameters, which have a great influence on the prediction level of the model. Nowadays, the commonly used parameter optimization methods at home and abroad
include gradient descent method, bootstrap method, Bayesian method and intelligent algorithm. The cross-validation method can obtain the optimal parameters
under certain specific premises, and can better solve the 'over-learning' and 'under-
learning' problems well, so that the accuracy of the prediction results can meet the
requirements. Therefore, the cross validation method is selected for parameter optimization. In this paper, the parameter 't' of the support vector machine is selected
as 2, and the cross validation method is used. The output of the selected parameters
is: bestc $= 64$, bestg $= 0.0625$. That is, the model we constructed, the final choice
of fitting parameters: '$-t$' is equal to 2, '$-c$' is equal to 64, '$-g$' is equal to 0.0625.
Parameter selection results in matlab printed contour map is shown in Fig. 4, 3D
view is shown in Fig. 5.

(5) Training SVM model

Fig. 4 Contour map of
rockburst cases

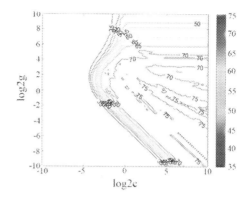

Fig. 5 3D picture of
rockburst cases

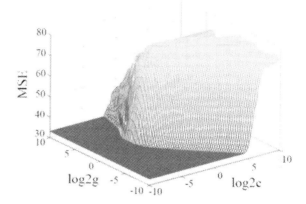

After the process of model training and simulation test, the prediction accuracy of
the final rockburst sample data is 89.2%, as shown in Fig. 6.

Fig. 6 Comparison of test
set and prediction results of
rockburst cases

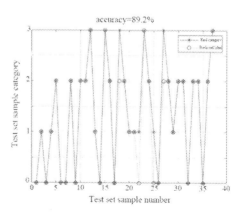

In the case of 185 samples, the learning accuracy of support vector machine can reach 84.5%, the prediction accuracy can reach 89.2%, and the prediction results have high reliability. This shows that the machine learning level of SVM is very high, and it is completely reasonable and feasible to use this model to predict rockburst.

4 Application

In order to verify the accuracy and applicability of the rockburst prediction model, Daxiangling Tunnel project [15] is selected as the verification object to predict the rockburst grade. The Daxiangling extra-long highway tunnel of Yalu Expressway in Sichuan Province is located in Daxiangling at the junction of Yingjing County and Hanyuan County, Ya'an City, Sichuan Province. It is a key and difficult control project of Yalu Expressway. In the large deformation section of Daxiangling tunnel, the surrounding rock is broken and the stability is extremely poor. It is easy to collapse after excavation. There is not only the interference in construction, but also the mutual influence and superposition on the formation disturbance. The poor geological conditions have promoted the rockburst in this area.

The surrounding rock grade of the tunnel is mainly grade III surrounding rock and grade IV surrounding rock. There are many rockburst phenomena in the construction, mainly minor and medium rockburst, accompanied by strong rockburst in some areas, which is harmful, as shown in Fig. 7. Therefore, reasonable prediction of rockburst intensity level is more important. We selected the data of 14 typical measuring points in the data sample of Daxiangling Tunnel, and listed the three indicators of the previous chapter, as shown in Table 2.

These 14 sets of data are used as test sets. For a set of data, the three independent variables are stress concentration factor SCF, strain energy storage index Wet, rock brittleness index $\sigma c/\sigma t$, and the dependent variable is rockburst intensity level. The fitting results are shown in Fig. 8 with an accuracy of 85.7%. The accuracy rate shows that the rockburst prediction model based on support vector machine has high

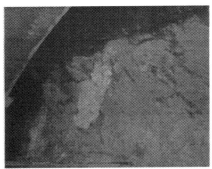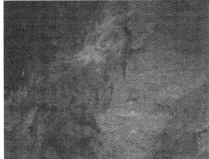

Fig. 7 Partial rockburst occurred in Daxiangling tunnel

Table 2 Prediction of rockburst Intensity of Daxiangling tunnel

Case No.	SCF	Wet	σ_c/σ_t	Rockburst intensity	Case No.	SCF	Wet	σ_c/σ_t	Rockburst intensity
1	0.43	1.7	45.9	Minor	8	0.57	3.8	25.6	Medium
2	0.42	2.4	29.9	Minor	9	0.43	2.4	29.9	Medium
3	0.56	1.9	34.3	Minor	10	0.56	1.9	34.3	Strong
4	0.60	3.4	28.3	Medium	11	0.51	5.2	50.9	Strong
5	0.53	3.6	21	Medium	12	0.60	3.4	28.3	Strong
6	0.66	4.1	21.5	Medium	13	0.56	4.6	31.3	Strong
7	0.52	4.3	17.8	Medium	14	0.49	4.7	49.6	Strong

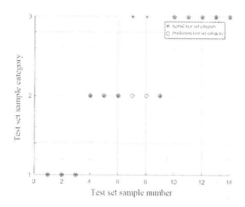

Fig. 8 Daxiangling tunnel rockburst monitoring point prediction results schematic diagram

applicability in Daxiangling Tunnel, which shows that this machine learning method is more reasonable in dealing with nonlinear problems such as rockburst.

5 Conclusion

Through research, the conclusions of this paper are as follows:

(1) Rockburst is a multi-factor and multi-mechanism geological phenomenon, so its prediction is a nonlinear problem. For a long time, experts and scholars at home and abroad have proposed many criteria to predict the intensity level of rockburst, but these methods have contingency and randomness. The method of machine learning is simple to operate and the model has strong generalization ability, which can further improve the reliability of the prediction model.

(2) In this paper, a representative case is collected to establish a sample database of rockburst data, and the main factors affecting the occurrence of rockburst are discussed. According to the engineering characteristics and the mechanism of rockburst, the stress concentration factor SCF, the strain energy storage index

Wet, and the rock brittleness index σ_c/σ_t are used as predictors. Through data processing methods such as cluster analysis, the most representative 185 samples are obtained to construct a rockburst case database.

(3) This paper introduces the basic working idea of support vector machine, and combines the cross validation method to optimize the parameters, and clarifies the advantages of support vector machine in solving nonlinear problems. The prediction results are analyzed, and the prediction accuracy is 89.2%, which is more accurate.

(4) The support vector machine rockburst prediction model is used to predict the rockburst of Daxiangling Tunnel, and the accuracy rate is 85.7%, which shows that the model has strong feasibility and applicability in practical engineering.

The rockburst prediction model is a meaningful attempt to use machine learning methods for disaster prediction. Although there are still many immature places in the application, it is believed that there will be broader development in this field in the future.

Acknowledgements Authors wishing to acknowledge the financial support from the High and New Technology Project from CCCC First Harbor Engineering Co. Ltd. (Grant No. 062021463), the Basic Research Program of Natural Science from Shaanxi Science and Technology Department (Grant No. 2022JM-191) and the Science and Technology Project of Gansu Province (21YF5FA002).

References

1. Liu CJ, Lu HJ (2005) Occurred mechanism and prevention countermeasures for rock outburst in deep mine roadway. Coal Sci Technol (11):30–32+67
2. Guo L (2004) The model to dynamically predict rockbursts proneness of hard rock at depth and its application. Central South University
3. Li ML, Li KG, Qin QC et al (2021) Discussion and selection of machine learning algorithm model for rockburst intensity grade prediction. Chin J Rock Mech Eng S1:2806–2816
4. Zhao HB, Feng XT (2004) Study of geotechnical engineering based on support vector machine. In: The 8th national academic conference on rock mechanics and engineering, Chengdu, China
5. Feng XT, Xiao YX, Feng GL et al (2019) Study on the development process of rockbursts. Chin J Rock Mech Eng 04:649–673
6. Adoko AC, Gokceoglu C, Li W et al (2013) Knowledge-based and data-driven fuzzy modeling for rockburst prediction. Int J Rock Mech Mining Sci 04:86–95
7. Zhang LW, Zhang DY, Li SC et al (2012) Application of RBF neural network to rockburst prediction based on rough set theory. Rock Soil Mech S1:270–276
8. Wang LS, Xu J, Xu LS et al (1999) Erlang mountain highway tunnel rockburst and rockburst intensity classification. Highway 02:41–45
9. Zhang ZL (2002) Study on rock burst and large deformation prediction of Xuefeng Mountain Tunnel of Shaohuai Expressway. Chengdu University of Technology
10. Lu RC (2014) Acoustic emission monitoring and numerical analysis about rock burst of Gaoligongshan Tunnel. Chongqing University
11. Zhou XT (2020) Prediction of rockbursts based on BP neural network. Xiangtan University
12. Bai MZ, Wang LJ, Xu ZY (2002) Study on a neutral network model and its application in predicting the risk of rock blast. China Saf Sci J 04:65–69

13. Wang DQ, Zhu JP, Liu XW et al (2018) Review and prospect of functional data clustering analysis. J Appl Stat Manage 01:51–63
14. Liu FY, Wang SH, Zhang YD (2018) Mathematical model and application survey of twin SVM. Meas Control Technol 08:10–15
15. Wen TX, Chen XY (2018) Forecast research on the rock burst liability based on the comprehensive evaluation H-PSO-SVM model. J Saf Environ 02:440–445

Application Practice of 3D Integration of Pipeline Information in Large-Scale and Complex Project Sites-Take Changsha Airport Addition and Alteration Project as an Example

Cheng Jiang, Xiaogang Dai, Yinqiang Huang, and Jinlei Li

Abstract Based on the practice of Changsha Airport Addition and Alteration Project, this paper refines, studies and summarizes the type, integration and interaction of pipeline information in large-scale and complex project sites, According to the whole life cycle of the Project, four types of pipeline information, including self-owned attribute, system attribute, management attribute and additive attribute, are sorted out. In actual engineering projects, there are two modeling ways, Method I: direct BIM design modeling, Method II: transformation and modeling after CAD design. The difference between Method I and Method II mainly sits on the difference in presentation form. One is three-dimensional and the other is two-dimensional. However, the base of both of them representing pipeline information is attributes and data. As long as these attributes and data can be extracted, information integration can be realized and applied to project management in the later stage. Based on the integrated research on the airport pipeline information, a corresponding BIM collaboration and management and control platform has been developed for the Project, and research results have been applied to the actual management of the project pipeline in an exploratory way, so as to achieve the project management requirements of the airport pipeline in terms of design, progress, cost, prefabrication and processing, and prepare for the later digital delivery and intelligent operation and maintenance management.

Keywords Site pipeline · BIM · Information integration · Project management application

C. Jiang (✉) · X. Dai
Hunan Airport Construction Command Department Hunan Airport Co., Ltd., Changsha, China
e-mail: zhongjun.wen@foxmail.com

Y. Huang · J. Li
China Southwest Architectural Design and Research Institute Co., Ltd., Chengdu, China

S. Wang et al. (eds.), *Proceedings of the 2nd International Conference on Innovative Solutions in Hydropower Engineering and Civil Engineering*, Lecture Notes in Civil Engineering 235, https://doi.org/10.1007/978-981-99-1748-8_34

373

1 Project Overview

The Changsha Airport Addition and Alteration Project (hereinafter referred to as "the Project") covers an area of more than 1100 ha, including an airfield area (including a runway and a taxiway), a parking area of 148 aircraft stands, a passenger terminal area, a freight terminal area, a working area, access roads and a central axis avenue outside the building line. The construction area is 1,048,800 m². In addition to a Passenger Terminal and a Ground Transportation Center (GTC), multiple supporting single buildings, municipal roads, viaducts, underpasses, municipal pipe networks, comprehensive pipe racks, civil air defense basements will also be built, covering more than 20 disciplines such as architectural, structural, municipal service road, landscape, etc. The total length of various site pipeline is up to thousands of kilometers.

2 Classification and Data Structure of Site Pipeline Information

The site pipelines of large-scale and complex projects generally include seven types of basic pipelines: water supply, sewage, rainwater, heat, gas, electricity and communication. In addition, there are special professional pipelines such as aviation oil at airport sites. Airport pipeline information is the basis and important link for building a "smart airport". These pipeline information are generated by several participating units by using professional design software. Therefore, whether it is the integration of original internal information generated by design software or the information exchange integration of external software/platforms, all kinds of pipeline information need to be classified, summarized, supplemented and confirmed on the integration platform according to the actual situation. Among them, the most important is to uniformly classify the types of pipeline information.

2.1 Type of Pipeline Information

According to the whole life cycle of the Project, the Project is divided into design, construction, operation and maintenance stages. In the Project, four types of pipeline information, including self-owned attribute, system attribute, management attribute and additive attribute, are sorted out according to the requirements for professional design software and its use in subsequent stages, as well as the commonness of pipeline information. See Fig. 1 for details.

In addition to the first three types of basic pipeline information, data can also be extended through additive attribute type. Additive attributes can be added through

Pipeline Attribute Category		
Self-owned attribute		
Attribute	**Value**	**Unit**
Pipe material	Copper	/
Pipe size	150	mm
Outside diameter of pipe	159	mm
Inside diameter of pipe	155	mm
Invert elevation of starting pipe	550.1	m
Invert elevation of end pipe	550.1	m
Pipe length	4.1	m
Pipe slope	0%	
System attribute		
Attribute	**Value**	**Remarks**
System type	Water supply	
Flow direction	P1→P2	
Management attribute		
Attribute	**Value**	**Remarks**
Installation unit		
Installation date		
Acceptance date		
Maintenance date I		
Maintenance date I_maintenance personnel		
Maintenance date II		
Maintenance date II_maintenance personnel		
Additive attribute		
Attribute	**Value**	**Remarks**
Additive attribute I		
Additive attribute II		
……		

Fig. 1 Classification of pipeline information

XDATA (AutoCAD), feature set (AutoCAD Civil 3D), attribute table (AutoCAD MAP 3D), plug-in provision, secondary development and other methods.

Fig. 2 Data structure of pipeline information in Revit software

2.2 Data Structure of Pipeline Information Library

According to the designed pipeline attribute type, and the built-in data structure of professional software, the data structure of pipeline information is supplemented and redefined, as shown in Figs. 2, 3 and 4.

3 Creation of 3D Model of Site and Pipeline

After the data structure of BIM pipeline model is established, the suitable BIM design software can be selected for 3D modeling. However, in actual engineering projects, due to the design habits of designers and their inadaptability to BIM software, most design units still use 2D method based on cad for preliminary design of sites and pipelines, and then submit them to modelers for conversion into 3D models through corresponding software. Therefore, there are also two modeling ways in the Project: direct BIM modeling and BIM model transformation after CAD design. The technical routes of these two modeling methods are as follows.

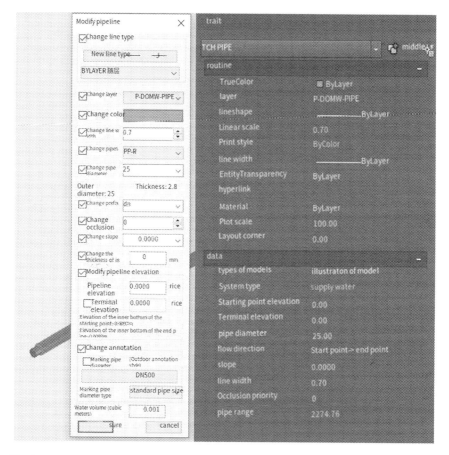

Fig. 3 Data structure of water supply and drainage pipeline information in Tangent software

Method I: direct BIM design modeling: The Project involves many design units. A small number of them use 3D BIM software (Revit, Civil 3D, OpenRoadcc, etc.) to carry out 3D site and pipeline design, and directly build 3D model of pipelines. The results are 3D models and 2D drawings. After proper processing, 3D models can be directly imported into the project BIM management and control platform for information integration and various applications. The technical route is shown in Fig. 5.

Method II: transformation and modeling after CAD design: In the Project, most design units still use 2D method based CAD for site and pipeline design, and the results are still 2D drawings. In order to integrate pipeline information into the BIM collaboration and management and control platform of the Project, participating or construction units need to conduct 3D processing on 2D drawing results, convert them into 3D models and supplement the missing pipeline information, import them into BIM collaboration and management and control platform of the Project, and

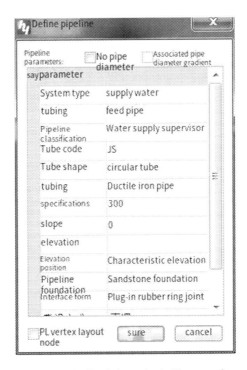

Fig. 4 Data structure of municipal pipeline information in Hongye software

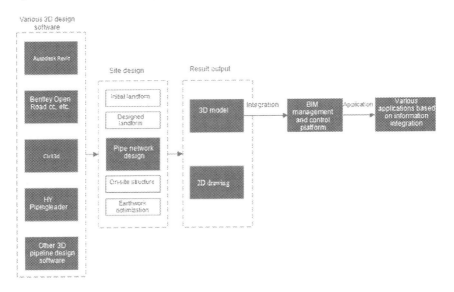

Fig. 5 Technical route of direct BIM design modeling

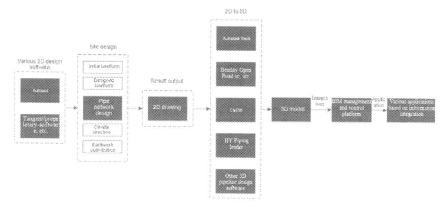

Fig. 6 Technical route of transformation and modeling after CAD design

integrate with models generated in Method I for information integration and various project management applications in the later stage. The technical route is shown in Fig. 6.

4 Integration and Interaction of Site Pipeline Information

4.1 Export of Site Pipeline Information

The created pipeline BIM model can export various geometrical information and attribute information of the required pipeline through "direct export", "plug-in export", "export.shp" and other methods according to the different application platforms/software in the later stage, such as SuperMap, Unreal4 Revit, and Revizto. The commonness of these plug-ins is that they are installed in Revit software. Users need to first open the Revit software and click "Export" button to export the required data. Some plug-ins also provide the function of model lightweight.

Although there are many ways to export pipeline information, data planning must be done in advance. Before selecting the integrated software/platform, developers of the other party should consult the design method and data structure of the native software in detail, and comprehensively consider the integration and fusion of the design software in the early stage and the integration platform in the later stage necessary for final seamless connection of pipeline data.

4.2 Integration of Site Pipeline Information

The difference between direct BIM modeling and 3D model transformation after CAD graphic design mainly sits on the difference in presentation form. One is three-dimensional and the other is two-dimensional. However, the base of both of them representing pipeline information is attributes and data. As long as these attributes and data can be extracted, information integration can be realized and applied to project management in the later stage. Therefore, these two design methods have been coexisting in the current design reality, and 2D graphic design is not completely denied because of the emergence of 3D design.

Meanwhile, the design software used in these two design methods has certain information integration capabilities. For example, for Tangent Water Supply and Drainage Software (AutoCAD plug-in), which is commonly used in CAD design, although design units often only use its 2D drawing function, the pipeline in it is a 3D entity from the 3D perspective, and the attributes of the pipeline, such as material, diameter and height, are also displayed in the feature bar. In addition, the software can automatically summarize all pipe types, length, and number of valves and fittings. This in itself is a certain degree of pipeline information integration. Although Tangent software is inferior to foreign BIM software (such as Revit) in 3D cooperation with other disciplines, it is better than foreign software in terms of convenience and localization of professional design. Therefore, at present, it still occupies a large market share in China.

For another example, some design units use Hongye Municipal Pipeline Software for site design. This is because the software is mainly aimed at municipal engineering. Compared with Tangent Water Supply and Drainage Software, it is more professional and comprehensive. Similarly, the software also can provide complete pipeline attribute and information integration, including 3D pipeline and tube well display, as well as quantity statistics and summary of pipeline, tube well and equipment.

BIM 3D software is oriented to 3D and multi-discipline collaboration at the beginning of design. Therefore, the concept of BIM 3D software is more advanced. For example, Revit developed by Autodesk and Openroad developed by Bentley, which are most commonly used in the building industry, have been widely promoted and used in China. Taking Revit as an example, in addition to the pipe material, pipe diameter, height, specification, system, length and other attributes of the pipeline, the software can flexibly add other types of attribute information to the pipeline, and conduct statistical analysis on the ancillary facilities of the pipeline and pipe fittings through BOM function to achieve the initial collection of pipeline information for design/construction/operation and maintenance. However, due to some problems existing in these software and poor localization and industry adaptability, they have not completely replaced 2D design software in China.

4.3 Interaction of Site Pipeline Information

After the information of various pipelines and sites is integrated into an unified platform, the pipeline information between the upstream and downstream of the same pipeline, between different pipelines, and between the pipeline and the site can be interacted through professional fusion and evaluation, so as to provide various professional analysis for the pipe network in the site [1].

(1) Condition query: Query various information of the pipeline according to user needs through the integration and interaction of upstream and downstream data of the whole pipeline; conduct all-round query operations on the pipeline and pipe points in terms of space, facilities, areas and fields;

(2) Pipeline statistics: Summarize and count the length, quantity, weight, amount and other information of the pipeline according to user needs through the integration and interaction of upstream and downstream data of the whole pipeline;

(3) Cross and vertical section analysis: Reflect the buried depth of various pipelines underground and the space among pipelines in a 3D visual and intuitive way through the interaction of spatial information among pipelines and the height information from the ground;

(4) Pipe break analysis: Analyze the pressure fluctuation and the degree of influence on upstream and downstream pipelines after a pipe break to determine the number of valves to be closed through the interaction of upstream and downstream pipelines;

(5) Collision analysis: Analyze the net distance among pipelines to determine whether the design is reasonable through the interaction of spatial information among pipelines;

(6) Pipe network flow direction and flow analysis: Realize the creation of pipe network flow direction model through the interaction of 3D pipe network scene model data, and query the specific flow direction and flow of the pipeline in a given area according to user needs;

(7) Pipe network connectivity reliability analysis: Calculate the connectivity probability from the source point to the sink point in the pipe network, and use it as the connectivity reliability index of the pipe network through the interaction of 3D network scene model data, so as to analyze and judge the degree of importance of the pipeline section in the pipe network, and check whether the two points in the pipe network are connected;

(8) Excavation analysis: Realize the ground excavation analysis and relevant pipeline data statistics under any excavation surface, and provide 3D ground excavation simulation of any area through the interaction of spatial information between pipeline section, pipe point and excavation surface;

(9) Early warning analysis: Provide early warning for pipeline sections that may have problems through the interaction between production time of pipe materials and environmental data;

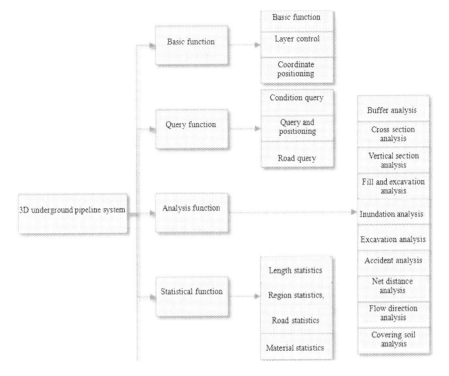

Fig. 7 Function module of underground pipeline system

(10) Accident influence scope analysis: Analyze the degree of influence of pipeline accidents on users around accident areas through the interaction between the spatial information of the pipeline route and the influencing radius of the destructive accident.

At present, all mature pipeline management software in the building market can realize the above functions, as shown in Fig. 7 Function Module of Underground Pipeline System [2].

4.4 Collaboration and Management and Control Platform of Site Pipeline Information

To get through the information integration and work collaboration of all links in the management of the Project, especially the collaborative design in the early stage, the Project strives to build a BIM-based airport pipeline information collaboration and control platform based on and centered on the pipeline BIM model. With B/S architecture and 3D model, the platform can track, collaborate and control the Project in the construction period from design, progress, cost, quality, prefabricated pipe fitting processing and other aspects, so as to improve project production efficiency, increase construction quality, control construction progress and construction cost.

In addition to meeting the current project applications and functions, the platform development also pays special attention to the following two aspects to ensure the unity and scalability of the platform [3].

Ensure Unity: Uniqueness, Standardization, Precision and Intelligence. The foundation of platforms is data, and the core of data is unify, including unity of structure, standard and interface. Therefore, the steps to unify data are to build a unified system and data standards for the generation and collection of upstream and downstream data sources, gather the data processed according to the standards into the library for unified management, establish an authoritative and unique site pipeline database for the realization of associated storage of attribute data, graphic data, and control data based on pipeline information, and provide intelligent tools for multi-source heterogeneous data collection and processing to improve data processing efficiency and accuracy. The database supports multiple data aggregation methods, and automatic parsing and loading of data in multiple formats (including word, jpg, Word, JPG, PDF, Excel, CAD, SHP, GDB, systems, and archives (paper/scanning)).

Face the Future: Exportable, Extensible, Connectable and Fusible. With the continuous progress of technology in the future, there will certainly be new platforms, new technologies and new languages. Therefore, it is necessary to ensure the exportability of original data and process data, or ensure the connectivity of the system, so as to facilitate good migration/integration into the new system. At the same time, with the progress of technology, VR, AR and other visualization technologies will be integrated in the later stage. For example, when mobile phones/AR helmets are used to observe underground virtual pipeline, the flow rate, temperature and other information of the liquid in the pipeline can be monitored in real time through the IoT technology, and pipeline information can be displayed in mobile phones/PCs/webpages and other platforms through cloud technology.

5 Application Practice of Pipeline Information in Project Management

At present, the Project has entered the construction stage. Based on the integrated research on the airport pipeline information, a corresponding BIM collaboration and management and control platform has been developed for the Project, and research results have been applied to the actual management of the project pipeline in an exploratory way, so as to achieve the project management requirements of the airport pipeline in terms of design, progress, cost, prefabrication and processing, and prepare for the later digital delivery and intelligent operation and maintenance management [4].

5.1 Application in Project Design Management

The Project is large in scale, involving more than ten major design units. Through the BIM collaboration and management and control platform built, the collaborative design mode based on BIM platform is realized for all design units. The combination of online and offline methods is adopted to improve the collaboration efficiency. With the BIM collaboration and management and control platform, the project department can coordinate the design interface, spatial relationship and technical cooperation scheme, find and resolve design conflicts in advance, and achieve the refined management of project design. Figure 8 shows the design collaboration interface. For some forward design contents, the competent design department innovates the management mode of design engineering with the help of this new collaboration and management and control platform, directly incorporates project management into the design process, advances the afterwards management of the old mode, and achieves event management. At the same time, it incorporates the user department into the design process by using the advantages of "What You See Is What You Get" (WYSIWYG) BIM design. The user department's interpretation of design results and review of functional requirements greatly improve the quality of design results and shorten the design cycle and the time required for confirmation.

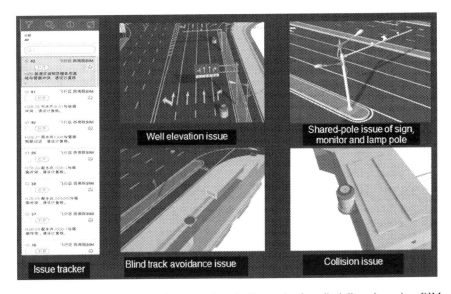

Fig. 8 Solve coordination issue between site pipeline and other disciplines by using BIM collaboration platform

5.2 Application in Project Cost Management

The Project covers a wide area of land, with complex and numerous types of pipelines, the total length of which is up to thousands of kilometers. Therefore, if the 2D design is adopted in quantity calculation, a large amount of original data should be exported from the 2D software for manual calculation. For example, when calculating the pipe length, inspecting well depth and trench earthwork quantities, it is necessary to manually input the well number, pipe diameter, pipe length, original ground elevation, design elevation, and pipe invert elevation into Excel tables, and calculate the quantities based on the data exported from the software and the quantities formula. However, it is easy to make mistakes when manually extracting elevation data from drawings, which will cause errors in the calculation of quantities.

After using BIM software to complete the site pipeline design, the pipeline information can be automatically integrated, the standardized data can be provided according to the requirements of the system, and the real quantities can be quickly and automatically calculated in the form of BOM, as shown in Fig. 9, which can be compared with the bill of quantities for bidding to avoid omissions and large deviations in the quantities. At the same time, the technology of automatically obtaining quantities from BIM software can be easily used for later project change management, and the change amount can be quickly calculated according to the change of quantities, which can be used for determining change level and speeding up the change process [5].

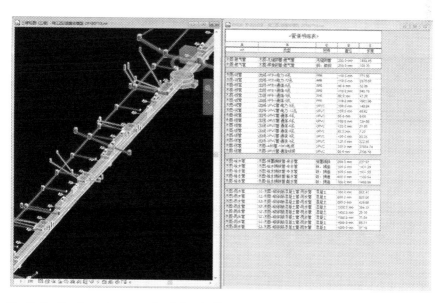

Fig. 9 Realize dynamic export of real site pipeline quantities by using BOM

5.3 Application in Project Progress Management

In the Project, the BIM collaboration and management and control platform integrates the 3D model of the pipe network, the quantity information, and the completion time information of project nodes. By formulating corresponding rules and automatically associating time parameters with components in the model, a BIM progress control model can be finally formed. By using these basic data and comprehensive data, the project department can schedule and control the construction schedule of the Project, so as to meet various functional requirements of construction schedule management, as shown in Fig. 10. The BIM model can store all the information in the whole life cycle of the airport pipeline, realizing the information coexistence and integration of multiple disciplines, the information sharing, transmission and query through the platform with centering on the pipe network model, and the whole process management of the construction progress [6].

BIM technology can also be used to simulate the method statement, which can be comprehensively rehearsed before the formal construction, exposing problems in the rehearsal process, and promoting construction units and relevant staff to take effective preventive measures in advance, which can largely ensure the construction progress and quality.

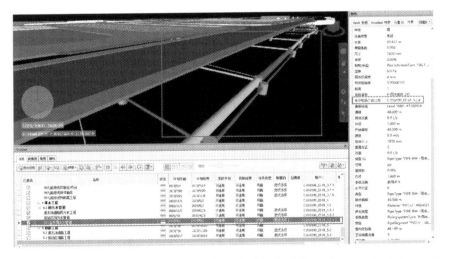

Fig. 10 Central axis avenue pipeline mounting schedule I of Changsha airport addition and alteration project

Fig. 11 Automatic segmentation and numbering of pipe sections

5.4 Application in Project Prefabrication and Processing Management

The Project has very high processing accuracy and quality requirements for prefabricated pipeline components. If there are mistakes in the design scheme or prefabricated components in the production process, it will have a serious impact on the later construction. In order to avoid this situation, the design scheme of prefabricated components shall be effectively verified according to the actual situation before production, and communication shall be made with relevant departments to accurately express the design intent, so that the design information can be accurately transmitted to the production link and ensure the consistency of information.

The traditional pipeline production and installation is to measure, divide, prefabricate, transport, store and assemble pipes on site according to 2D construction drawings and on-site construction conditions. The traditional program has many problems, such as complicated calculation process, troublesome material quality management and low manufacturing accuracy, which greatly affects the accuracy of final components. The integration of pipeline information based on BIM can well solve the above problems, as shown in Figs. 11 and 12. BIM technology can generate accurate pipe section splitting before construction based on digitized 3D visual model data, attribute parameters and spatial relations of pipelines and pipe fittings, and count the size, specification and quantity in the form of BOM, so as to achieve the project objective of accurate production and control of prefabricated pipeline components.

5.5 Application in Digital Delivery and Intelligent Operation and Maintenance Management of Project

In the Project, the BIM collaboration and management and control platform is used to integrate various information and data of the airport pipeline, including drawings, design data, construction data, operation and maintenance data, based on and centered on the pipeline 3D model data, as shown in Fig. 13.

With the continuous progress of airport construction, various information and data of the airport are being improved and accumulated. At the later completion stage, the

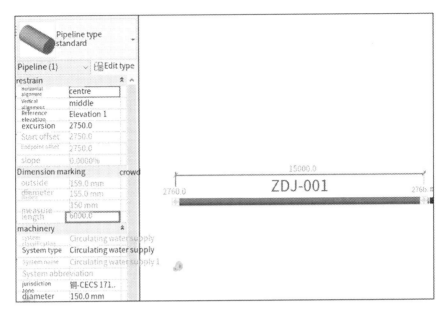

Fig. 12 Standard length of pipeline

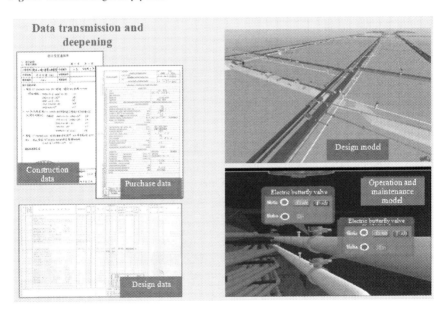

Fig. 13 Pipeline information and data integration based on 3D model

platform will submit a BIM completion model (LOD500 in depth) that contains the same information as the field for digital delivery. If the operation and maintenance data is added to the platform, the platform can completely provide a comprehensive and accurate 3D digital baseboard for the intelligent operation and maintenance of the airport, which will continue to play a role in the long-term operation and transformation of Changsha Airport in the future, and lay a solid foundation for building a safe, smart, green, humanistic airport.

6 Summary

In summary, this paper investigates and analyzes the whole process of pipeline information integration from design → operation and maintenance, points out that the two design methods have native pipeline information integration, analyzes the platforms and software that can be used for pipeline information integration on the market, sorts out and formulates pipeline attribute tables, attribute attachment methods, export and import platforms and other workflows, and carries out the application practice of airport pipeline information integration in the design and construction stages based on the research. It covers design management, cost management, prefabrication processing management, digital delivery and intelligent operation and maintenance management, which has reference value for realizing airport/large-scale project pipeline information integration and digital application.

References

1. General Introduction of Underground Pipeline Integrated Management System. https://blog.csdn.net/wk2133/article/details/113948633
2. Skyline 3D Pipeline System. https://blog.51cto.com/speciallist/5717023?articleABtest=1
3. SuperMap 3D Pipeline. https://blog.csdn.net/supermapsupport/article/details/52869056
4. Eastman C, Teicholz P, Sacks R, Liston K (2012) BIM handbook: a guide to building information, modelling for owners, managers, designers, engineers and contractors, 2nd edn. Wiley, New York
5. Forgues D, Poirier E, Staub-French S (2016) The impact of BIM on collaboration: an action-cognition perspective
6. Zhao D, McCoy AP, Bulbul T, Fiori C, Nikkhoo P (2015) Building collaborative construction skills through BIM-integrated learning environment. Int J Constr Educ Res 11:1–24

Seismic Interferometry by Multi-Dimensional Deconvolution Free from Point-Spread Function

Lu Bin, Wang Ji, Niu Huan, Li Bo, and Ji Guangzhong

Abstract Seismic interferometry (SI) can be used to reconstruct a pseudo-acquisition from response of a passive source, while the reconstructed data can be used to recover a portion of the model space that is different from that recovered by the inversion of original measurements. The SI by crosscorrelation requiring the seismic wave field to be evenly distributed, which limits the scope of application of this method. SI by multi-dimensional deconvolution (MDD) broken through the limitation that the wave field must be evenly distributed, but there are still some limitations in some special practices. Interferometric point-spread function (PSF), like the correlation function, is a necessary condition of the MDD method, but in some practices it cannot be derived from the field data. A new MDD method is proposed in this paper that free from PSF, and theoretically proved that it is equivalent to usual MDD. We demonstrate the effectiveness of this method with two numerical examples of one-sided illumination, and the source blurring phenomenon in the results of SI by crosscorrelation is effectively eliminated. The first numerical example is far-field one-side illumination, which can also be treated with the usual MDD, and the comparison of the results shows that the two MDD methods are equally effective. The second example is the near-field one-side illumination, and only the MDD method proposed in this paper can be used because the PSF cannot be obtained.

Keywords Passive source · Multi-dimensional deconvolution · Point-spread function · Seismic interferometry · Noise imaging

L. Bin (✉) · L. Bo
Wenzhou University of Technology, Wenzhou 325035, China
e-mail: lubin2000@163.com

L. Bin · W. Ji · N. Huan
The Xi'an Research Institute, CCTEG, Xi'an 710077, China

J. Guangzhong
Anhui University of Science and Technology, Huainan 232001, China

© The Author(s) 2023
S. Wang et al. (eds.), *Proceedings of the 2nd International Conference on Innovative Solutions in Hydropower Engineering and Civil Engineering*, Lecture Notes in Civil Engineering 235, https://doi.org/10.1007/978-981-99-1748-8_35

1 Introduction

Passive seismic is seismic imaging using sources of opportunity [1]. This method does not use standard air guns, vibrators or dynamite, but only deploys an array of seismic acquisition equipment to record the seismic response of opportunistic source. Sources of opportunity include microseisms, ambient-noise, industrial noise, etc. Because sources of opportunity exist for a long time relative to artificial sources, they are suitable for time-lapse seismic assessments [1]. The main industrial applications of passive seismic time-lapse assessments include: reservoir monitoring [2], stress redistribution around the coal longwall mining panel [3–5], geological stability of train tunnel [6]. The seismic response of passive source cannot be directly used for imaging, because the passive source often has some obvious shortcomings, such as the uncertainty of source location and of shot time, and multi-sources mixing, etc. In order to use the passive sources to imaging the target medium, their seismic responses need to be reconstructed. Reconstructing the seismic response of a passive source into the seismic response between geophone pairs is a method known as seismic interferometry (SI). This method is also called Green's function extraction method, or virtual source method, because one of the geophones appears to be a "virtual" source for reconstructing seismic response.

SI is a rapidly developing research field in recent years, which generally reconstructs the Green's function between two geophones by the cross-correlation method. Compared with the SI by cross-correlation, the deconvolution interferometry proposed by Snieder et al. [7] has some advantages; for example, it can compensate for the properties of the source wavelet, and there is no need to assume that the medium is lossless [8]. Therefore, this method has a good performance for recovering the impulse response from noise records by a long and complicated source-time function. Nakata et al. [9] applied the cross-coherence method to the SI of traffic noise to retrieve both body waves and surface-waves. By using only the phase information and ignoring amplitude information, the method effectively removes the source signature from the extracted response and yields a stable structural reconstruction even in the presence of strong noise. These methods belong to 1D SI.

All of the methods mentioned above require the wave field to be evenly distributed. The Green's function between two points could be recovered using the cross-correlation function of the ambient noise measured at these two points, because the ambient-noise is more uniform from all directions [10, 11]. However, when the industrial noise is used as the source, one-sided illumination can lead to severe distortions of the retrieved Green's function, which is proportional to a Green's function with a blurred source. These limitations can be partially solved by MDD methods [12]. The source blurring of cross-correlation can be quantified by the point-spread function (PSF), which can be obtained from the correlation function of the data [13]. The source of the Green's function obtained by the correlation method can be deblurred by deconvolving the correlation function for the PSF [8]. MDD greatly expands the scope of application of SI. However, for a type of thread sources in tunnels, such as the belt conveyor in the coal working face tunnels or subway trains, it is difficult to get PSF from the field data, because the geophones and the thread sources in the same tunnel. In the studies conducted by Wapenaar et al. [14], similar problems were encountered when trying to apply MDD for deblending data from simultaneous sources, where the source and receiver arrays even coincided. The results indicated that in the cases where the sources and receivers were very close together, it was not possible to invert the PSF using conventional MDD-methodology. The aforementioned researchers believed that this problem had occurred due to the fact that the spatial spectrum of the PSF was too broad. They reasoned that the problem could potentially be mitigated by extending the overburden, or possibly by including an additional spatial filter inside their formulation. Lu found that another PSF could be obtained from the data of the adjacent tunnel, and a new MDD formula based on the new PSF is derived [15].

In this paper, we propose a new SI by MDD method which free from PSF, which is suitable for practices where PSF cannot be obtained from field data. This method is similar to a multi-dimensional whitening filter, which can be proved to be partly equivalent to usual MDD method.

2 Green's Function Representation and SI by MDD

Figure 1a shows the illustration of SI by MDD for the situation of direct-wave interferometry. In an arbitrary inhomogeneous anisotropic dissipative medium, a volume V was defined, which enclosed by a surface S with outward pointing normal vector $\mathbf{n} = (n_1, n_2, n_3)$. Surface S was coincided with the surface S_{rec} and a hemisphere S_1. For the practices discussed in this paper, S_{rec} stands for tunnels in which geophone array $\mathbf{x}$ is arranged, and $\mathbf{x}_B$ is a geophone in an adjacent tunnel. A straightforward crosscorrelation of responses at $\mathbf{x}$ and $\mathbf{x}_B$ gives a set of impulse response (Green's function) at one receiver $\mathbf{x}_B$ of virtual sources at the position of $\mathbf{x}$. The mass density in V is ρ. The sources are outside of V, and the i-th source is named $\mathbf{x}_S^{(i)}$. We define the Fourier transform of $G(\mathbf{x}_B, \mathbf{x}_S, t)$ as

$$\hat{G}(\mathbf{x}_B, \mathbf{x}_S, \omega) = \int_{-\infty}^{\infty} \exp(-j\omega t) G(\mathbf{x}_B, \mathbf{x}_S, t) dt,$$

with j the imaginary unit and ω the angular frequency. The convolution-type Green's function representation is given by [13]

$$\hat{G}(\mathbf{x}_B, \mathbf{x}_S, \omega) = -\oint_S \frac{1}{j\omega\rho(\mathbf{x})} \Big(\partial_i \overline{\hat{G}}(\mathbf{x}_B, \mathbf{x}, \omega) \hat{G}(\mathbf{x}, \mathbf{x}_S, \omega)$$
$$-\overline{\hat{G}}(\mathbf{x}_B, \mathbf{x}, \omega) \partial_i \hat{G}(\mathbf{x}, \mathbf{x}_S, \omega)\Big) n_i d\mathbf{x} \tag{1}$$

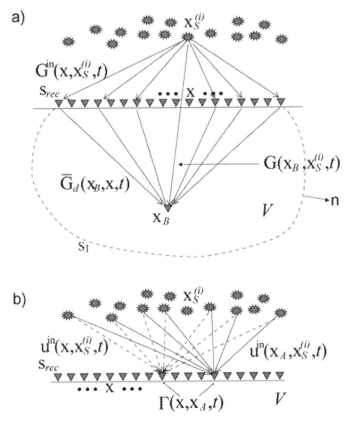

Fig. 1 **a** Illustration of interferometry by MDD for the situation of direct-wave interferometry. **b** Illustration of the point-spread function $\Gamma(\mathbf{x}, \mathbf{x}_A, t)$ (Eq. 7) (Modified from [13])

(Einstein's summation convention applies to repeated lower case Latin subscripts). The notation $\widehat{\overline{G}}$ is introduced to denote a reference state with possibly different boundary conditions at S and/or different medium parameters outside S (but in V the medium parameters for $\widehat{\overline{G}}$ are the same as those for $\hat{G}$). The bar is usually omitted because $\hat{G}$ and $\widehat{\overline{G}}$ are usually defined in the same medium throughout space. ∂_i denotes that the differentiation is carried out with respect to the components of n_i.

The convolution-type Green's function representation (Eq. (1)) is a basic expression for SI by MDD in open systems [16]. $\hat{G}(\mathbf{x}, \mathbf{x}_S, \omega)$ is writed as the superposition of an inward- and outward-propagating part at $\mathbf{x}$ on S_{rec}, according to $\hat{G}(\mathbf{x}, \mathbf{x}_S, \omega) = \hat{G}^{in}(\mathbf{x}, \mathbf{x}_S, \omega) + \hat{G}^{out}(\mathbf{x}, \mathbf{x}_S, \omega)$. After a series of simplifications of the integral of Eq. (1), convolution-type representation is obtained [13]:

$$G(\mathbf{x}_B, \mathbf{x}_S, t) = \int_{S_{rec}} \overline{G}_d(\mathbf{x}_B, \mathbf{x}, t) \otimes \hat{G}^{in}(\mathbf{x}, \mathbf{x}_S, t) d\mathbf{x} \qquad (2)$$

where, $\otimes$ denotes temporal convolution.

The passive source is regarded as a series of discrete sources $\mathbf{x}_S^{(i)}$, and the wavelet of each discrete source is $\omega^{(i)}(t)$. The convolution of Green's function and wavelet $\omega^{(i)}(t)$ was used to obtain the corresponding wave displacement [13]:

$$u^{in}\left(\mathbf{x}, \mathbf{x}_S^{(i)}, t\right) = G^{in}\left(\mathbf{x}, \mathbf{x}_S^{(i)}, t\right) \otimes \omega^{(i)}(t) \tag{3}$$

$$u\left(\mathbf{x}_B, \mathbf{x}_S^{(i)}, t\right) = G\left(\mathbf{x}_B, \mathbf{x}_B^{(i)}, t\right) \otimes \omega^{(i)}(t) \tag{4}$$

Then, by substituting Eqs. (3) and (4) into Eq. (2), the following was obtained [13]:

$$u\left(\mathbf{x}_B, \mathbf{x}_S^{(i)}, t\right) = \int_{S_{rec}} \overline{G}_d(\mathbf{x}_B, \mathbf{x}, t) \otimes u^{in}\left(\mathbf{x}, \mathbf{x}_S^{(i)}, t\right) d\mathbf{x} \tag{5}$$

In order to solve the above formula in the terms of a least square, the incident wave $u^{in}\left(\mathbf{x}_A, \mathbf{x}_S^{(i)}, t\right)$ was cross-correlated on both sides, and $\mathbf{x}_A$ represented one point in $\mathbf{x}$. Subsequently, the following was obtained [13]:

$$C(\mathbf{x}_B, \mathbf{x}_A, t) = \int_{S_{rec}} \overline{G}_d(\mathbf{x}_B, \mathbf{x}, t) \otimes \Gamma(\mathbf{x}, \mathbf{x}_A, t) d\mathbf{x} \tag{6}$$

where, the PSF is defined as

$$\Gamma(\mathbf{x}, \mathbf{x}_A, t) = \sum_i u^{in}\left(\mathbf{x}, \mathbf{x}_S^{(i)}, t\right) \otimes u^{in}\left(\mathbf{x}_A, \mathbf{x}_S^{(i)}, -t\right)$$

$$= \sum_i G^{in}\left(\mathbf{x}, \mathbf{x}_S^{(i)}, t\right) \otimes G^{in}\left(\mathbf{x}_A, \mathbf{x}_S^{(i)}, -t\right) \otimes W^{(i)}(t) \tag{7}$$

and the correlation function (also known as virtual shot gather) is defined as

$$C(\mathbf{x}_B, \mathbf{x}_A, t) = \sum_i u\left(\mathbf{x}_B, \mathbf{x}_S^{(i)}, t\right) \otimes u^{in}\left(\mathbf{x}_A, \mathbf{x}_S^{(i)}, -t\right)$$

$$= \sum_i G\left(\mathbf{x}_B, \mathbf{x}_S^{(i)}, t\right) \otimes G^{in}\left(\mathbf{x}_A, \mathbf{x}_S^{(i)}, -t\right) \otimes W^{(i)}(t) \tag{8}$$

with $W^{(i)}(t) = \omega^{(i)}(t) \otimes \omega^{(i)}(-t)$, and Σ denotes the superposition of all the sources.

$\Gamma(\mathbf{x}, \mathbf{x}_A, t)$, the interferometry PSF, as defined in Eq. (7) is the crosscorrelation of the inward-propagating wavefields at $\mathbf{x}$ and $\mathbf{x}_A$, summed over the sources [17], see Fig. 1b. When the medium is lossless and the wavefield is equipartitioned, the PSF would approach a temporally and spatially band-limited delta function [13]. When these assumptions are violated, Eq. (6) shows that the PSF blurs the source of Green's function in the spatial directions [7]. MDD involves inverting Eq. (6). This removes the distorting effects of the PSF $\Gamma(\mathbf{x}, \mathbf{x}_A, t)$ from the correlation function $C(\mathbf{x}_B, \mathbf{x}_A, t)$ and yields an improved estimate of the Green's function $\overline{G}_d(\mathbf{x}_B, \mathbf{x}, t)$ [13].

Then, through the discretization of the integral in Eq. (6), the following was obtained [13]:

$$C\left(\mathbf{x}_B, \mathbf{x}_A^{(l)}, t\right) = \sum_k \overline{G}_d\left(\mathbf{x}_B, \mathbf{x}^{(k)}, t\right) \otimes \Gamma\left(\mathbf{x}^{(k)}, \mathbf{x}_A^{(l)}, t\right) \qquad (9)$$

For all $\mathbf{x}_A^{(l)}$ on the receiver surface S_{rec}.

In the frequency domain, the convolution becomes the following multiplication:

$$\hat{C}\left(\mathbf{x}_B, \mathbf{x}_A^{(l)}, \omega\right) = \sum_k \widehat{\overline{G}}_d\left(\mathbf{x}_B, \mathbf{x}^{(k)}, \omega\right) \hat{\Gamma}\left(\mathbf{x}^{(k)}, \mathbf{x}_A^{(l)}, \omega\right) \qquad (10)$$

Then, by solving this system of equations, the required Green's function was obtained [13].

3   Practices in Which PSF Cannot Be Obtained from Field Data

SI by MDD, based on Eq. (10), needs two sets of data, the one is the result of SI by correlation function $C\left(\mathbf{x}_B, \mathbf{x}_A^{(l)}, t\right)$ and the other one is $\Gamma\left(\mathbf{x}^{(k)}, \mathbf{x}_A^{(l)}, t\right)$. The matrix equation can be solved per frequency component via such as weighted least-square inversion to obtain $\overline{G}_d(\mathbf{x}_B, \mathbf{x}, t)$ [8]. In practices, $C(\mathbf{x}_B, \mathbf{x}_A, t)$ can always be obtained, but the PSF is not always available. In this section we introduce two practices, in where the PSF cannot be obtained from field data.

Working face is any place where coal is extracted during a mining cycle, at the same time where is the primary disaster zone, geological hazards include caving of roof, gas outburst, pressured water outburst, etc. Figure 2a shows a top view of common longwall working face. After the coal is mined by the mining machine, it is transported to crusher by scraper conveyors. After reduced in size, the coal in crusher is loaded onto the conveyor belt and transported to the surface. Belt conveyor is installed in the belt tunnel, and it is about 0.5 m from the coal wall. The belt conveyer is in the belt tunnel from the beginning to the end, so the seismic ray can cover the whole working face [5].

For the foundation below the subway tunnel, accumulated deformation of soil under the subway loading should not be ignored. For example, when the deformations of foundation exceed a certain range, the safety of adjacent buildings and underground pipe networks will be endangered, causing a series of environmental and geotechnical problems [6]. Figure 2b is a cross-sectional view of subway train tunnels and utility tunnels under urban roads. Geophones are arranged in these tunnels to monitor the

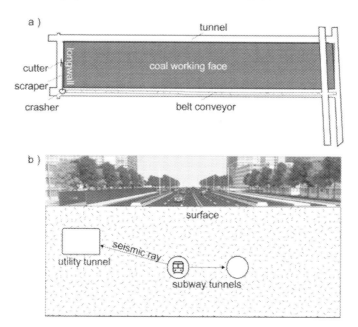

Fig. 2 **a** Top view of coal working face. **b** A cross-sectional view of the stratum below the road, including subway tunnels and utility tunnel

stratum through which seismic waves pass by, using the trains' noise as sources. Geophones can only be arranged in these tunnels. Because it almost coincides with the distribution of passive sources, it is difficult to obtain a high signal-to-noise ratio PSF in the data of the train tunnel, so the usual MDD method is not suitable for such practice.

Figure 3a shows 10 s of the response of belt conveyor's noise along the belt tunnel in a coal mine working face, with a total of 72 traces. The trace spacing is 10 m. Seven of them are bad traces, which are 19–21, 26 and 31–33. The belt conveyor is composed of belt and belt support, the spacing between each belt support is about 3 ~ 4 m, and the belt support is installed on the tunnel floor. It can be seen from the data that the belt conveyor is started at about 2 s. Figure 3b shows the PSF of the 36th trace of the data shown in Fig. 3a. It is found that only the autocorrelation function of the 36th trace has a relatively high signal-to-noise ratio, while the cross-correlation function with other traces is completely covered by noise, so the PSF is invalid. Therefore, the usual MDD method is not applicable to this practice.

Fig. 3 **a** Response along the belt tunnel of belt conveyor's noise in a coal mine working face. **b** PSF of the 36th trace of the response of belt conveyor's noise

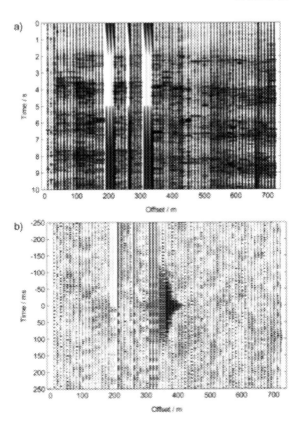

4 SI by MDD Free from PSF

The PSF does not needed to be calculated for the proposed new MDD methods which, like multi-dimensional whitening filter, can extract PSF from the cross-correlation function $C(\mathbf{x}_B, \mathbf{x}_A, t)$ implicitly. Because the data process only uses the results of SI by cross-correlation and not need calculate PSF explicitly from seismic response, the method is named as MDD free from PSF.

Equation (7) defines the PSF in the time domain. Transforming the PSF to the frequency domain, according to

$$\hat{\Gamma}(\mathbf{x}, \mathbf{x}_A, \omega) = \sum_i \hat{G}^{in}\left(\mathbf{x}, \mathbf{x}_S^{(i)}, \omega\right)\left\{\hat{G}^{in}\left(\mathbf{x}_A, \mathbf{x}_S^{(i)}, \omega\right)\right\}^* \hat{W}^{(i)}(\omega) \qquad (11)$$

Here we discuss how the alternative to the PSF is obtained from the result of SI by cross-correlation.

$\hat{C}(\mathbf{x}_B, \mathbf{x}, \omega)$ is cross-correlation of response at geophones at $\mathbf{x}_B$ and $\mathbf{x}$, in frequency domain. $\mathbf{x}$ is coordinate of geophones of virtual sources. Transforming the Eq. (8) to the frequency domain, according to

$$\hat{C}(\mathbf{x}_B, \mathbf{x}_A, \omega) = \sum_i \hat{G}\left(\mathbf{x}_B, \mathbf{x}_S^{(i)}, \omega\right) \left\{\hat{G}^{in}\left(\mathbf{x}_A, \mathbf{x}_S^{(i)}, \omega\right)\right\}^* \hat{W}^{(i)}(\omega) \qquad (12)$$

Similarly, the $\hat{C}(\mathbf{x}_B, \mathbf{x}, \omega)$ according to

$$\hat{C}(\mathbf{x}_B, \mathbf{x}, \omega) = \sum_i \hat{G}\left(\mathbf{x}_B, \mathbf{x}_S^{(i)}, \omega\right) \left\{\hat{G}^{in}\left(\mathbf{x}, \mathbf{x}_S^{(i)}, \omega\right)\right\}^* \hat{W}^{(i)}(\omega) \qquad (13)$$

$\hat{C}(\mathbf{x}_B, \mathbf{x}, \omega)$ and $\hat{C}(\mathbf{x}_B, \mathbf{x}_A, \omega)$ are cross-correlated, according to

$$\left\{\hat{C}(\mathbf{x}_B, \mathbf{x}, \omega)\right\}^* \hat{C}(\mathbf{x}_B, \mathbf{x}_A, \omega) = \sum_i \hat{G}^{in}\left(\mathbf{x}, \mathbf{x}_S^{(i)}, \omega\right) \left\{\hat{G}^{in}\left(\mathbf{x}_A, \mathbf{x}_S^{(i)}, \omega\right)\right\}^*$$

$$\left\{\hat{G}\left(\mathbf{x}_B, \mathbf{x}_S^{(i)}, \omega\right)\right\}^* \hat{G}\left(\mathbf{x}_B, \mathbf{x}_S^{(i)}, \omega\right) \left\{\hat{W}^{(i)}(\omega)\right\}^* \hat{W}^{(i)}(\omega)$$

$$= \hat{\Gamma}(\mathbf{x}, \mathbf{x}_A, \omega) \left\{\hat{G}\left(\mathbf{x}_B, \mathbf{x}_S^{(i)}, \omega\right)\right\}^* \hat{G}\left(\mathbf{x}_B, \mathbf{x}_S^{(i)}, \omega\right) \left\{\hat{W}^{(i)}(\omega)\right\}^* \qquad (14)$$

The right-hand side of the above equation contains 3 components, and the first part is just the PSF. The second part is

$$\sum_i \left\{\hat{G}\left(\mathbf{x}_B, \mathbf{x}_S^{(i)}, \omega\right)\right\}^* \hat{G}\left(\mathbf{x}_B, \mathbf{x}_S^{(i)}, \omega\right) \qquad (15)$$

In the absence of multiples, this part has zero phase and has no effect on the total phase.

The third part is the last term on the right side of the equation, which is the conjugate of the source wavelet $\sum_i \hat{W}^{(i)}(\omega)$. $\hat{W}^{(i)}(\omega) = \hat{\omega}^{(i)}(\omega)\left\{\hat{\omega}^{(i)}(\omega)\right\}^*$ has zero phase, and the conjugate of them still has zero phase, which will not affect the phase of Eq. 13, according to

$$\sum_i \left\{\hat{W}^{(i)}(\omega)\right\}^* = \sum_i \hat{W}^{(i)}(\omega) \qquad (16)$$

so it will not affect the phase of Eq. 14, but this term will have a certain impact on the total amplitude spectrum.

In general, $\left\{\hat{C}(\mathbf{x}_B, \mathbf{x}, \omega)\right\}^* \hat{C}(\mathbf{x}_B, \mathbf{x}_A, \omega)$ is an alternative of the point spread function $\hat{\Gamma}\left(\mathbf{x}^{(k)}, \mathbf{x}_A^{(l)}, \omega\right)$ in terms of phase. The amplitude spectrum will be affected by the zero-phase correlation wavelet. Compared with PSF, the amplitude spectrum of $\left\{\hat{C}(\mathbf{x}_B, \mathbf{x}, \omega)\right\}^* \hat{C}(\mathbf{x}_B, \mathbf{x}_A, \omega)$ is narrower, which is equivalent to that the PSF is filtered by the source wavelet $\sum_i \hat{W}^{(i)}(\omega)$.

$\left\{\hat{C}(\mathbf{x}_B, \mathbf{x}, \omega)\right\}^* \hat{C}(\mathbf{x}_B, \mathbf{x}_A, \omega)$ can be written as $\hat{P}$. The inverse filter can be obtained per frequency component via weighted least-squares inversion, according to

$$\hat{F} = \left(\hat{P} + \varepsilon^2 \mathbf{I}\right)^{-1} \tag{17}$$

where the ε^2 is a stabilization parameter, $\mathbf{I}$ is the identity matrix. ε^2 is frequency dependent. Applying the inverse filter for each frequency component and transforming the result to the time domain is equivalent with the deconvolution in the time domain.

5 Numerical Example 1: Thread Source One-Side Illumination

The numerical example in this section is a general one-side illumination, the source is a thread source, such as the train [18–21] or the highway [9]. With the development of transportation infrastructure, this kind of passive source is getting more attention [22–25]. In this example, there is a certain distance between the passive source and the geophones (Line1 shown in Fig. 4). In the example in the next section, the passive sources and the geophones which as the virtual sources are almost coincident (shown in Fig. 6).

Consider the configuration in Fig. 4, which consists of a set of thread sources and two sets of geophone arrays. The length of geophone array is 500 m, and the thread source is composed of a group of independent noise sources, with a total of 51 sources and trace spacing of 10 m. The source spacing is 10 m too. It is 300 m from the thread source to Line1 and 300 m from Line1 to line2.

The source time function of each noise source is represented by a random sequence of standard normal distribution, which is uncorrelated with each other. The longer the random sequence used, the higher the signal-to-noise ratio of the shot gather obtained by cross-correlation. Because it propagates in the homogeneous medium, the signal from each source point arrives at each geophone only with time delay and energy diffusion attenuation.

Figure 5a shows the result of SI by cross-correlation, and the random sequence length is 100,000. The virtual source is at the 26th geophone in Line1 (the middle trace), and this results in a great signal-to-noise ratio in the image. In the results, the source blurring is very obvious due to the one-side illumination, especially in the middle traces. Figure 5b shows the PSF of trace 26 in Line1, which is the cross-correlation function between trace 26 and all the traces in Line1. Figure 5c is the result obtained by using the usual MDD method, and compared with Fig. 5a, the source blurring completely disappears. Figure 5d is the result obtained by using

Fig. 4 Configuration of 'virtual source method' for far-field illumination from one-side. Two sets of geophone arrays are respectively in two tunnels (the triangles at Line1 and Line2)

noise

Line1

Line2

MDD method free from PSF proposed in this paper; the source blurring is also well eliminated, and the signal-to-noise ratio is greater than that of Fig. 5c.

6 Numerical Example 2: Near Field Thread Source One-Side Illumination

In Fig. 6, the thread passive source and geophones in Line1 almost coincide with each other, which simulates the belt conveyor or train source in an underground tunnel, because the geophones as virtual sources can only be installed in the tunnel. This is a common problem for all thread sources in the tunnel. Section 3 briefly describes such practices.

In this example, the thread source is composed of 51 independent point sources with intervals of 10 m. The source time function of each source is represented by a random sequence with length of 100,000, which is uncorrelated with each other. The two groups of geophones Line1 and line2 are composed of 51 geophones respectively with a trace spacing of 10 m. The distance between Line1 and line2 is 300 m. In actual, most of the coal working face width is about 300 m. This numerical example can be directly used to study the interferometry imaging of working face with belt conveyor as the source.

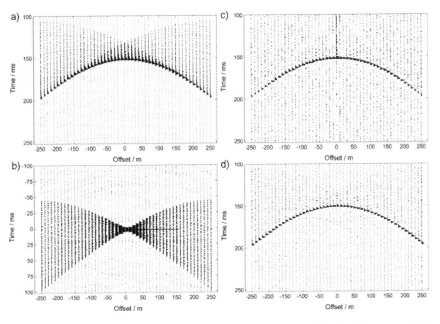

Fig. 5 **a** Result of SI by cross-correlation. **b** PSF of trace 26 in Line1. **c** Result of SI by usual MDD method. **d** Result of SI by MDD free from PSF

Fig. 6 Configuration of 'virtual source method' for near-field illumination from one-side. Two sets of geophone arrays are respectively in two tunnels (the triangles at Line1 and Line2)

Figure 7a is the result of SI by cross-correlation, which if the cross-correlation function between the response of trace 26 (middle channel) of Line1 and the responses of all traces of Line2. It is obvious that there are a series of spurious events before the true event, and these spurious events are caused by one-side illumination. Compared with the source blurring in Fig. 5a, the source blurring in Fig. 7a are some spurious events. True and spurious events can be easily distinguished in the

Fig. 7 **a** Result of SI by cross-correlation. **b** Result of SI by MDD free from PSF

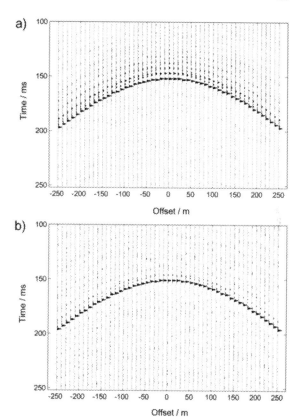

numerical synthesis example, but it is difficult to distinguish them in the actual field data. Figure 7b is the result of SI by MDD free from PSF proposed in this paper, which largely eliminates the spurious events on the seismogram.

In the numerical simulation, the distance of geophone between the source and Line1 is an important influencing factor. If the positions of 51 source points and 51 geophones completely coincide with each other, the proportion of overlapped source points is too large for a certain geophone, and the contribution of the remaining 50 source points can be ignored. Under the above conditions, there is no source blurring in the cross-correlation results, which is inconsistent with the actual observation results. In practices, the belt conveyor or train causes vibration on the tunnel floor, and the geophone is installed on the tunnel wall, so there is a distance of 2–3 m between the source point and the geophone at the same position.

For practices where PSF cannot be obtained in the field data, the usual MDD method cannot be used to eliminate source blurring, while the MDD method free from PSF proposed in this paper can be used to solve this problem. In Fig. 7b, the spurious events still have some residuals, but the true event is clearer compared to Fig. 7a.

7 Conclusions

We propose a new MDD method free from PSF as an alternative to the usual MDD. Using this method, MDD processing can be carried out directly from the results of SI by cross-correlation. The main advantage of the new MDD method is that the source of the Green's function obtained by the correlation method can be deblurred for practices where PSF cannot be obtained, while the conventional MDD method must calculate PSF first. In addition, for practices where PSF can be obtained from field data, the results of the new method show that the new method and the usual method show similar processing effects.

The new method is similar to a 3-D whitening filter. Each geophone in Line1 is taken as the virtual source, and virtual shot gathers is obtained by cross-correlation. The virtual shot gathers are arranged as a 3D array for whitening. MDD needs to set stabilization parameters ε^2 which can only be determined by multiple attempts. The results of SI by MDD are given by subjective evaluation, so the results obtained by different people will be slightly different.

In practices, the SI by cross-correlation and by MDD method should be used together. The research shows that both 1D deconvolution and MDD can improve the resolution of shot gather, and at the same time, the method itself may cause spurious events. There is no spurious event in the result of SI by cross-correlation, so it can be used to determine which are spurious events in the result of SI by MDD. Although the seismic interference of such sources will still face many problems in practical applications, the use of linear sources in tunnels is getting more and more attention. This new MDD method proposed in this paper provides an attractive solution.

Acknowledgements This project was funded by the National Natural Science Foundation of China (Grant No. 41974209, No. 42074148, No. 51978531).

References

1. Duncan P (2005) Is there a future for passive seismic. First Break 23:111–115
2. Issa NA, Lumley D, Pevzner R (2017) Passive seismic imaging at reservoir depths using ambient seismic noise recorded at the Otway CO_2 geological storage research facility. Geophys J Int 209(3):1622–1628
3. Luo X, King A, Van de Werken M (2009) Tomographic imaging of rock conditions ahead of mining using the cutter as a seismic source—a feasibility study. IEEE Trans Geosci Rem Sens 47(11):3671–3678
4. Hosseini N, Oraee K, Shahriar K et al (2011) Studying the stress redistribution around the longwall mining panel using passive seismic velocity tomography and geostatistical estimation. Arab J Geosci 6(5):1407–1416
5. Lu B, Feng J (2017) Coal working face imaging by seismic interferometry-using conveyer belt noise as source. J Seism Explor 26:411–432
6. Yan CL, Tang YQ, Wang YD et al (2012) Accumulative deformation characteristics of saturated soft clay under subway loading in Shanghai. Nat Hazards 62:375–384 (in Chinese)

7. Snieder R, Sheiman J, Calvert R (2006) Equivalence of the virtual source method and wave-field deconvolution in seismic interferometry. Phys Rev E 73:066620
8. Wapenaar K, Van der Neut J, Ruigrok E (2008) Passive seismic interferometry by multidimensional deconvolution. Geophysics 73:A51–A56
9. Nakata N, Snieder R, Tsuji T (2011) Shear wave imaging from traffic noise using seismic interferometry by cross-coherence. Geophysics 76(6):SA97–SA106
10. Shapiro N, Campillo M (2004) Emergence of broadband Rayleigh waves from correlations of the ambient seismic noise. Geophys Res Lett 31(7):1615–1619
11. Shapiro NM, Campillo M, Stehly L (2005) High-resolution surface-wave tomographyfrom ambient seismic noise. Science 307:1615–1618
12. Schuster G, Zhou M (2006) A theoretical overview of model-based and correlation-based redatuming methods. Geophysics 71(4):SI103–SI110
13. Wapenaar K, van der Neut J, Ruigrok E et al (2011) Seismic interferometry by crosscorrelation and by multidimensional deconvolution: a systematic comparison. Geophys J Int 185(3):1335–1364
14. Wapenaar K, Van der Neut J, Thorbecke J (2012) On the relation between seismic interferometry and the simultaneous-source method. Geophys Prospecting 60:802–823
15. Lu B (2021) Multi-dimensional deconvolution for near-field thread seismic sources in tunnels. J Environ Eng Geophys 26(4):305–313
16. Slob E, Wapenaar K (2007) GPR without a source: cross-correlation and cross-convolution methods. IEEE Trans Geosci Rem Sens 45:2501–2510
17. van der Neut J, Ruigrok E, Draganov D et al (2010) Retrieving the earth's reflection response by multi-dimensional deconvolution of ambient seismic noise. In: EAGE, Extended Abstracts. EAGE, Houten, pp P406-1–P406-4
18. Ditzel A, Herman G, Holscher P (2001) Elastic waves generated by high-speed trains. J Comput Acoust 9(3):833–840
19. Chen QF, Li L, Li G et al (2004) Seismic features of vibration induced by train. Acta Seismologica Sinica 17(6):715–724. (in Chinese)
20. Li L, Peng WT, Li G et al (2004) Induced by trains: a new seismic source and relative test. Chin J Geophys-Ch 47(4):680–684
21. Li WJ, Li L, Chen QF (2008) Research on the movement of vibration source of train by means of SSA. Chin J Geophys-Ch 51(4):1146–1151
22. Xu SH, Guo J, LI PP et al (2017) Observation and analysis of ground vibrations caused by the Beijing–Tianjin high-speed train running. Prog Geophys 32(1):421–425. (in Chinese)
23. Zhang HL, Wang BL, Ning JY et al (2019) Interferometry imaging using high-speed-train induced seismic waves. Chin J Geophys-Ch 62(6):2321–2327
24. Cao J, Chen JB (2019) Solution of Green function from a moving line source and the radiation energy analysis: a simplified modeling of seismic signal induced by high-speed train. Chin J Geophys-Ch 62(6):2303–2312
25. Wang XK, Chen JY, Chen WC et al (2019) Sparse modeling of seismic signals produced by high-speed trains. Chin J Geophys-Ch 62(6):2336–2343

Numerical Simulation of Seismic Liquefaction for Treasure Island Site

Hua Lu(ID)**, Ziyun Lin, Xudong Zhan, Yanxin Yang**(ID)**, and Di Wu**

Abstract By conducting numerical simulations of Treasure Island site in 1989 Loma Prieta earthquake, the Hilbert-Huang transform was used to study soil liquefaction. The UBC3D-PLM constitutive model was used to simulate the dynamic behavior of sand, the non-liquefied soil was simulated by Hardening Small Strain model, and the seismic information recorded in the PEER database was used for deconvolution analysis by one-dimensional equivalent linear analysis method, and the within motion was applied. Rayleigh damping is used in the numerical simulation process, and the damping ratio is 5%. Based on the Hilbert-Huang transform analysis, it is concluded that the UBC3D-PLM constitutive model can effectively simulate the dynamic behavior of soil. The amplification of the peak acceleration was from the bottom of Sandy fill to the ground surface, and the low-frequency content was shifted due to liquefaction in the process of transmission to the surface, and the high-frequency content was identified at the surface.

Keywords Liquefaction · Dynamic analysis · UBC3D-PLM · Hilbert-Huang transform

1 Introduction

Liquefaction induced by earthquake is one of the important research topics in geotechnical earthquake engineering. During liquefaction of sand or sandy soil, the pore water pressure of saturated sand increases rapidly under dynamic loads, resulting in a decrease in the effective stress. When the effective stress approaches to zero, the soil bearing capacity reduces and the soil will liquefy and behave like liquid.

Liquefaction will cause severe damages to structures and foundations. For example, in 1995 Kobe earthquake [1], the soil liquefied due to the seismic motions,

H. Lu · Z. Lin · X. Zhan · Y. Yang (✉) · D. Wu
School of Architecture and Transportation Engineering, Guilin University of Electronic Technology, Guilin 541000, Guangxi, China
e-mail: yanxinyangswjtu@foxmail.com

© The Author(s) 2023
S. Wang et al. (eds.), *Proceedings of the 2nd International Conference on Innovative Solutions in Hydropower Engineering and Civil Engineering*, Lecture Notes in Civil Engineering 235, https://doi.org/10.1007/978-981-99-1748-8_36

resulting in widespread lateral spreading and the collapse of the road surface. The liquefaction in 1999 Chi-chi earthquake [2], caused the settlement of buildings and severe lateral spreading.

A variety of site liquefaction evaluation methods have been proposed. In 1970, Seed simplified method [3, 4] was proposed to identify sand liquefaction phenomenon. In 1982, Tanimoto and Noda proposed the liquefaction potential index method to identify liquefaction [5]. Davis and Berrill proposed the energy method to identify sand liquefaction [6]. With the development of finite element method, numerical simulation has become an effective an effective method for liquefaction analysis, several constitutive models simulating sand liquefaction were proposed. Wobbes et al. [7] verified the accuracy of the finite element method (FEM) for simulating sand liquefaction based on shaking table test, and the results showed that the method was feasible. Subasi et al. [8] analysis of three soil profiles with different compactness with finite element method, and the simulation results yielded similar results to the empirical method. When adopting numerical simulation, the appropriate constitutive model should be selected for soil dynamic analysis, and the variation of pore water pressure and the accumulation of plastic strain during dynamic loading should be considered, which are the key factors to predict sand liquefaction.

In this paper, finite element method was used to analyze the liquefaction of the site under the excitation of earthquake, and the UBC3D-PLM constitutive model was used to simulate the liquefiable soil. To analyze the liquefaction characteristics, the surface and subsurface motions recorded in the dynamic analysis were analyzed by the Hilbert-Huang transform.

2   Case Study

In this paper, Treasure Island site was analyzed. The site is a man-made island located northwest of Yerba Island in the San Francisco area. In the 1989 Loma Prieta earthquake, liquefaction and lateral spreading were observed on the island. Bedrock motions recorded at Yerba Island were available in the PEER database [9].

Soil profile shown in Fig. 1 was used in the analysis, based on the site investigation, the Sandy fill shown in Fig. 1 was suspectable to liquefaction. Below the Sandy fill layer, there exists the Young Bay mud and Old Bay mud in sequence. The Rock fill was constructed adjacent to the seaside as used as supporting structure.

3   UBC3D-PLM

UBCSAND was developed by Beaty and Byrne [10] for prediction of liquefaction behavior of sand, which has been implemented in Plaxis. P. V. UBC3D-PLM is a 3-D sand liquefaction model developed based on UBCSAND by Puebla et al. [11], which greatly improved the accuracy of calculation results under cyclic loading and

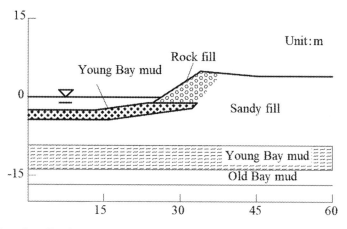

Fig. 1 The soil profile of Treasure Island

earthquake motions. The model requires four primary parameters as inputs. The constant volume friction angle, φ_{cv} φ_{cv}, the peak friction angle, φ_p φ_p, and cohesion, c, were evaluated from direct shear tests on material. In the absence of the test results, the SPT blow count of the liquefiable soil, $(N_1)_{60}$, can be used to estimate the input parameters.

3.1 Seismic Input

During the 1989 Loma Prieta earthquake, seismic waves were recorded on Yerba Buena Island, available in the PEER database. Conducting one-dimensional equivalent linear analysis, the input motion applied at the bottom of the model in the finite element analysis from the deconvolution analysis is shown in Fig. 2. The characteristics of input motion denoted as YBI 090 is shown in Table 1.

3.2 Numerical Model

The mesh of finite element model based on the soil profile of Treasure Island is shown in Fig. 3. Three monitoring points are used to obtain the dynamic response of the site. Point A is located at the surface of the sandy layer (the elevation is 0.00 m), point B is located at the bottom surface of the sandy layer (the elevation −8.65 m), and point C is located at the bottom of the model. The Mohr-Coulomb constitutive model is used to model the behaviors of all layers of soil under static stress and the Hardening Small Strain constitutive model is used to model the behavior non-liquefiable soil in the dynamic analysis. Table 2 shows the material parameters as

Fig. 2 Input motion (YBI 090 from deconvolution analysis)

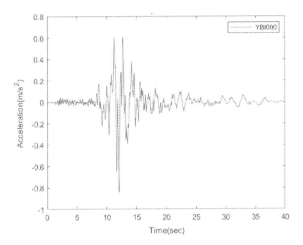

Table 1 The para maters of YBI 090

No	Site	Earthquakes	Time(s)	M_w	$PGA(g)$	Source
1	Yerba Island	Loma Prieta (1989)	40	7.0	0.06	Power [12]

used in the Mohr-Coulomb model for modelling. The generation of the initial stresses is generated via the gravity loading process. Where, the left and right boundaries are set to normal fixed and defined as undrained and the bottom boundary is set to completely fixed and defined as undrained condition. In the dynamic analysis phase, Sandy fill was replaced by UBC3D-PLM sand liquefaction model, and the parameters were evaluated based on the literature [13]. The drainage condition is set to undrained-A, which represents the effective-stress analysis in Plaxis P. V. and the material parameters are shown in Table 3. The x_{min} and x_{max} boundary conditions are set to the free field boundary to absorb seismic waves and prevent seismic waves from being bounced at the boundary, and the bottom boundary is set as the compliance basis. At the same time, in order to avoid complete loss of shear strength of soil at the x_{min} and x_{max} and large deformation in the analysis results, 1 m-wide drainage boundaries using Mohr-Coulomb constitutive model are set at x_{min} and x_{max}, and the soil parameters in the drainage area was the same as the adjacent soil. To reduce the impact of boundaries on the analysis results, the model must be wide enough [14], so the model width was set to 100 m. In order to eliminate the influence of high-frequency components in seismic motion, the material damping ratio, ξ, was set to $\xi = 5\%$ and Rayleigh Damping was set based on the method by Hudson et al. [15].

The motion was applied at the bottom of the model, converted to a specified line displacement in Plaxis P. V. The x-direction displacement components have to be divided by a factor of two, considering that the signal that downward going waves needs to be absorbed.

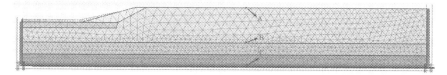

Fig. 3 Finite element analysis model

Table 2 Material parameters as used in the Mohr-Coulomb model for modelling

Parameters	Rock fill	Sandy fill	Young Bay mud	Old Bay mud
Saturated unit weight (kN/m^3)	28.1	18.4	16.5	19.2
Dry density (kg/m^3)	23	14.11	13.1	15.23
Cohesion (kPa)	1	1	5	5
Internal friction angle ($^\circ$)	36	33	25	25
Shear wave velocity (m/s)	500	152	186	336
Poisson's ratio	0.3	0.3	0.45	0.45

Table 3 Main parameters assued in UBC3D-PLM for dynamic analysis

Parameters	Description	$(N_1)_{60} = 8.45$ [16]
φ_p ($^\circ$)	Peak friction angle	31.29
φ_{cv} ($^\circ$)	Friction angle at constant volume	30.45
c (kPa)	Effective cohesion	1
$(N_1)_{60}$	SPT blow count of the liquefiable soil	8.45

4 The Results of Element Analysis

The step-by-step construction stages of finite element analysis are as follows, Phase 1. In the initial phase, the initial static stress is generated by gravity load process and the steady groundwater seepage is used to generate the static pore pressure. Phase 2. In this stage, plasticity analysis was used to simulate the static conditions of the model to remove the unbalanced static stress. Phase 3. The dynamic analysis was performed in this phase by using the UBC3D-PLM constitutive model for the liquefiable soil, and the line displacement was activated.

4.1 Dynamic Response

The acceleration time histories at different elevations (Point A, B and C) are shown in Fig. 4. It can be seen from Fig. 4a that the recorded value of the signal at the bottom

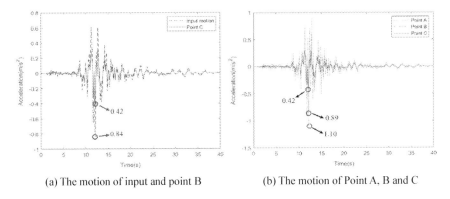

(a) The motion of input and point B   (b) The motion of Point A, B and C

Fig. 4 Time history of acceleration

of the model (point C) is the average value of the input signal, as the coefficient of 0.5 was specified when the signal was as input of within motion and a complaint boundary condition was used. It can be seen from Fig. 4b the PGA at the ground surface is 1.10 m/s². From the point C to point A the peak value amplified from 0.84 to 1.10 m/s².

4.2 Excess Pore Water Pressure Ratio

The time history curve of excess pore water pressure at point B is plotted in Fig. 5, and according to Fig. 5, the soil liquefied at the time 11.95 s.

Fig. 5 Excess pore water pressure versus time

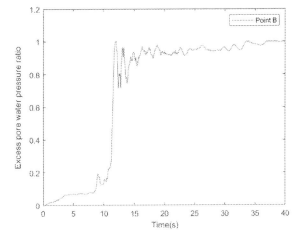

4.3 *Hilbert-Huang Transform*

The Hilbert-Huang transform is a technique for non-stationary signal processing. The signal is decomposed in the form of intrinsic mode function (IMF) by empirical mode decomposition (EMD) [17, 18]. The two constraints of natural mode function are as follows: (1) the number of extreme points and zeros should be equal or not more than one difference; (2) The upper and lower envelope should be locally symmetric with respect to the time axis, and then the corresponding Hilbert spectrum can be obtained by applying Hilbert transform to each IMF.

In this paper, the Hilbert-Huang transform was conducted for analyzing the nonlinear response of the site. The Hilbert spectrum of the recorded motion of Point A, B and C. Figure 6 shows the distribution of the two acceleration time history signals at monitoring points A, B and C in the time-frequency-amplitude spectrum. For the bottom of the model (point C), the maximum amplitude was 0.23 corresponding with the time of 12.1 s, and the corresponding frequency was 0.8 Hz. For the bottom of the sand layer (point B), the maximum amplitude occurred at 12.4 s with a value of 0.24, corresponding with a frequency of 0.6 Hz. For the surface corresponding to the top of the Sandy layer (point A), the maximum amplitude occurred at the time of 13.0 s with a peak value of 0.55 and a corresponding frequency of 3.2 Hz. The differences between the three Hilbert-Huang spectra shows that the amplification of the amplification occurs from a depth of 8.65 underground to the surface, and the low-frequency content is shifted due to liquefaction in the process of transmission to the surface, at the same time, the high-frequency content is identified at the surface.

5 Conclusions

In this paper, the finite element software was used to simulate the liquefaction within the Treasure Island site, in which UBC3D-PLM was selected for the sand liquefaction constitutive model, and the calculation results were analyzed by Hilbert transform.

The UBC3D-PLM constitutive model can simulate the dynamic behavior of cyclic and liquefiable soil and generate pore water pressure. For the motion YBI 090, the sand layer liquefies in 11.95 s, and the PGA is amplified from 0.84 to 1.10 m/s^2 on the surface. The amplification of the peak value is found from the depth of 8.65 m underground to the surface, and the low-frequency content is shifted due to liquefaction in the process of transmission to the surface, and the high-frequency content is identified at the surface.

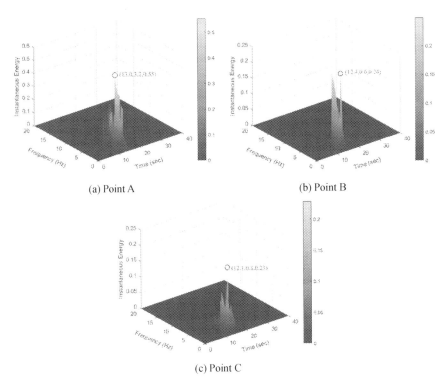

(a) Point A (b) Point B

(c) Point C

Fig. 6 Hilbert spectrum

Acknowledgements The research was funded by the Natural Science Foundation of Guangxi Grant (No. 2022GXNSFBA035569).

References

1. Miyachi Y, Kimura K (2021) Liquefaction and the associated lateral migration of the Hanshin District in the 1995 Hyogoken-Nanbu earthquake (Kobe earthquake), Central Japan. In: International geological congress. VSP, pp 409–422
2. Chu DB (2006) Liquefaction-induced lateral spreading in near-fault regions during the 1999 Chi-Chi, Taiwan earthquake. J Geotech Geoenviron Eng 132:1549–1565
3. Seed HB (2021) Simplified procedure for evaluating soil liquefaction potential. J Soil Mech Found Divs 97:1249–1273
4. Seed HB (1975) Evaluation of soil liquefaction potential during earthquakes. University of California, Berkeley
5. Tanimoto K (2011) Prediction of liquefaction occurrence of sandy deposits during earthquakes by a statistical method. Proc Jap Soc Civ Eng 1976:79–89
6. Davis RO (1982) Liquefaction potential of saturated sands. Earthq Eng Struct Dyn 10(1):59–68
7. Wobbes E (2017) Modeling of liquefaction using two-phase FEM with UBC3D-PLM model. Procedia Eng 175:349–356

8. Subasi O (2022) A numerical study on the estimation of liquefaction-induced free-field settlements by using PM4S and model. KSCE J Civ Eng 26:673–684
9. Seyhan E, Stewart JP, Ancheta TD (2014) NGA-West2 site database
10. Beaty MH, Byrne PM (2011) UBCSAND constitutive model version 904ar. Technical report, Itasca UDM Web Site
11. Puebla H (1997) Analysis of CANLEX liquefaction embankments: prototype and centrifuge models. Can Geotech J 34:641–657
12. Power M, Egan J (1998) Analysis of liquefaction-induced damage on treasure island. In: Liquefaction TLH (ed) The Loma Prieta, California, earthquake of October 17, 1989 . U.S. Geological Survey Professional Paper 1551-B, pp 87–120
13. Daftari A (2014) Prediction of soil liquefaction by using UBC3D-PLM model in Plaxis. Int J Civ Environ Eng 8(6)
14. Elsäcker WV (2016) Evaluation of seismic induced liquefaction and related effects on dynamic behaviour of anchored quay walls. Delft University of Technology, Master
15. Hudson M, Idriss I, Beirkae M (1994) QUAD4M user's manual
16. Hatanaka M (1996) Empirical correlation between penetration resistance and internal friction angle of sandy soils. Soils Found 36:1–9
17. Huang NE (1999) A new view of nonlinear water waves: the Hilbert spectrum. Annu Rev Fluid Mech 31:417–457
18. Huang NE (2014) Hilbert-Huang transform and its applications. World Scientific, Singapore

Analysis of Mode and Response Spectrum of an OBS Anti-trawl Sinking Coupling Frame Based on ANSYS Workbench

Le Zong, Kaiben Yu, Yuexia Zhao, Shengqi Yu, Lichuan Zhao, Jun Ran, Zhiguo Yang, and Baohua Liu

Abstract Three-dimensional solid modeling of an OBS anti-trawl sinking coupling frame was carried out using the ANSYS, and modal simulation analysis of the overall structure of the anti-trawl sinking coupling frame was made through the ANSYS Workbench. The natural frequencies and mode shape contour diagrams of the first six modes were extracted. The displacement response spectrum was analyzed, and the equivalent stress and directional displacement contour diagrams were obtained. Simulation results show that the natural frequency of the anti-trawl sinking coupling frame evades the working frequency band of the OBS, which verifies the effectiveness of the anti-trawl structure design.

Keywords OBS · Anti-trawl sinking coupling frame · ANSYS workbench

1 Introduction

An ocean bottom seismometer (OBS) is a geophone put on seabed for three-component data acquisition, and ocean bottom seismometers are widely used for study on the deep crust-mantle structure of the earth [1, 2]. With the development in the fields of OBS apparatus development and deep detection technology in China, China has carried out study on application of OBS in shallow-water continental crust areas, such as Bohai Sea, on the basis of using an OBS for detection of the deep structure of a deep sea, and has conducted seabed detection mainly with an OBS

L. Zong (✉) · K. Yu · Y. Zhao · L. Zhao · J. Ran · Z. Yang · B. Liu
National Deep Sea Center, Ministry of Natural Resources, 69 Wenhai East Road, Jimo District, Qingdao 266237, China
e-mail: Zongl@ndsc.org.cn

L. Zong · K. Yu · S. Yu · Z. Yang · B. Liu
Laboratory for Marine Geology, Qingdao National Laboratory for Marine Science and Technology, 1 Wenhai Road, Jimo District, Qingdao 266237, China

S. Yu
Qingdao Innovation and Development Base, Harbin Engineering University, 1777 Sansha Road, Huangdao District, Qindao 266000, China

© The Author(s) 2023
S. Wang et al. (eds.), *Proceedings of the 2nd International Conference on Innovative Solutions in Hydropower Engineering and Civil Engineering*, Lecture Notes in Civil Engineering 235, https://doi.org/10.1007/978-981-99-1748-8_37

carried by a simple cross-frame-type sinking coupling frame [3]. However, owing to growing marine engineering works and frequent fishing activities in the coastal shallow-water areas, the apparatuses put on seabed in shallow-water areas were often damaged by impact or dragged and displaced by fishing boat carried trawls, in particular, the apparatuses were fished out by fishing boat carried trawls and lost, so that valid seismic measurement data could not be obtained finally [4]. In this paper, study on and design of anti-trawl structure for OBS sinking coupling frame were carried out in the environment for placement of OBS in a shallow sea based on the existing work experience, and an engineering sample was made and used in the sea.

Since an OBS has a given working frequency band, and the natural frequency of the OBS anti-trawl sinking coupling frame and the coupling between OBS and sinking coupling frame and that between sinking coupling frame and seabed surface must be taken into full consideration for the purpose of making the acquired data reliable in acquisition of seismic data of seabed, it is crucial to carry out modal analysis of sinking coupling frame. In this paper, based on the structural and technical features of the OBS, modeling of an anti-trawl sinking coupling frame was carried out using the SolidWorks, and modal analysis of the anti-trawl sinking coupling frame was conducted using the large-scale simulation analysis software ANSYS Workbench. The natural frequencies and mode shape contour diagrams of various modes of the sinking coupling frame were obtained. Based on comparative analysis of the natural frequencies of the sinking coupling frame and the working frequency range of the OBS, the structure of the sinking coupling frame was optimized, so resonance was avoided. This provides a theoretical basis for the rationality of the structural design of the sinking coupling frame.

2   Basic Theory of Modal Analysis

Modes are the natural vibration characteristics of a structural system, and the natural frequencies and natural mode shapes of a structure are the bases for analysis of the dynamic response and other dynamic characteristics of the structure. An anti-trawl sinking coupling frame is a vibration system with N degrees of freedom. According to mechanics theory, the general equation of motion of an object in vibration mechanics is as follows:

$$Mx + Cx + Kx = F(t) \tag{1}$$

where M is the mass matrix, C is the damping matrix, K is the stiffness coefficient matrix, x is the displacement vector, and F is the excitation force vector.

The damping value of the vibration system of the anti-trawl sinking coupling frame structure is very small and it has quite little influence on the self vibration frequency of the system, so the influence of the damping value can be neglected in modal analysis of the anti-trawl sinking coupling frame structure to calculate the natural frequencies and mode shapes. Consequently, the natural frequencies and mode shapes of the structure can be studied by solving for undamped vibration under the condition of no external load, and Eq. (1) is simplified as:

$$M\ddot{x} + Kx = 0 \tag{2}$$

It can be known from vibration theory that the analytical solution of free vibration of the system is:

$$x = A \sin(\omega_n t + \phi) \tag{3}$$

where A is the principal mode shape, i.e., the mode shape of the system; ω_n is the natural frequency of the system; and φ is the phase angle.

When Eq. (3) is substituted into Eq. (2), the following can be obtained:

$$(K - \omega_n^2 M)A = 0 \tag{4}$$

Equation (4) is a multi-degree-of-freedom dynamic characteristic equation, with non-zero solutions existing, it has n eigenvalues, i.e., ω1, ω2, ..., ωn, and ωi is the natural frequency of the ith mode of the system [5].

3 Composition of the Anti-trawl Sinking Coupling Frame

The OBS anti-trawl sinking coupling frame designed in this paper was mainly composed of such parts as seabed coupling panel, guardrail, side support bracket, OBS release opening, rope storage box, chassis, and anti-trawl shield, and the OBS was mounted on the chassis.

The OBS anti-trawl sinking coupling frame employed a frustum-shaped streamline design with relatively high strength, which could effectively prevent the sinking coupling frame platform and the carried OBS from being damaged by various fishing boat carried trawls, and flow nets. The design diagrams are shown in Fig. 1 and the real object pictures are shown in Fig. 2.

In the design, three-dimensional modeling of an OBS anti-trawl sinking coupling frame was carried out using the ANSYS, and then finite element analysis of the sinking coupling frame was conducted with the modal analysis types in the software ANSYS Workbench [6], as shown in Fig. 3. The material of the sinking coupling frame was austenitic stainless steel of a given model number, and the material properties are shown in Table 1.

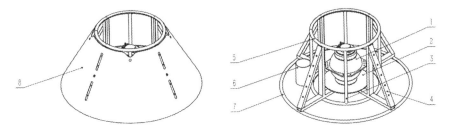

Fig. 1 Diagrams of the OBS anti-trawl sinking coupling frame. 1-OBS, 2-Seabed coupling panel, 3-Guardrail, 4-Side support bracket, 5-OBS release opening, 6-Rope storage box, 7-Chassis, 8-ANTI-trawl shield

Fig. 2 Real object pictures of OBS anti-trawl sinking coupling frame

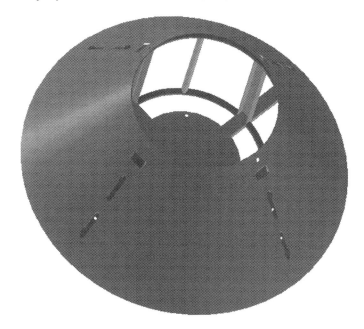

Fig. 3 Finite element analysis model of the anti-trawl sinking coupling frame

Table 1 Material properties

Density (g cm³)	Poisson's ratio	Elastic modulus (MPa)
8030	0.3	2×10^{11}

4 Analysis of the Anti-trawl Sinking Coupling Frame

To prevent the OBS anti-trawl sinking coupling frame from resonance under the undersea conditions, the natural frequencies and mode shapes of the sinking coupling frame need to be calculated. In this paper, modal analysis of the sinking coupling frame was carried out using the constraint modal analysis method under the condition that the sinking coupling frame was only subjected to gravity without any external force applied. The bottom of the sinking coupling frame, which came into contact with the seabed surface, was fixed with full constraints [7], and finite element gridding of the sinking coupling frame was carried out using the free gridding method. The density of grids was increased according to the mean value, and 182, 481 grids in total were generated, with the number of nodes of 366, 244. Figure 4 shows the schematic diagram of gridding.

After fixation constraints were applied to the bottom of the anti-trawl sinking coupling frame, the natural frequencies and natural mode shapes of the OBS anti-trawl sinking coupling frame were calculated through simulation. The mode shapes of the first six modes are shown in Fig. 5, and the natural frequencies of various modes of the sinking coupling frame and corresponding maximum deformation values are shown in Table 2.

The mode shapes of low-order modes determined the dynamic characteristics of a structure, and had higher influence on the dynamics of the structure than high-order modes [8]. The working frequency of the OBS ranged from 0.0083 to 100 Hz. In

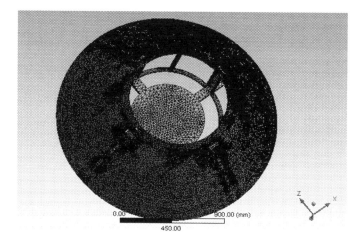

Fig. 4 Schematic diagram of gridding

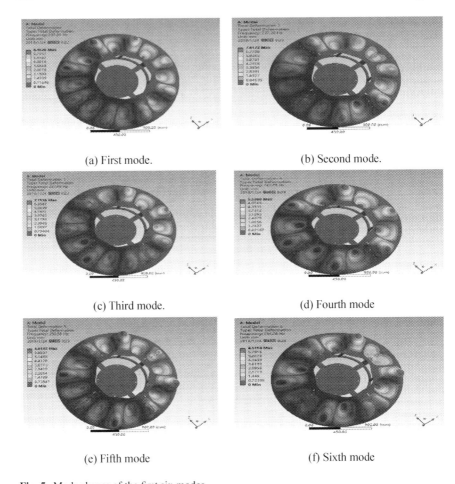

<table>
<tr><td>(a) First mode.</td><td>(b) Second mode.</td></tr>
<tr><td>(c) Third mode.</td><td>(d) Fourth mode</td></tr>
<tr><td>(e) Fifth mode</td><td>(f) Sixth mode</td></tr>
</table>

Fig. 5 Mode shapes of the first six modes

Table 2 Natural frequencies of various modes of the sinking coupling frame

Mode	Frequency (Hz)	Maximum deformation value (mm)
1	237.09	6.4526
2	237.38	7.6172
3	237.48	7.1536
4	242.25	5.5968
5	250.56	6.6192
6	254.76	6.5158

theory, the designed resonance frequency of the anti-trawl sinking coupling frame should evade 100 Hz. Moreover, there might be mutual influences between modes. According to general experience, a frequency range 1.5–2 times the frequency of vibration source needed to be considered, so all resonance frequencies below 200 Hz should be avoided. It can be known from the analysis that the deformation of the sinking coupling frame was mainly shield deformation. The natural frequency of the first mode of the shield was 237.09 Hz, and it effectively evaded the working frequency band of the OBS, avoiding the resonance phenomenon.

5 Analysis of the Seismic Response Spectrum

Spectral analysis is an extension of modal analysis, and the spectra represent the response of a single-degree-of-freedom to time-related load. Spectral analysis can make the modal analysis results associated with a spectral curve, to calculate the displacement and stress of the structure from the perspective of frequency domain, thus to determine the dynamic response of the structure to a random load [9].

In this paper, analysis of mechanical strength of the anti-trawl sinking coupling frame was carried out using the single-point response spectrum analysis method, with fixation constraints applied to the bottom of the anti-trawl sinking coupling frame. The equivalent stress contour diagram and displacement contour diagram of the sinking coupling frame was obtained through simulation analysis, as shown in Figs. 6 and 7.

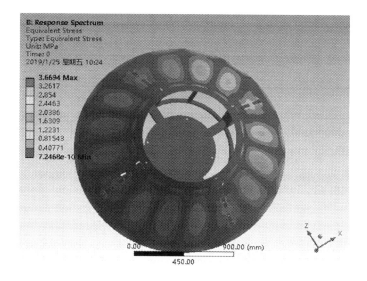

Fig. 6 Equivalent stress contour diagram of the anti-trawl sinking coupling frame

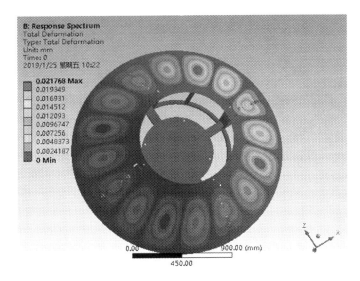

Fig. 7 Displacement contour diagram of the anti-trawl sinking coupling frame

As can be discerned from Fig. 6, the maximum stress of the anti-trawl sinking coupling frame is 3.6694 MPa, occurring in the shield of sinking coupling frame, and it is less than the yield limit of austenitic stainless steel, so the strength of the sinking coupling frame meets the requirement. It can be discerned from Fig. 7 that the maximum deformation value of the sinking coupling frame is 0.021768 mm, which is very small. It can be known from summarizing the above analyses that this sinking coupling frame structure meets the requirements.

6 Conclusions

Through analyses of an OBS anti-trawl sinking coupling frame with the above methods, the following conclusions could be made:

(1) The vibration frequencies of an OBS anti-trawl sinking coupling frame can be determined. Comparing the natural frequencies of the sinking coupling frame structure obtained from analysis with finite element analysis software with the working frequency range of the OBS can achieve avoidance of resonance phenomenon. This method demonstrates that it is feasible to avoid resonance through finite element analysis.

(2) Through finite element modal analysis, the natural frequencies of various modes of the OBS anti-trawl sinking coupling frame effectively evaded the working frequency band of the OBS and would not influence the results recorded by the OBS, and resonance phenomenon was avoided. This demonstrates the rationality of the structural design of the sinking coupling frame.

(3) According to the spectral analysis, the shield of sinking coupling frame underwent relatively large stress and had relatively great displacement, but the overall stiffness of the sinking coupling frame was high and the structural design met the requirements.

Acknowledgements This work was supported by the NSFC-Shandong Joint Fund for Marine Science Research Centers (Grant No. U1606401), the Taishan Scholar Project Funding (Grant No. tspd20161007), and the Shandong Provincial Key Research and Development Program Major Scientific and Technological Innovation Project (Grant 2019JZZY010802). The authors would also like to express gratitude to the editors and the anonymous reviewers for their constructive comments and suggestions.

References

1. Pan J, Liu BH, Hua QF et al (2012) Forward modeling based on OBS and first break recognition research. Adv Geophys 27(6):2437–2443
2. You QY, Liu FT, Ran CR et al (2003) Development of high-frequency and micro-power ocean bottom seismometer. Adv Geophys 3:173–176
3. Li ZH, Zheng YP, Zhi PY, Liu BH et al (2015) New progress in study on deep seismic detection and crustal structure in the southeastern Bohai Sea - OBS 2013 profile data processing and analysis. Adv Geophys 30(3):1402–1409
4. Yu KB, Liu ZC, Wei ZX et al (2012) Development of seabed-based anti-trawl ADCP in shallow-water areas. Ocean Technol 31(1):41–44
5. Yu WQ, Wu FQ (2017) Basic course for ANSYS workbench and detailed explanation of engineering analysis. Tsinghua University Press, Beijing
6. Wang ZW (2011) Study on method of data exchange between solidworks and ANSYS. Coal Mine Mach 32(9):248–250
7. Shaomao HU (2014) Anti-seismic analysis of telecommunication cabinet based on ANSYS workbench. Mech Electric Eng Technol 43(4):77–79
8. Cui P, Wang JH, Li B (2009) Optimization design of an OBS anti-trawl sinking coupling frame based on ANSYS. Machinery 47(10):36–38
9. Cui Q (2013) Dynamic analysis of steel structure tower based on ANSYS. Coal Mine Mach 34(4):132–133

Reliability Analysis on Horizontal Bearing of Pile Foundation in Sloping Ground Based on Active Learning Kriging Model

Hao Liang, Chang Liu, and Xiuqing Yan

Abstract The uncertainty of pile and soil and slope effect are two of the major factors affecting the horizontal bearing capacity of piles of transmission tower in sloping ground. In order to analyze the influence of the two factors on the reliability of pile, this paper proposes a reliability analysis method for horizontal bearing of pile foundation in sloping ground based on proxy model. Firstly, the analytical model of horizontal bearing of the pile foundation in sloping ground was derived, and corresponding performance functions were constructed. Secondly, by combining Kriging model method with the performance functions, the reliability analysis method of pile foundations in sloping ground is established. Finally, taking a typical transmission line project in mountainous area as an example, the horizontal bearing reliability of pile foundation was analyzed. The results show that the proposed analysis method can quickly converge to the horizontal bearing limit state of pile. Slope effect has more significant influence on horizontal deformation than that of material yield. Among the uncertainty parameters, the bearing capacity of pile foundation is sensitive to the dispersion degree of horizontal force, pile diameter and the elastic modulus of foundation pile.

Keywords Pile foundation in sloping ground · Kriging model · Reliability analysis theory · Horizontal bearing characteristics · Transmission line

1 Introduction

During the construction of transmission lines in mountainous areas, a large number of transmission towers are located in steep terrain and complex geological conditions. Unilateral soil of the pile foundation located on the slope is missing, and the internal

H. Liang (✉)
Southwest Jiaotong University, Chengdu 610031, China
e-mail: liangboju@my.swjtu.edu.cn

C. Liu · X. Yan
Southwest Electric Power Design Institute Co., Ltd. of China Power Engineering Consulting Group, Sichuan, Chengdu 610021, China

S. Wang et al. (eds.), *Proceedings of the 2nd International Conference on Innovative Solutions in Hydropower Engineering and Civil Engineering*, Lecture Notes in Civil Engineering 235, https://doi.org/10.1007/978-981-99-1748-8_38

force and deformation of the pile foundation under horizontal load are complicated [1]. Therefore, more attention should be paid to the safety of horizontal bearing capacity of pile foundation in sloping ground under special terrain conditions such as mountain area.

Numerous studies have been conducted on the horizontal bearing characteristics of pile foundation in sloping ground. Cheng et al. [2] analyzed the influence of distance between the foundation edge and the slope and slope ratio on the horizontal bearing deformation performance of pile foundations through model tests and numerical simulation. Sivapriya et al. [3] studied the key factors affecting the horizontal bearing capacity of pile foundations according to model test. Yin et al. [4] revealed the influence mechanism of slope spatial effect on displacement and internal force of pile foundation under horizontal load. The above research mainly focuses on the influence mechanism of slope effect through deterministic analysis method. However, in fact, the horizontal bearing performance of pile presents obvious uncertainties due to the strong random characteristics of pile and soil [5].

The uncertainty is one of the safety hazards which affect the horizontal bearing capacity of pile. In that case, reliability analysis method has been used to evaluate the horizontal bearing performance of pile. Yin et al. [6] derived the limit state formula of pile-column bridge pile and analyzed the reliability of its horizontal deformation by checking point method. Based on response surface and Monte Carlo theory, Chan et al. [7] proposed a new reliability analysis method and conducted reliability evaluation for horizontal bearing of pile foundations. However, current studies are mainly based on traditional reliability theories such as check point method, response surface and Monte Carlo. It is difficult to ensure the solution efficiency and convergence when the limit state equation of pile foundation is high-dimensional and complex. In addition, the research object is mostly pile foundation in level ground, while the horizontal bearing reliability of pile in sloping ground is less considered.

This paper establishes a theoretical model of horizontal bearing capacity of transmission tower pile foundation in sloping ground. An efficient analysis method of reliability is proposed by introducing Kriging model, which can greatly improve the calculation efficiency on the premise of accuracy. On this basis, the reliability of the transmission tower pile foundation in sloping ground in an actual project is evaluated, and the influence of different factors on the reliability is analyzed.

2 Theoretical Model of Horizontal Bearing of Pile Foundation in Sloping Ground

2.1 Analysis of Pile Foundation Deformation Mechanism Considering Slope Effect

As shown in Fig. 1, pile foundation in sloping ground affected by load includes the slope affected section (l_1) and the embedded section (l_2). Due to the free face in front of the foundation pile, the soil on both sides of the pile foundation is asymmetrical, which weakens the horizontal resistance of soil in slope affected section, namely 'slope effect'. The length of slope affected section is:

$$l_1 = \lambda d \tan \theta \tag{1}$$

where λ is reduction coefficient of slope effect($\lambda \in [3, 5]$ [3, 5]); d is the pile diameter; θ is grade of slope.

The differential equation of deformation deflection of pile foundation considering slope effect, can be written as follows:

$$EI \frac{d^4 x}{dy^4} + C(y) Bx = 0 \tag{2}$$

where EI is the bending stiffness of pile; x is the horizontal displacement of pile; y is the distance from the calculation point of deformation to the pile top; $C(y)$ is the resistance coefficient of pile side; B is the calculative width of pile; when $d \leq 1$ m, $B = 0.9(1.5d + 0.5)$; otherwise, $B = 0.9(d + 1)$.

The m method is used to estimate resistance coefficient of pile side foundation:

$$C(y) = \begin{cases} m \frac{y^2}{l_1}, & 0 < y \leq l_1 \\ m(y - l_1), & l_1 \leq y \leq H \end{cases} \tag{3}$$

where m is the proportional coefficient of foundation resistance.

Fig. 1 Force analysis model of pile foundation in slope section

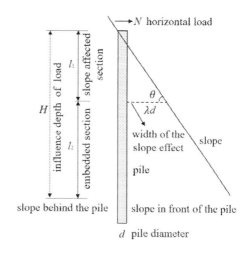

2.2 Theoretical Analysis Model

The central difference theory is used to solve the deflection differential equation of pile in sloping ground. Firstly, the influence depth of load is equally divided into 'n' parts, and two additional virtual nodes are set at both tips of pile. Then the boundary conditions need to be determined. For the boundary conditions of transmission tower pile, free pile top and embedded pile end can be considered. The difference schemes of boundary conditions are as follows:

$$\text{pile top} \begin{cases} x_2 - 2x_1 + 2x_{-1} - x_{-2} = \frac{2Ls}{EI} \\ x_1 - 2x_0 + x_{-1} = \frac{Ms^2}{EI} \end{cases} \tag{4}$$

$$\text{pile end} \begin{cases} x_{N-1} - x_{N+1} = 0 \\ x_N = 0 \end{cases} \tag{5}$$

where x_i is the horizontal displacement at different nodes; L and M are transverse load and bending moment respectively; s is the distance between nodes.

The controlling difference equation at different nodes of transmission tower pile in sloping ground is:

$$\text{slope affected section } x_{i-2} - 4x_{i-1}$$
$$+ \left(6 + \frac{Bmi^2s^6}{EIl_1}\right)x_i - 4x_{i+1} + x_{i+2} = 0 \tag{6}$$

$$\text{embedded section } x_{i-2} - 4x_{i-1}$$
$$+ \left(6 + \frac{Bm(is - l_1)s^4}{EI}\right)x_i - 4x_{i+1} + x_{i+2} = 0 \tag{7}$$

By combining the control difference equation of each node of pile foundation with the boundary condition equation, the horizontal displacement matrix equation of pile foundation in sloping ground can be obtained as follows:

$$\begin{bmatrix} a_{0,-2} & a_{0,-1} & \cdots & a_{0,N+1} & a_{0,N+2} \\ a_{1,-2} & a_{0,-1} & \cdots & a_{1,N+1} & a_{1,N+2} \\ \cdots & \cdots & \cdots\cdots & \cdots \\ a_{N,-2} & a_{N,-2} & \cdots & a_{N,N+1} & a_{N,N+2} \\ b_{1,-2} & b_{1,-1} & \cdots & b_{1,N+1} & b_{1,N+2} \\ \cdots & \cdots & \cdots\cdots & \cdots \\ b_{4,-2} & b_{4,-1} & \cdots & b_{4,N+1} & b_{4,N+2} \end{bmatrix} \cdot \begin{bmatrix} x_0 \\ x_1 \\ \cdots \\ \\ x_{N-1} \\ x_N \end{bmatrix} = \begin{bmatrix} 0 \\ 0 \\ \cdots \\ 0 \\ 2Ls/EI \\ Ms^2/EI \end{bmatrix} \tag{8}$$

where $a_{i,j}$ is coefficient of each node in the difference equation of foundation pile, and i and j represent the calculation node from pile top to pile end and the serial number of each node in the recurrence relationship, respectively. $b_{k,h}$ is the coefficient of each node in the boundary condition, $k = 1, 2, 3, 4$, representing different boundary equations, and h is the serial number of each node in the corresponding boundary.

By solving the above matrix equation, the horizontal displacements of transmission tower pile foundation under horizontal load can be obtained. In addition, the difference scheme of rotation angle, bending moment and shear of pile in sloping ground can be expressed as:

$$
\begin{cases}
\gamma_i = (x_{i+1} - x_{i-1})/2s \\
M_i = (x_{i-1} - 2x_i + x_{i+1})EI/s^2 \\
L_i = (-x_{i-2} + 2x_{i-1} - 2x_{i+1} + x_{i+2})EI/2s^2
\end{cases}
\tag{9}
$$

where γ_i, M_i and L_i are the rotation angle, bending moment and shear at different nodes of the pile foundation respectively.

By substituting the calculated horizontal displacement into Eq. (9), the corresponding rotation angle, bending moment and shear can be obtained.

3 Reliability Analysis Method of Pile Foundation Based on Kriging Model

3.1 Horizontal Bearing Performance Function of Pile Foundation in Sloping Ground

The failure modes of pile under lateral load mainly include excessive horizontal displacement and material yield. When excessive horizontal displacement is considered, the performance function is:

$$
g_d(s) = d(s) - D_l
\tag{10}
$$

where $g_d(s)$ is the performance function of the horizontal displacement; s is system random variables; $d(s)$ is the horizontal displacement value; D_l is the horizontal displacement limit.

When material yield is considered, the performance function is:

$$
g_M(s) = M(s) - M_l
\tag{11}
$$

where $g_M(s)$ is performance function of material yield of pile; $M(s)$ is the bending moment value; M_l is flexural capacity.

3.2 Active Learning Kriging Mode

The performance function of pile foundation includes complex process of horizontal bearing analysis. Using traditional reliability theory may lead to imprecise solution and excessive calculation scale. Kriging model technology can replace complex analytical process by fitting physical process, making it possible to evaluate performance functions efficiently and accurately.

In Kriging model, complex system is regarded as Gaussian static random process. In that case, the optimal linear unbiased estimation and mean square deviation of the system response at the unknown input point can be deduced as:

$$\hat{y}(x) = f^T(x)\hat{\beta} + r^T(x)R^{-1}\left(Y - F\hat{\beta}\right) \tag{12}$$

$$\sigma^2 = \sigma_z^2\left(1 + u^T\left(F^T R^{-1} F\right)^{-1} u - r^T R^{-1} r\right) \tag{13}$$

where $\mathbf{r}$ is correlation function vector between the unknown input point and training sample; $u = F^T R^{-1} r - f$.

The accuracy of initial Kriging model is greatly affected by selected samples. Therefore, it is necessary to update the model through active learning function. In this paper, efficient global reliability analysis method (EGRA) which is widely used in the reliability field, is adopted for model optimization.

3.3 Reliability Analysis Method of Pile Foundation in Sloping Ground

By combining Kriging model with the Monte Carlo simulation process (MCS), the failure probability of pile foundation can be calculated efficiently. According to the statistics of Kriging model response corresponding to numerous pile and soil random variables, failure probability is:

$$P_f \approx \hat{P}_f = \sum_{i=1}^{n_{mc}} I(g(s_i) \leq 0) \Big/ n_{mc} \tag{14}$$

where P_f is failure probability; $\hat{P}_f$ is simulated failure probability; n_{mc} is the number of samples; $I(\cdot)$ is the fault indication function, and its expression is:

$$I(g(s_i)) = \begin{cases} 0 \ g(s_i) > 0 \\ 1 \ g(s_i) \leq 0 \end{cases} \tag{15}$$

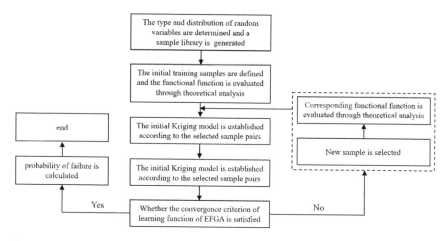

Fig. 2 Flow chart of reliability analysis of foundation pile in sloping ground

To sum up, the specific reliability analysis process of pile foundation in sloping ground based on Kriging model is shown in Fig. 2.

4 Reliability Analysis of Pile Foundation in Sloping Ground

4.1 Calculation Conditions and Parameters

A UHV DC transmission line passes through the mountainous area of Ganzi Prefecture. The slope of this area varies from 25° to 55° with an average gradient of 40°. The transmission towers in this area adapt pile-type foundation using C25 concrete. The pile length H is 15 m, the pile diameter d is 2.5 m, the elastic modulus E is 28 GPa, and the coefficient of variation of pile parameters are 0.025. The average proportional coefficient of soil resistance m is 60 MN/m^4. The horizontal load L and bending moment M acting on the pile top is 1000 kN, 1500 kN m respectively. The coefficient of variations of L, M and m are 0.25. The flexural bearing capacity of pile is 8295 kN m, and the displacement limit of pile top is 10 mm.

4.2 Analysis of Calculation Result

Failure Probability of Horizontal Bearing. Based on the above calculation conditions and parameters, failure probability of pile is calculated through the reliability analysis process shown in Fig. 2.

Table 1 Comparison of calculation results of different reliability analysis methods

Failure mode	Analysis method	Number of numerical analyses	Failure probability (‰)	Coefficient of variation	Relative error
Material yield	Kriging	234	3.81	0.0114	2.1%
	MCS	2×10^6	3.73	0.0116	–
Excessive horizontal displacement	Kriging	311	5.01	0.01	0.8%
	MCS	2×10^6	4.97	0.01	–

Assuming that the uncertainty parameters (s = [H, d, E, m, L, M]) obey normal distribution and are independent of each other, the sample library is firstly generated using the normal distribution theory, and the sample size is 105. Secondly, 50 training samples are selected by Latin hypercube sampling and substituted into the analytical model to obtain the corresponding displacement and bending moment of pile, so as to evaluate the performance functions. According to the sample pairs composed of input samples and performance function values, the initial Kriging model is constructed, and the model is optimized with EFGA method until it met the convergence criterion and the failure probability is finally determined. The comparison of results based on Kriging model method and Monte Carlo method is shown in Table 1.

The failure probability calculated by MCS can be regarded as exact result. As can be seen from Table 1, compared with material yield, the probability of excessive displacement is higher. The failure probabilities simulated by Kriging method and MCS method are basically the same, and the maximum relative error is only 2.1%. In addition, 2×106 numerical analysis processes are needed in MCS to get failure probability, while Kriging method needs 311 numerical analysis processes at most. The comparison indicates that the reliability analysis method based on Kriging model can significantly reduce the calculation cost and solve the complex limit state equation of pile foundation efficiently on the premise of accuracy.

Influence of Uncertainty Parameters on Reliability of Pile Foundation. In this section, the influence of the dispersion degree of random variables on the reliability of pile foundation are analyzed based on the proposed method.

Figure 3 shows the failure probability of transmission tower pile foundation in sloping ground under different coefficients of variation.

It can be seen from Fig. 3 that the reliability of pile is greatly affected by the variability, and the influence is also different under different failure modes. When the pile foundation fails due to material yield, the variability of the horizontal force and pile diameter has the greatest influence on the reliability, followed by bending moment, proportional coefficient of soil resistance and pile length. Compared with the above parameters, the influence of variability of elastic modulus on the pile is almost negligible. Among them, the failure probability keeps increasing approximately linearly with the increase of the variation coefficient of horizontal force, pile diameter and bending moment. When pile fails due to excessive displacement, the

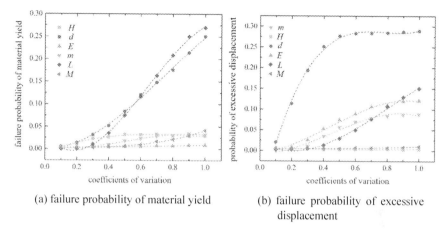

(a) failure probability of material yield   (b) failure probability of excessive
                                                      displacement

Fig. 3 Influence of variability on pile foundation reliability under different failure modes

variability of pile diameter has the greatest influence on the reliability of pile foundation, followed by the horizontal force, elastic modulus and proportional coefficient
of soil resistance, while the variability of bending moment and pile length have the
least influence on the horizontal deformation.

Influence of Slope Effect on Reliability of Pile Foundation. In order to reveal the
influence of slope effect on the reliability of pile foundation, this section calculates the failure probability and corresponding average safety margin of pile in
sloping ground under different reduction coefficient and slope gradient through the
constructed Kriging model.

Transmission tower piles located in level ground, 25°, 35° and 45° slopes are
selected respectively, and their failure probabilities are respectively calculated under
the reduction coefficients of 3, 3.5, 4, 4.5, 5 and no reduction, as shown in Fig. 4.

As shown in Fig. 4, when reduction coefficient or slope gradient is not considered, the failure probability of pile foundation is almost zero. Under the condition
of constant reduction coefficient, the increase of slope leads to the decrease of reliability. When the slope or reduction coefficient is large, the failure probability increases
significantly with the increase of slope. For the same slope gradient, failure probability increases with the reduction coefficient, and the increase amplitude is also
affected by the gradient. The reliability decreases rapidly under the condition of
large slope. In addition, the maximum failure probability of excessive displacement
and material yield caused by the slope effect are 42.13% and 8.80%, respectively,
indicating that the slope effect has a greater influence on the horizontal deformation.

The value of performance function is the safety margin. Figure 5 shows the
average safety margin of pile foundation under different reduction coefficient and
slope gradient.

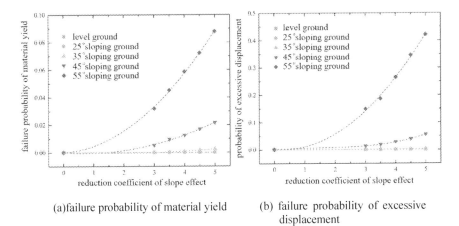

(a)failure probability of material yield   (b) failure probability of excessive displacement

Fig. 4 Influence of slope effect on foundation pile reliability under different failure modes

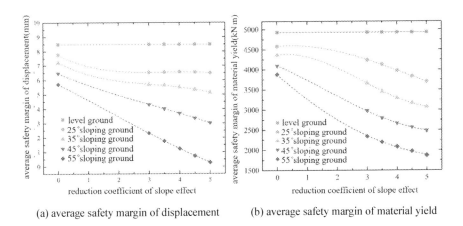

(a) average safety margin of displacement   (b) average safety margin of material yield

Fig. 5 Average safety margin of each failure mode under slope effect

As can be seen from Fig. 5, the change of reduction coefficient does not affect the average safety margin for pile in level ground while the increase of the reduction coefficient reduces the safety margin of pile in sloping ground. The reduction amplitude of safety margin of displacement increases with the slope gradient, while the change amplitude of safety margin of bending moment is not affected by the gradient. The influence of slope gradient on the safety margin is similar to the reduction coefficient. The safety margin decreases with the slope gradient. The average safety margin of horizontal displacement and bending moment is minimized to 0.29 mm and 1867.91 kN m when the gradient and reduction factor are 55° and 5, respectively.

5 Conclusion

In this paper, an active learning reliability analysis method of pile foundation in sloping ground is proposed. Horizontal bearing reliability of transmission tower pile foundation under actual conditions is evaluated using proposed method, and the influence mechanism of random factors and slope effect on the reliability is further analyzed. The main conclusions are as follows:

(1) Compared with traditional method, the proposed Kriging method can rapidly significantly improve the computational efficiency.
(2) The displacement of pile in sloping ground is more likely to occur than that of the material yield when the horizontal load is carried.
(3) Horizontal bearing capacity of pile in sloping ground is greatly affected by the dispersion degree of elastic modulus of pile, horizontal force and pile diameter.
(4) Slope effect decreases horizontal bearing safety margin of transmission tower pile, which may greatly increase the failure probability; slope effect has more significant influence on horizontal deformation under two failure modes.

References

1. Sawwaf EM (2006) Lateral resistance of single pile located near geosynthetic reinforced slope. J Geotech Geoenviron Eng 132(10):1336–1345
2. Cheng LY, Xu XC, Chen SX et al (2014) Model test and numerical simulation of horizontal bearing capacity and impact factors for pile foundations in slope. Rock Soil Mech 35(9):2685–2691
3. Sivapriya SV, Gandhi SR (2020) Soil-structure interaction of pile in a sloping ground under different loading conditions. Geotech Geol Eng 38(2):1185–1194
4. Yin PB, He W, Zhang JR et al (2018) Study on spatial effect of slope and horizontal bearing behavior of piles in sloping ground. Chin Civil Eng J 51(4):94–101
5. Zhao MH, Zeng ZY, Su YH (2007) Improved response surface method and its application to reliability analysis of piles under inclined loads. Rock Soil Mech 28(12):2539–2542
6. Yin PB, He W, Cheng Y et al (2015) Reliability analysis of horizontal displacement of a pile beneath column-supported bridge located in mountainous area. Chin J Geotech Eng 37(1):120–124
7. Chan CL, Low BK (2009) Reliability analysis of laterally loaded piles involving nonlinear soil and pile behavior. J Geotech Geoenviron Eng 135(3):431–443

Force Analysis of Anti-slip Pile Bodies on Soil Mudstone Slopes Under Heavy Rainfall

Xiaobin Liu

Abstract The soil mudstone slope anti-slip piles are used as the object of study to analyse the pile forces under the action of heavy rainfall. Introduction to the project, description of the general topographic elevation of the area; extraction of seepage potential energy from soil mudstone, calculation of rainfall and rainfall infiltration, assumption of vertical wall backs and horizontal fill behind the wall, calculation of lateral pressure on the rock supported by anti-slip piles, analysis of its damage mode according to the morphology of the bedrock surface and the conditions of the rock outwash structural surface, and calculation of slope stability under the action of strong rainfall. Analysis of the results: setting the length of the anti-slip piles at 15.5–17.5 m is most reasonable under the effect of heavy rainfall.

Keywords Heavy rainfall · Earthy mudstone · Slopes · Anti-slip piles · Force analysis · Piles

1 Introduction

Slopes refer to all geological bodies on the surface of the earth's crust that have a lateral protruding surface. With China's infrastructure construction in full swing, various projects such as highway embankment filling and graben excavation, deep foundation excavation, earth and rock dam filling, mining excavation and other projects are also in full swing, and at the same time the number of slopes has also increased. The formation and development of landslides are not independently influenced by a single factor. The deformation and stability of earthy mudstone is a function of time. Just after construction, the deformation of the slope is not obvious. As one of the three major global geological hazards, the occurrence of landslides is often related to the success or failure of the entire project, accompanied by huge economic losses and, in serious cases, even casualties. Slopes are sensitive to rainfall factors and extreme rainfall factors can cause sudden changes in slope displacement [1, 2]. The

X. Liu (✉)
Chongqing Branch of Zhonglianchuang Design Co., Ltd., Chongqing, China
e-mail: 249186700@qq.com

© The Author(s) 2023
S. Wang et al. (eds.), *Proceedings of the 2nd International Conference on Innovative Solutions in Hydropower Engineering and Civil Engineering*, Lecture Notes in Civil Engineering 235, https://doi.org/10.1007/978-981-99-1748-8_39

stability analysis methods of slopes are all based on the geotechnical elastic-plasticity theory. When the slope is overstressed, the strength of the soil is insufficient to maintain the elastic phase and enters the plastic phase, at which time the deformation will continue to develop until destabilisation and damage, so the stability problem of the slope can be considered as a strength problem. However, the stability of slopes also has to consider internal and external factors. For a slope that has been excavated or is being excavated, the internal factors have already been determined, such as physical factors like gravity and water content. Not many studies have considered the combined effects of creep factors and extreme rainfall factors. The design of slope stabilisation measures in engineering is only calculated with short-term stability in mind, while the engineering geology and the performance of the support structure are subject to change during long-term use due to natural phenomena such as rainfall. Geometric factors such as slope height and slope gradient. The external factors such as load and boundary condition changes are also crucial for slope stability analysis, and the infiltration of water is one of the points that cannot be ignored. In slope engineering, the influence of water on slopes is mainly manifested in the infiltration of groundwater and rainfall infiltration, and most of the above-mentioned accidents occur after continuous strong rainfall, so it is also crucial to study the influence of seepage field on slope stability.

2 Project Overview

The soil-rock combination slope refers to the intrusion range of the underground function area including both soil and rock layers. The project is located in the southwest of S. Taking the S area as an example, through the regional geological boreholes included in the geological cloud platform developed under the auspices of the Geological Survey, it can be found that the distribution of strata in the area is dominated by rock formations, mainly sandstone and mudstone, and the nearby mountains show a long monopoly shape with a general northwest direction. The terrain is low and undulating, with an approximate elevation of 210–280 m. The alluvial ditches are mostly "V" shaped, mostly dendritic in development, and are Quaternary floodplain, alluvial, loess-like chalk and gravelly soils, with a grain size of 200 mm. The overlying soils are often fill, clay and sand, and the rock below the soils is strongly weathered for about 1 m, while the lower part of the rock is moderately weathered and weakly weathered in that order. The lithology is mostly shale, sandstone, mudstone and sandy mudstone, with shale, sandstone and mudstone being the main ones. The landslide of the new project is semi-circular in plan and bending in profile, with a maximum height difference of 60 m, a trailing edge width of 114 m and a leading edge width of 288 m, covering an area of 3.11×106 m^2, with an average thickness of 14 m and a volume of 33.64×106 m^3. The rock type and distribution characteristics, mudstone interbedded structure, except for a group of slopes with parallel orientation and dip close to upright unloading fissures developed on the slopes of the two banks near the river, the rock is generally relatively intact. Three distinct fissures are visible at the back edge boundary, with the longest fissure 162.07 m and the deepest 44.3 m. As the leading edge of the landslide, the road at the site is severely deformed, with severe arching of the road surface and significant cracking, with the maximum crack width exceeding 22 cm. The left side boundary: the left side of the landslide is bounded by an alluvial ditch, the road is severely arched and obvious fissures can be seen, the fissures are about 6–12 cm wide, locally up to 17–22 cm. right side boundary: the right side of the landslide is bounded by a ridge, the right side of the mountain body is fractured and severely damaged, the maximum fissures can be 22–35 cm. according to the borehole, the stratigraphy of the slope body from top to bottom is: the Holocene landslide accumulation layer, the Upper Pleistocene of the Quaternary The slide body is mainly composed of the Middle Pleistocene Loess of the Quaternary, which is light brown in colour and hard in texture, with occasional calcareous nodules.

3 Force Analysis of Anti-slip Piles on Earthy Mudstone Slopes

3.1 Extraction of Seepage Potential Energy from Earthy Mudstone

When the soil is completely dry, it can be regarded as a soil-air two-phase body, and when the soil air is completely filled with fluid, then the object of analysis can be regarded as a soil–water two-phase body, and the flow of water between the pores is called saturated seepage. Under natural conditions, the lower part of the landslide is the slip-resisting section [3, 4]. Due to the action of various external forces making the leading edge of the slope the height of the prograde increases, the sliding resistance gradually decreases, resulting in the leading edge of the slope deformation and the slope surface cracks. For groundwater infiltration channels elongate deformation cracks, continuous rainfall formation of surface water can also be through the infiltration of deeper slopes, a combination of these unfavourable conditions lead to landslide body front edge deformation damage and landslide body front rear edge support weakened or even open, the front edge of the sliding body rear edge deformation caused by the weakening of sliding resistance. Rainfall infiltration is also a kind of vertical infiltration problem. The study of rainfall infiltration is mainly to investigate the relationship between rainfall and infiltration recharge, i.e. how much rainfall is converted into groundwater infiltration, especially in the seepage analysis of slope problems, due to the slope, it is more important to clarify the relationship between rainfall and groundwater infiltration, for which a rainfall infiltration model needs to be established. In this regard, the calculation of rainfall and rainfall infiltration is given by:

$$L = \int_0^E \frac{|E - 1|^2}{\varepsilon} \tag{1}$$

$$H = \int_0^E G \times \varepsilon \tag{2}$$

In Eqs. (1) and (2), E represents the depth of accumulation of rainfall per unit area per unit time, G represents the rate of infiltration and ε represents the duration of rainfall. From Eqs. (1) and (2), it is possible to derive the depth of water accumulation per unit time, i.e. by subtracting the rainfall infiltration from the rainfall volume. In short, the steepness of the leading edge of the landslide eventually leads to a concentration of stresses located near the foot of the slope, and the slope produces an adjustment in the redistribution of stress internal forces due to the stress concentration phenomenon, resulting in the deformation of the slide and a weakening of the support for the middle section of the slide. Typical moisture content profiles are, from top to bottom: saturated zone, transitional zone, conductive zone, wet zone, the boundary of the wet zone is called the wetting front. And according to Darcy's saturated seepage theory, the maximum infiltration rate in any direction for a slope soil of the same material can be found as follows:

$$\beta_{\max}(\varepsilon) = -l \frac{\mu(1 + l)^2}{\varpi} \tag{3}$$

In Eq. (3), l represents the saturated permeability, μ represents the pressure head and ϖ represents the water content of the topsoil. The essence of seepage is the flow of liquids such as water in a pore medium. When the pore space is fully filled with water it is called saturated seepage and when water is only occupying part of the pore space it is called unsaturated seepage. According to the theory of soil water potential, the total potential energy of water in both saturated and unsaturated soils can be expressed by a formula:

$$\eta = \eta_1 + \eta_2 + \eta_3 \tag{4}$$

In Eq. (4), η_1 represents the gravitational potential, η_2 represents the pressure potential and η_3 represents the substrate potential. The matrix potential is the absorption of water by the porous medium thereby reducing the free energy of water, the magnitude of which is a negative value, which is mainly caused by the matrix suction (adsorption + curved lunar surface force). The matrix potential in saturated soils is generally 0. The surface tension in the void is created by the presence of unbalanced forces on both sides of the liquid surface. The central slip zone of a landslide is mostly a geologically existing soft zone (surface) that softens with water, decreases in strength and cannot withstand the thrusts generated by the sliding body at the trailing edge and deformation occurs. The middle section of the landslide moves downwards as the plastic zone expands, and the overall sliding of the landslide begins. Eventually the damage process of front traction, back loading, mid-slip zone shear connection and landslide sliding is formed. When the soil is not saturated, that is, when there is both air and water in the pore space, at this time the soil is a three-phase soil, water and gas, the study of the unsaturated soil seepage law is unsaturated seepage theory. Soil water infiltration can be broadly divided into vertical infiltration problems and lateral infiltration problems according to the direction of infiltration, lateral infiltration including embankment seepage problems, groundwater infiltration problems, vertical infiltration including rainfall infiltration, the earliest research on vertical infiltration problems is Klarman and Burdeman's research on dry soil under waterlogging conditions, through the experiment to obtain the relationship between soil moisture content and soil depth. In addition, landslides often occur at the trailing edge of dangerous rock collapse and human activities (coal mining, construction, etc.) lead to the accumulation of the trailing edge of the landslide body, trailing edge tension cracks, the accumulation of load on the trailing edge of the landslide body gradually reaches and exceeds its ability to resist slippage, resulting in instability of the trailing slide at the initial stage, but the front of the landslide body produced a thrust, the slide body to the trailing edge deformation landslide centre compression deformation, is a gradual loading process, resulting in drum-shaped tension cracks.

3.2 Calculation of Lateral Pressure on Rock Supported by Anti-slip Piles

In the lateral pressure calculations for soil-rock combination slopes, the soil pressure is calculated according to the Coulomb or Rankine active earth pressure formulae and the rock pressure is calculated by converting the equivalent internal friction angle also according to the earth pressure formulae. As the cohesion of the soil is generated by the combined water and capillary pressure, its strength is very low and mutual slip can be generated between the soil particles, therefore the damage of the soil under the action of external forces is shear damage along the soft structural surface between the mineral particles, so it satisfies Moore a Coulomb's theorem. If the shear strength of the anti-slip pile is inadequate, the pile is highly susceptible to cracking under landslide thrusts, and even gradually expanding into pile shear or fracture damage [5, 6]. This form of damage is permanent damage to the structure itself, indicating that the pile has lost its ability to resist the slope slide. Considering the lateral pressure generated by the soil layer when the soil layer will not slide along the soil-rock interface, the mathematical expression of the lateral pressure of the soil layer can be calculated according to the active earth pressure based on the flat section assumption, assuming that the back of the wall is vertical and the fill behind the wall is horizontal, the mathematical expression of the lateral pressure of the soil layer is specified as follows:

$$V_{\lambda_1} = \frac{1}{2}\phi_1\sigma_1 S_{\lambda_1} \tag{5}$$

In Eq. (5), ϕ_1 represents the soil weight, σ_1 represents the height of the soil part of the retaining wall and S represents the active earth pressure coefficient. However, in the case of rock, although it has the same three-phase nature as soil, the mineral particles form a strong polymerisation through crystalline and cemented connections, so that the rock is much stronger than the soil and is in an unplastic state. This damage may be due to a misjudgement of the slip surface, resulting in a low slip resistance due to insufficient anchorage depth, or due to an unreasonable design of the anti-slip pile cross-section and reinforcement, resulting in a pile with insufficient bending and shear resistance. If the anti-slip pile is placed in a poorly anchored soil stratum, the pile will be susceptible to tilting or tipping under landslide thrust before damage occurs but the anchorage is insufficient to cause the overall instability of the anti-slip pile. In terms of rock microstructure, rocks are composed of a variety of mineral grains, pores and cements [7, 8]. It has the same three-phase nature as soil and therefore borrows many of the theories of soil mechanics for the study of rock mechanics. However, the strength of the rock is considerably higher than that of the soil due to the strong polymerisation of the mineral particles through crystalline and cemented joints. The formula for the active earth pressure ensemble for a standard combination of loads is:

$$R = \frac{|1 - k|^2}{\delta^2} \times \sqrt{\|k + n\|} \tag{6}$$

In Eq. (6), k represents the cohesion of the soil, δ represents the standard value of the mean surface load and n represents the angle of internal friction of the soil. In this case, the pile body is usually not structurally damaged, and the anchorage capacity of the strata can be improved by means of grouting or adding anchor cables to limit the deformation of the pile top to achieve the purpose of secondary reinforcement. If the shear strength of the soil between the piles is low and the soil flows out from between the piles, then the anti-slip piles are intact but the stability of the slope is reduced and the anti-slip piles fail. If there is an outwardly inclined hard structural surface, the calculation can be made according to the active rock pressure formula, in which the soil layer should be considered as the surface uniform load, the specific expression formula for the inclination of the outwardly inclined structural surface of the side slope is as follows:

$$V = 1 + \frac{2g \sin\theta \cos\gamma}{\phi_1\sigma_1 \sin(\theta + \gamma)} \tag{7}$$

In Eq. (7), g represents the cohesive force on the surface of the outwardly dipping structure, θ represents the angle of internal friction on the surface of the outwardly dipping structure and γ represents the angle of friction between the rock and the back of the retaining wall. As the excavation area of the slope is a shallow stratum, the rock layer is assumed to be in a linearly elastic natural stress field without consideration of tectonic stresses. The residual sliding force of the soil layer is calculated as follows:

$$\varphi_z = W_{z-1}\xi_{z-1} + C_z - \frac{U_z}{Q} \tag{8}$$

In Eq. (8), W represents the slip force per unit width of the $z - 1$ th calculation bar due to gravity and other external forces, and ξ represents the slip force per unit width of the z th calculation bar due to gravity and other external forces, C indicates the transfer coefficient of calculation strip $z - 1$ to calculation strip z, Q indicates the angle of internal friction between the rock slope and the fill, and 3 indicates the length of the sliding zone at the bedrock face. Normally, the piles are not damaged at this point and the original anti-slip piles can continue to be used when appropriate measures are taken to manage the newly formed landslide. Anti-slip piles have a significant effect on slope management and are widely used in slope engineering, but their structural design parameters largely determine the reinforcement effect of the slope. Therefore, the rock excavation unloading is in the elastic recovery stage, and the corresponding rock pressure acting on the support structure is the elastic deformation pressure. For the soil-rock combination slope supported by anti-slip piles, as the piles are applied first and then the rock and soil in front of the piles are excavated, the elastic deformation of the rock between the piles is completed soon after excavation, but as the anti-slip piles limit the elastic deformation of the rock afterwards, there is a residual elastic deformation pressure acting on the anti-slip piles.

3.3 Calculation of Slope Stability Under the Effect of Heavy Rainfall

The infiltration of rainfall is a dynamic process, the infiltration of water will change the moisture content of the soil, the moisture content of the soil changes with height, the surface soil is the first to reach saturation, when the moisture content of the soil increases, first of all the soil capacity increases, the increased mass can be seen as an additional load may cause significant changes in the force of the slope. The soil-rock combination slope is composed of the upper soil layer and the lower rock layer with different properties. The damage mode should be analysed according to the morphology of the bedrock surface and the conditions of the outwardly inclined structural surface of the rock layer, which mainly includes possible damage modes such as internal circular sliding damage of the soil layer, sliding damage along the soil-rock interface, flat sliding of the rock layer and composite damage of the rock and soil layer [9, 10]. Secondly, the shear parameter of the soil decreases significantly with changes in water content, and the reduction in shear strength causes slope damage to occur in a less stressed state, a phenomenon that can also be explained by the extended Moore Coulomb strength criterion, i.e. the presence of pore water pressure in the

Fig. 1 Structural
relationship between soil and
rock layers

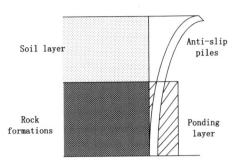

water-bearing soil leads to a reduction in inter-skeleton stress in the pore medium
and a consequent reduction in shear strength, which occurs when rainfall reaches a
certain level of damage. The structural relationship between the soil body and the
rock layer is shown in Fig. 1.

Due to the low strength of the soil, the slope stability of the soil layer can be
calculated using the circular sliding method. When the bedrock face dips outwards,
the soil layer may also produce sliding damage along the bedrock face, when the
transmission coefficient method can be used to calculate the stability of the soil layer.
Usually when the safety factor for such landslides is greater than 1.25, the soil slope
can be judged to be in a stable state. If, under special circumstances, the coefficient
of safety of a landslide is greater than 1.15, the soil slope can be judged to be in a
stable state. If the cohesion of the slope is equal to zero, the coefficient of safety of
the landslide can be expressed as:

$$A = \frac{\tan m}{\tan n} \tag{9}$$

In Eq. (9), m indicates the inclination of the sliding surface and n indicates the
slope angle. With the continuation of rainfall, especially the continuous heavy rainfall
weather that occurs in the southern rainy season, the road surface becomes water-
logged and the groundwater level can be too high, which is the initial condition for
seepage field analysis, so it is also essential to study the coupled analysis of seepage
stress under different groundwater level conditions. After the end of rainfall, the
water accumulated in the slope will continue to seep out and the water table will fall
with time. At this time, different bedrock inclination angles will affect the rate of
dissipation of residual water, so the stability of the slope flow-solid coupling under
different bedrock angles should also be considered. Then the calculation formula for
the sliding force is:

$$P = x \frac{\tan m}{\tan n} \sum (\psi - u)^2 \tag{10}$$

In Eq. (10), x represents the dip angle in the direction of the slide infiltration
pressure, ψ represents the structural correction factor and u represents the single
wide infiltration pressure of block u. When there is an outward sloping structural

surface of the rock layer, the wedge-shaped rock block is prone to sliding damage along the outward sloping structural surface, and at the same time drives the upper overburden layer to deform outward. The thickness of the upper soil layer is 8.2 m and the bedrock face is gently sloping and slightly inclined inwards, so the soil layer will not produce sliding damage along the bedrock face. When rainfall occurs, as the rainfall continues, the water content of the upper layer of the soil gradually increases, and the negative pore water pressure gradually dissipates. 8 days after the strong rainfall occurs, the pore water pressure within the more moist soil of the surface layer has increased to between -50 and -20 kN/m^2. In calculating the stability of slopes, the linear rupture surface method is often used in order to visualise and simplify the analysis of such slope planes. Usually a slope containing permeable gravel, sand, gravel or sandy soils will form a linear rupture surface when it breaks. In this case, the equation for the shear strength of the soil is as follows:

$$d = \alpha + r \sum \frac{(1 - \alpha)}{r^2} + D \qquad (11)$$

In Eq. (11), α represents the effective stress factor, r represents the pore water pressure and D represents the material strength. The maximum negative pore water pressure inside the slope also rose to -277 kN/m^2, and the eight days of rainfall caused the negative void is pressure within the slope to drop by 20% as a percentage and 30% as a peak. As the permeability coefficient of the lower mud layer of the slope is much smaller than the permeability coefficient of the soil above the rock and the intensity of the rainfall, the negative pore water pressure of the mud rock above the water table does not change much as seen from the cloud map. Rainwater collects at the toe of the slope causing the water content at the toe to rise gradually and the soil to become saturated until the precipitation is connected to groundwater, at which point a large saturated seepage zone appears on the slope. The criteria for slope instability damage include excessive displacement change, in order to analyse the change in slope displacement field after the end of rainfall.

4 Case Studies

4.1 Construction Preparation

TJ-22 earth pressure gauges (range 0–0.1 MPa, diameter 8 cm) were buried behind the wall and placed on the centreline of the anti-slip piles and baffles at distances of 2 m, 4 m, 6 m, 8.5 m and 11.5 m respectively from the top line of the wall. The maximum height of the excavated slope is 25 m, and the maximum slope technology is grade 3. The slope of this section is a typical overburden bedrock type slope, the upper layer of soil is a thicker gravel-bearing miscellaneous fill with a thickness ranging from 3–30 m, which varies greatly, and the lower bedrock is a sandy mudstone layer, and the dip of the rock layer at the project site is 10–30°. All three levels of slope slopes are cut at a rate of 1:1 and the slope reinforcement scheme is proposed to use anchor frame beam protection. The reinforcement gauge is a GXR series vibrating string type reinforcement gauge corresponding to the main reinforcement. The rebar meter is connected to the main reinforcement cut into sections and tied together with the main reinforcement using sleeves and other strengths, and the test leads are carefully led to the top of the pile and protected.

4.2 Force Analysis

Controlling the length of the anti-slip pile within a reasonable range can effectively enhance the safety factor of the construction, therefore, this force analysis on the anti-slip pile body of the soil mudstone slope, based on and a series of changes in the pile displacement and pile shear stress parameters, the optimum pile body length was selected. The variables were set to three types: anti-slip pile length 13.5–15.5 m, anti-slip pile length 15.5–17.5 m, and anti-slip pile length 17.5 m and above. The specific results are shown in Figs. 2 and 3.

As can be seen from Figs. 2 and 3, the pile displacement is minimum and the pile shear stress is maximum when the length of the anti-slip pile is 15.5–17.5 m. For the upper part of the pile, the shear stress is positive under the action of the landslide, for the lower part of the pile, the passive soil pressure in front of the pile is greater than the active soil pressure behind the pile and the shear stress is negative. When the pile length is short, the absolute value of shear stress in the upper part of the pile is larger than the absolute value of shear stress in the lower part, while when the pile length

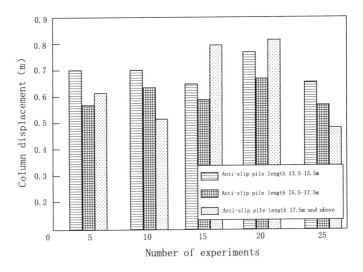

Fig. 2 Variation in pile displacement (m)

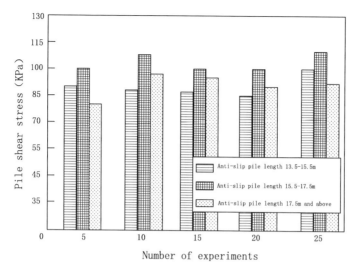

Fig. 3 Variation in pile shear stress (KPa)

is long, the absolute maximum value of shear stress in the upper part of the pile and the absolute maximum value of shear stress in the lower part are almost equal. With the increase in the length of anti-slip piles, the displacement of each point of the pile roughly shows a gradually decreasing trend, in the local range with the increase in the length of anti-slip piles the displacement of each point of the pile has a slight increase, indicating that too long anti-slip piles do not have a significant effect on increasing the stability of the landslide body. Therefore, under the effect of heavy rainfall, the length of anti-slip piles should be set at 15.5–17.5 m, which is the most reasonable.

5 Conclusion

After rainfall occurs, the water content of the soil within a certain range of the slope surface layer increases. The increase of water in the pore space increases the soil weight on the one hand, and on the other hand, the increase of pore water pressure leads to the decrease of matrix suction of the pore medium eventually reduces the shear strength of the soil, so the increase of soil weight and the decrease of effective stress of the slope are the main reasons for the damage of unsaturated soil slopes under rainfall infiltration conditions. The long-term stability state of anti-slip pile slopes under different strong rainfall types shows different response laws. The displacement of the slope, which gradually stabilises under conventional creep, does not change abruptly due to rainfall under low intensity long duration rainfall conditions, but the rate of change of the slope displacement under heavy rainfall is much greater than that under conventional creep, and the displacement of the monitoring point changes more significantly due to heavy rainfall. Due to the constraints of the study, the reinforcement effect and soil arch effect under the double-row and multi-row arrangement were not analysed in this study, which can be further explored in the future.

References

1. Buddo IV, Shelokhov IA, Misyurkeeva NV et al (2021) Transient electromagnetic sounding in 2D, 3D, and 4D modes: sequence of geological exploration activities. Geodyn Tectonophys 12(3):715–730
2. El-Hadidy M (2021) The relationship between urban heat islands and geological hazards in Mokattam plateau, Cairo, Egypt. Egypt J Remote Sens Space Sci 24(3P2):547–557
3. Gerzsenyi D, Albert G (2021) Geological hazards of the Gerecse Hills (Hungary). J Maps 17(2):730–740
4. Brighenti F, Carnemolla F, Messina D et al (2021) UAV survey method to monitor and analyze geological hazards: the case study of the mud volcano of Villaggio Santa Barbara, Caltanissetta (Sicily). Nat Hazard 21(9):2881–2898
5. Rızaoğlu T (2021) An overview of the impacts of geological hazards on production. Multidiscip Aspects Prod Eng 4(1):153–165
6. Ercilla G, Casas D, Alonso B et al (2021) Offshore geological hazards: charting the course of progress and future directions. Oceans 2(2):393–428
7. Zeng L, Yu HC, Liu J et al (2021) Mechanical behaviour of disintegrated carbonaceous mudstone under stress and cyclic drying/wetting. Constr Build Mater 282(6):122656
8. Cheng L, Liu J, Ren Y et al (2022) Study on long-term uniaxial compression creep mechanical behavior of rocksalt-mudstone combined body. Int J Damage Mech 31(2):275–293
9. Wang Y, Cong L, Yin X et al (2021) Creep behaviour of saturated purple mudstone under triaxial compression. Eng Geol 288(3):106159
10. Shi HC (2019) Simulation study on vulnerability of tunnel structure under insufficient lining thickness. Comput Simul 36(7):230–233,375

Water Environment Simulation and Ecological Restoration in Sanhekou Reservoir Basin

Yuqiang Zheng, Ying Wang, Feng Gao, and Xin Zhang

Abstract Sanhekou Reservoir is an important water source for the Han Wei River Diversion Project, which plays a very important role in supplying water to Guanzhong area. Taking Sanhekou reservoir basin as the research object, MIKE21 model is used to simulate the water environment of the reservoir basin. MIKE21 hydrodynamic module is used to analyze the flow field characteristics of the reservoir study area in high flow year, normal flow year and low flow year. MIKE21 water quality module is used to simulate the water pollution of the reservoir, and analyze the maximum concentration value of pollutant migration in 6, 12 and 24 h under four working conditions, as well as the pollution peak value of 5 and 20 km sections. The results show that the overall velocity variation range of the study area in the wet, normal and dry years is 0–0.139 m/s, 0–0.102 m/s and 0–0.096 m/s respectively. In the four working conditions, the maximum pollutant concentration of working condition 1 is 4.76 mg/L under 6 h of pollution leakage, and the peak value of pollutants under the four working conditions of 5 km section is larger than that of 20 km section. In view of the ecological environment problems in Sanhekou basin, reasonable ecological restoration suggestions and measures are proposed.

Keywords Sanhekou reservoir · MIKE21 model · Water environment simulation · Ecological restoration

1 Introduction

Water resources are important resources for people's survival and development. With the rapid development of social economy, people's demand for water resources is increasing, which makes water resources become more and more scarce. At the same

Y. Zheng · Y. Wang (✉) · F. Gao
School of Water Resources and Hydropower, Xi an University of Technology, Xi an 710048, China
e-mail: wangying@xaut.edu.cn

X. Zhang
Shaanxi Yinhan Jiwei Engineering Construction Co., LTD., Hanzhong 723000, China

© The Author(s) 2023
S. Wang et al. (eds.), *Proceedings of the 2nd International Conference on Innovative Solutions in Hydropower Engineering and Civil Engineering*, Lecture Notes in Civil Engineering 235, https://doi.org/10.1007/978-981-99-1748-8_40

time, it also causes increasingly serious water environment pollution, which not only threatens people's water safety, but also affects the sustainable development of the country [1]. As one of the important water resources closely related to people's production, life and social development, the reservoir has the functions of flood control, water storage and irrigation, water supply, power generation, fish farming, tourism, etc. [2]. Due to the pollution of rivers, lakes and groundwater, reservoirs have become an important guarantee of urban drinking water supply in China [3]. Therefore, reservoir water security, water quality and other water resources issues have been highly valued by people [4].

Sanhekou Reservoir is one of the two important water sources of the Han Wei River Diversion Project, which plays an important role in water supply in Guanzhong area. At present, there is little research on the water environment of Sanhekou Reservoir. Therefore, this paper uses MIKE21 model to simulate the water environment of Sanhekou Reservoir basin. At the same time, it puts forward some reasonable ecological restoration measures and suggestions for the ecological water environment of the basin, and provides theoretical reference for the water environment management and protection of the reservoir basin.

2 Overview and Methods of the Study Area

2.1 Overview of the Study Area

The dam site of Sanhekou Reservoir is located about 2 km downstream of Sanhekou Village, Daheba Township, Foping County. The total storage capacity of the reservoir is 710 million m^3, and the regulating storage capacity is 660 million m^3. It is a multi-year regulating reservoir. The normal pool level of the reservoir is 643 m, and the dead water level is 558 m. Sanhekou Reservoir is located on Ziwu River, a tributary of Hanjiang River. The reservoir area is composed of Wenshui River, Puhe River and Jiaoxi River. The controlled drainage area above the dam site is 2186 km^2, accounting for 72.6% of the whole basin. Figure 1 shows the distribution of Sanhekou River Basin.

2.2 Model Building

MIKE21 is one of the more commonly used two-dimensional mathematical models and can be used to simulate a variety of water environments. With the continuous development of the simulation over the years, it has been widely used in the simulation of water flow and sediment in rivers and lakes [5].

The hydrodynamic module of MIKE21 was used to analyze the flow field of the watershed in the study area. Three typical years of 5% guaranteed rate flow rate in

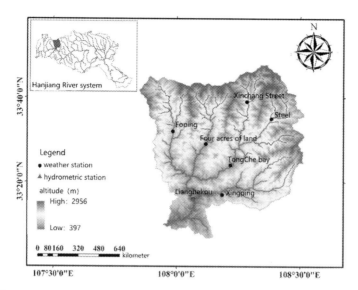

Fig. 1 Sanhekou river basin study area

wet year, 50% guaranteed rate flow rate in normal year and 95% guaranteed rate flow rate in dry year are selected. The hydrodynamic calculation time was set as 1 year, the inlet flow was input into the model according to relevant literature, and other parameters were set as follows: dry water depth of 0.005 m, submerged water depth of 0.05 m and wet water depth of 0.1 m. Water quality module of MIKE21 was used to simulate water pollution, and its degradation coefficient was set at 0.25/day, diffusion coefficient at 1 m^3/s, and other parameters were set according to the situation of the study area.

2.3 Calculation Condition Setting

It is assumed that a pollutant leakage accident occurs at the backwater end of any tributary of Sanhekou River, causing similar nitrogen pollutants to enter the reservoir. The flow rate is set according to the upstream incoming water in different hydrological years (5% guaranteed rate wet year), and the summer flow rate is 144.6 m^3/s as a comparison. Two types of pollutant leakage flow were selected, with different duration and different leakage flow as comparative analysis. The specific working conditions are shown in Table 1. The initial concentration of pollutants was assumed to be 0.5 mg/L (Class II).

Table 1 Conditions set

Working condition	Flow (m³/s)	Contaminant leakage (m³/s)	The last time (min)
1	144.6	2	5
2		0.2	10
3	48.2	2	5
4		0.2	10

2.4 Model Validation

The hydrodynamic model is verified by the measured and simulated water level in front of the dam in the reservoir area, and the results are shown in Fig. 2. It can be seen from the figure that there is a certain deviation between the measured value and the simulated value, but the overall trend is consistent, so the simulation effect is good.

The validation of water quality model selects the measured value of COD and the simulated value to compare and evaluate the accuracy of the model. The distribution comparison between the measured COD scattered points of 10 months and the calculated value change curve of the state variable is selected, and the verification results are shown in Fig. 3. It can be seen from Fig. 3 that the measured COD values fluctuate slightly around the calculated values, but the basic trend remains the same. It can be seen that the simulation of water quality model can ensure the accuracy of calculation.

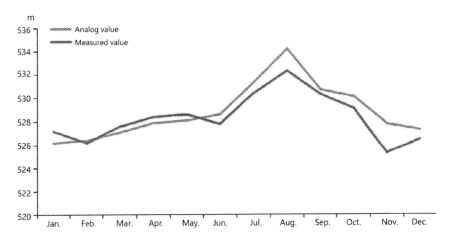

Fig. 2 Verification diagram of water level in front of the dam in Sanhekou reservoir area

Fig. 3 Verification of COD in reservoir

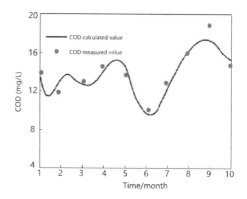

3 Results and Discussion

3.1 Hydrodynamic Flow Field Analysis

The hydrodynamic module is used to analyze the flow field in the study area of the reservoir in wet, normal and dry years, and the results are shown in Fig. 4.

Figure 4 shows the flow field diagram at the junction of three rivers on the left and the flow field diagram at the junction of Puhe River and Wenshui River on the right. As can be seen from Fig. 4a, under the hydrological calculation of 5% guarantee rate in a wet year, the overall velocity of the study area ranges from 0 to 0.139 m/s, in which the phenomenon of a larger velocity occurs in some areas where the river channel suddenly narrows. It can be observed from the figure on the left that the area of water at the intersection of water bodies is larger, the riverbed is wider, and the flow speed is lower. Because of the complicated riverbed topography, the flow velocity decreases from the bank to the center of the main channel. The downstream channel shape of the junction is narrow first and then wide, and the velocity changes from large to small, which accords with the law of reality. As shown in the figure on the right, Jiaoxi River on the left, Wenshui River and Puhe River on the right river confluence. Due to the change of the terrain on the right bank of Jiaoxi River and the shrinkage of the river, the direction of the flow on the left bank shifts to the right and quickly passes through the narrow area of the river.

As shown in Fig. 4b, the overall velocity in the study area is basically between 0 and 0.102 m/s in normal water years, and the velocity increases in some areas. The velocity at the intersection of Wenshui River and Puhe River is between 0–0.044 m/s. Before the intersection, the velocity at the area with large velocity on the right bank of Puhe River exceeds 0.040 m/s. After the confluence of the two rivers, they continue to flow downstream. After the confluence with Jiaoxi River at the third estuary, the river channel becomes wider, and the flow velocity is between 0.024 and 0.032 m/s. The flow continues to converge downward in front of the dam, and the flow velocity continues to decrease. Compared with the hydraulic calculation

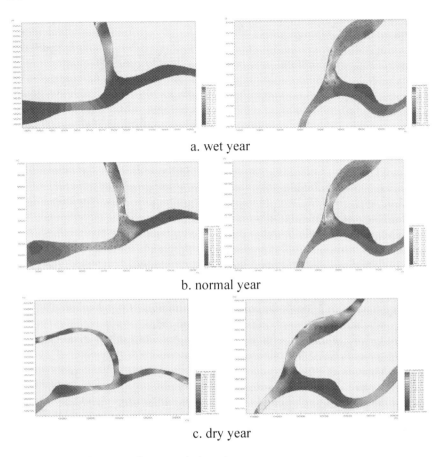

a. wet year

b. normal year

c. dry year

Fig. 4 Flow velocity vector diagram at the interchange

condition of 5% assurance rate in high water year, the overall flow rate is decreasing due to the decrease of upstream inflow.

As shown in Fig. 4c, under the hydrological calculation condition of 95% guarantee rate in a dry year, the flow velocity shows an obvious decreasing trend, and the flow velocity changes obviously. The overall flow velocity of this section is roughly between 0 and 0.096 m/s. From the front of the dam to the junction of the three rivers, the flow velocity is small, ranging from 0.015 to 0.045 m/s, while the flow velocity is mostly in the narrow flow area, ranging from 0.072 to 0.096 m/s. The velocity distribution of the whole study area is basically consistent with the results of the previous two hydraulic calculation conditions.

As can be seen from Fig. 5, in wet years, the overall river channel in front of the Sanhekou Dam becomes wider and the flow velocity of the water decreases, entering an area with slow flow velocity. In normal water years, the water flow in front of the dam tends to be static, and the overall velocity is less than that in wet water years.

However, in dry years, the flow velocity gradually becomes smaller after the flow is collected, and the flow velocity tends to 0 in some areas before the dam. On the whole, the three hydraulic calculation conditions show a decreasing trend, and the velocity changes regularly.

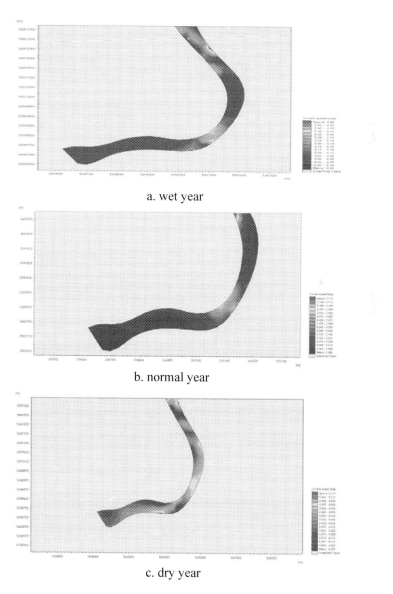

a. wet year

b. normal year

c. dry year

Fig. 5 Vector diagram of velocity in front of dam

3.2 Water Pollution Simulation

Time Node Analysis. As shown in Fig. 6, the change process of pollutant content under four working conditions at three different moments was analyzed. It can be seen from the figure that the change trend of pollutant under four working conditions is basically the same. After the occurrence of pollutants for 6 h, the peak value of pollutants in each working condition is 4.76 mg/L, 2.70 mg/L, 3.64 mg/L and 2.23 mg/L respectively, among which the content of pollutants in working condition 1 is the highest, and the content of pollutants in working condition 4 is the lowest. After 12 h of pollution occurrence, the peak value of pollutants in each working condition is as follows: 3.46, 1.76, 2.75 and 1.81 mg/L, among which the content of pollutants in working condition 1 is the highest and the content of pollutants in working condition 2 is the lowest. After the occurrence of pollution for 24 h, the peak value of pollutants in the four working conditions changed as follows: 2.31, 1.50, 1.86 and 1.23 mg/L, among which the maximum pollutant content was in the first working condition and the minimum was in the fourth working condition.

According to the analysis of the first and second conditions, when the flow of the river is constant, the pollutant leakage determines the pollutant diffusion rate. The more the leakage, the shorter the duration of the peak, the less the leakage, the longer the duration of the peak. At the same time, the closer to the pollution point, the greater the pollutant content. When the pollutant content is unchanged, the longer the time of pollution occurrence, the farther the peak occurs from the pollution leakage point. According to the analysis of the first and third conditions, when the pollutant leakage is certain, the river flow determines the pollutant diffusion rate. When the upstream

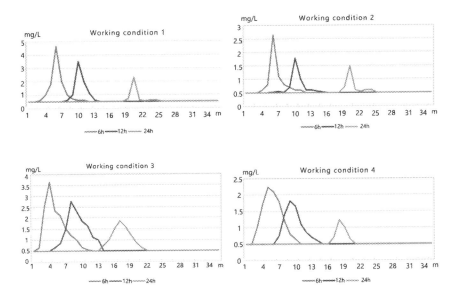

Fig. 6 Analysis results of working conditions

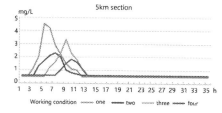 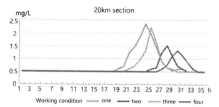

Fig. 7 Analysis results of spatial section

water flow is larger, the diffusion rate of pollutants is faster and the residence time of pollutants is shorter. The larger the incoming water flow is, the slower the peak time is, and the farther the peak location is from the leakage point.

Analysis of Different Sections. It can be seen from the analysis in Fig. 7 that the peak concentration of pollutants under Condition 1 is 4.36 mg/L at a distance of 5 km from the accident, the peak time is 6 h after the leakage, and the pollution duration is 6 h; The peak concentration of pollutants under condition 2 is 2.32 mg/L, the peak time is 8 h after leakage, and the pollution duration is 8 h; The peak concentration of pollutants under Condition 3 is 3.65 mg/L, the peak time is 10 h after the leakage, and the pollution duration is 8 h; The peak concentration of pollutants under Condition 4 is 1.89 mg/L, the peak time is 11 h after the leakage, and the pollution duration is 6 h.

When it is 20 km away from the accident, the peak concentration of pollutants under Condition 1 is 2.35 mg/L, the peak time is 24 h after the leakage, and the pollution duration is 10 h; The peak concentration of pollutants under condition 2 is 1.52 mg/L, the peak time is 29 h after the leakage, and the pollution duration is 7 h; The peak concentration of pollutants under Condition 3 is 2.21 mg/L, the peak time is 25 h after the leakage, and the pollution duration is 9 h; The peak concentration of pollutants under Condition 4 is 1.31 mg/L, the peak time is 31 h after the leakage, and the pollution duration is 6 h.

To sum up, it can be seen from the comparison between different working conditions that at the same section, the pollution duration is affected by the incoming flow and pollutant content. The greater the inflow, the shorter the pollution duration; The greater the pollutant leakage, the longer the duration.

4 Countermeasures for Watershed Ecological Restoration

4.1 Ecological Status of the Basin

Ziwu River Basin is located in the Qinling Mountains and consists of Wenshui River, Puhe River and Jiaoxi River. The main ecological and environmental problems in the reservoir basin: the domestic sewage and garbage generated by the residents around the reservoir area are discharged into the basin at will, which affects the water quality of the basin; In agriculture, due to the unreasonable use of pesticides and fertilizers by residents, the utilization rate of pesticides and fertilizers is low, so that some pesticides and fertilizers will flow into the reservoir with rainfall, irrigation and surface runoff, causing pollution; In terms of livestock breeding, the main pollution source comes from the excreta of pigs, cattle, chickens, ducks, etc., which are free range raised by farmers and livestock. These pollutants are directly discharged without treatment, and eventually some of them will enter the water, causing the total nitrogen, total phosphorus and other indicators of the reservoir to exceed the standard; At the same time, the construction of water conservancy projects also affects the normal reproduction of aquatic organisms, resulting in the loss of biodiversity, the decline of river water self purification capacity and buffer capacity against interference, resulting in the imbalance of ecological environment development and water environment problems.

4.2 Ecological Restoration Measures and Suggestions

Habitat Construction. The original stone body in the river channel is used to build a small permeable rockfill dam. The water flows through the rock crevices or overflow from the dam crest, so it does not affect the connectivity of the river channel. The water surface is raised to about 1 m, creating a deep water environment favored by fish, and landscape cold water fish and benthos are put into the river channel. Construct the water area conditions required for fish and other aquatic animals to complete the whole life process in the river, such as spawning area, feeding place and recreation area. Build a small ecological landscape community, allocate aquatic plants and landscape fish, and increase the flexibility of river landscape.

The tributary of Puhe River is left with a small stone quarry, which is in a stepped shape and has a small slope. In areas with different heights and flow velocities, pave permeable materials such as sand gravel to build habitats and build small water ecosystems. On the premise of not affecting the water circulation, create an environment suitable for fish reproduction and reproduction, and increase the abundance of aquatic organisms. The original ecological restoration and water ecological restoration shall be carried out according to the principle of sustainable development and ecological priority to create conditions for the growth of aquatic plants and the construction of ecosystems, so as to gradually complete the restoration of water resources. The aquatic animal and plant communities are conducive to building a perfect ecological food chain, restoring the natural restoration capacity of the water area, and comprehensively improving the water quality of the basin.

Fish dropping. The construction of Sanhekou Reservoir has resulted in the reduction of fish exchange and genetic diversity due to the blocking effect. As the water flow slows down and some water loving species on the dam gradually decrease, these fish will be forced to go up to the upper reaches of the reservoir area or tributaries to find a new living environment. The reservoir area will be dominated by sedentary fish groups, such as carp, carp, fish, culter, catfish, black plum, wheat head fish, and artificially released salmon, fish, grass carp, etc. Drifting spawning fishes, such as common edible fishes; Fish that produce sticky eggs, such as carp, gill, whale and catfish, should be protected for reproduction. The water temperature of Sanhekou Reservoir is stratified, and the number of fishes adapted to the life in the upper and middle waters of the reservoir area may increase. There are 32 species of fish in Ziwu River, mainly Cyprinidae. The spawning period of fish is from April to July. After the reservoir is built, the average temperature drop of the discharged water is the most obvious in May September, with the maximum temperature drop of 3.75 °C. The water temperature fluctuation is delayed about one month. The discharged low temperature water will affect the life and reproduction of downstream fish.

Comprehensive management of reservoir area

(1) In areas where agricultural cultivated land is concentrated, it is recommended to change the traditional agricultural cultivation mode and apply fertilizer reasonably and accurately to the cultivated land near the river. During the crop fertilization period, heighten the field boundary to prevent fertilizer from flowing into the river. It is prohibited to use pesticides with high toxicity in the upstream catchment area, and it is recommended to use pesticides with low residue and low toxicity. It is prohibited to burn agricultural straw directly to pollute the air. It is encouraged to break the compost in the straw cultivated land to reduce pollution and improve soil fertility.

(2) The pollution of upstream rural residential areas is mainly feces, garbage and untreated sewage. This directly affects the water quality of the downstream reservoir, and corresponding countermeasures should be taken. The impact of domestic garbage on the water quality of the reservoir is obvious. It is necessary to speed up the construction and management of local household garbage disposal sites and implement the resource utilization and harmless treatment of household garbage. Build drainage collection and treatment system with village and town households. Now, due to the loose distribution of residents living in the reservoir area, it is difficult to build large-scale sewage treatment facilities. Therefore, in order to treat various social sewage technologies and meet the discharge standards, it is necessary to build a concentrated and single type household sewage treatment device.

(3) The pollution load brought by aquaculture cannot be ignored. Research shows that a certain amount of fish will have a large number of nitrogen and phosphorus pollutants, which will flow into the water body in dissolved form, and the rest of the fish residues will be deposited in the mud in solid form, which may become endogenous pollution, causing eutrophication of the water body, and even more serious harm to human and livestock breeding. Optimize the cultivation mode of livestock and poultry and fish breeding industry, appropriately reduce the scale of backward industries, centralize the treatment of untreated wastewater pollutants discharged from livestock farms to meet the discharge standard, and stipulate that wastewater shall not be discharged at will. It is recommended to strengthen the construction of biogas projects, strengthen the promotion of

the ecological cycle model of "livestock manure biogas crops", and strive to achieve the goal of cleaner production. Promote the construction of new clean energy treatment projects, strengthen the promotion of the "livestock manure biogas crop" ecological cycle model, and strive to achieve the goal of more clean, harmless and efficient.

5 Conclusion

(1) The MIKE21 hydrodynamic module simulation shows that the flow velocity range in the study area of Sanhekou Reservoir basin is 0–0.139 m/s in high flow years, 0–0.102 m/s in normal flow years and 0–0.096 m/s in low flow years. In the dam area, the flow velocity will decrease, and some areas are close to the still water state.

(2) By using MIKE21 water quality module to simulate water quality pollution, under four working conditions, the maximum pollutant peak value at 6, 12 and 24 h is working condition 1, which is 4.76 mg/L, 3.46 mg/L and 2.31 mg/L respectively. In the comparison between 5 km section and 20 km section, the pollutant concentration in 5 km section under four working conditions is relatively high. Combined with the actual situation, the threat degree of the reservoir area under specific working conditions is obtained, which provides a reference for the downstream dam monitoring and early warning.

(3) For the ecological environment problems in the Sanhekou reservoir basin, reasonable ecological restoration measures and suggestions are put forward to build a perfect watershed ecosystem, restore the living and activity space of organisms, and provide effective guarantee for the reservoir water environment treatment and protection, and water quality safety.

Acknowledgements We would like to express our sincere gratitude to the editors and reviewers who have put considerable time and effort into their comments on this paper. This work was supported by the Shaanxi Natural Science Foundation (grant number 2019JLM-63).

References

1. Li ZS (2022) Simulation study on important of water environment in Yanghe river basin based on MIKE SHE model. Yellow River 44(2):100–105
2. Yang PC (2022) Eutrophication evaluation and water quality analysis of Youche reservoir. Technol Econ Changjiang 6(1):36–40
3. Li D (2018) Water pollution risk simulation and prediction in a drinking water catchment. J Zhejiang Univer (Agric Life Sci) 44(1):75–88
4. Zhang EZ (2022) Analysis of reservoir water resources issues and guarantee measures for governance—taking Dongiao reservoir in Jiangning district, Nanjing city as an example. China Res Compr Utilization 40(1):137–139

5. Mu C (2019) Application of MIKE model in urban and basin hydrological-environmental simulation. J Water Resour Water Eng 30(2):71–80

The Variation of Hydrological Regime According to the Daily Operation of a Complementary Hydro-Photovoltaic Reservoir and Its Impact

Chonglin Wang, Sizhen Liang, Jingjie Feng, Ran Li, and Gaolei Zhao

Abstract Daily operation of a hydropower station is conducted to meet the energy requirement. The hydraulic parameters of the downstream are significantly affected by the dam operation, which has a negative impact on the aquatic system. When the multi energy complementary method is used, such as hydro-photovoltaic (hydro-PV) combined power generation, the problem will worsen. Hydropower station A (HSA) on River X was selected to investigate the impact of daily operation. HSA is a part of hydro-PV complementary power generation. The spawning and breeding period of typical fish, April to July, was selected as the study period. According to various scheduling, the changes of hydrological regime were analyzed. The results show that the maximum flow variation was 334 m³/s, and the variations in water surface width and velocity during reservoir operation were between natural conditions. The maximum daily water level variations under the two operating scenarios were 1.6 m and 3.5 m respectively. The remarkable change of water level may have a negative impact on aquatic organisms. Considering the daily variation limit of 1.2 m under natural condition, the relationship between the allowable daily variation of reservoir outflow and the reference base flow was proposed. The results in this paper serve as a technical reference for studying changes in the hydrological regime and lessening their impacts on aquatic organisms in hydro-photovoltaic complementary development.

Keywords Reservoir · Daily operation · Hydrological regime · Daily water level variation · Hydro-photovoltaic complementary

C. Wang · S. Liang · J. Feng (✉) · J. Feng (✉) · R. Li · G. Zhao
State Key Laboratory of Hydraulics and Mountain River Engineering, Sichuan University, Chengdu 610065, Sichuan, China
e-mail: fengjingjie@scu.edu.cn

465

S. Wang et al. (eds.), *Proceedings of the 2nd International Conference on Innovative Solutions in Hydropower Engineering and Civil Engineering*, Lecture Notes in Civil Engineering 235, https://doi.org/10.1007/978-981-99-1748-8_41

1 Introduction

The development of hydropower station can manage runoff, compensate for dryness, control floods, and increase the assurance of irrigation and water supply [1–3]. Reservoirs have assumed a significant role in advancing socioeconomic development [4]. Dams have major impacts on river hydrology, primarily through changes in the timing, magnitude, and frequency of flow, producing flow regime differing significantly from the pre-impoundment natural [5–8]. The change of hydrological characteristics with time and space, such as discharge, water depth, water surface width, and velocity, are called hydrological regime [9]. The hydrological regime is a significant indicator of the ecosystem stability and plays important role in maintaining the potential of aquatic organisms [10, 11]. To achieve the "30·60" carbon peaking and carbon neutrality objective, China has accelerated the building of hydro-wind-photovoltaic (PV) integrated clean energy base and encouraged the construction of hydropower station [12]. As a result of frequent and quick changes in power demand, hydropower can be used to regulate short-term variability in the power grid [13, 14].

In the hydro-PV complementary system, the daily regulation of hydropower station is operated according to the changes in photovoltaic output [15–17]. This activity makes outflow changes more significant and frequent [18]. In the affected basin, the combined operation of hydro-PV exacerbates intraday fluctuation in water levels, velocity, water level width, and other hydrological characteristics [19]. The ecosystem may be impacted by changes to the hydrological regime [20, 21]. In addition, the daily peak-shaving operation of the hydropower station also significantly altered hydrological parameters [22]. The daily operation of the hydropower station affects the survival of aquatic organisms [23–25] and even leads to the reduction of fish populations [26]. For example, the daily operation of Itiquira power station resulted in approximately 400 adult fishes being stranded and dying when discharge flow rapidly changed over 18 h [27].

Statistical analysis and numerical simulation methods are mainly used to assess the impact of reservoirs on hydrological regime. The statistical analysis method mainly uses the observed historical data to examine the changes of hydrological indicators. The indicators of hydrologic alteration (IHA) [28, 29] and the range of variability approach (RVA) [30] are two commonly used statistical analysis methods. IHA includes 33 indicators, and RVA is a subdivision of IHA. The effects of Kerr Reservoir on the Roanoke River's hydrological regime were assessed using the IHA and the RVA [31]. For three different scenarios, the overall degree of hydrological changes assessed by traditional RVA was 0.39, 0.42 and 0.40 respectively. The construction of the Kerr Reservoir significantly altered the natural hydrologic regime downstream. The RVA was also employed by Ban et al. [32] to analyze the hydrological regime downstream of the Three Gorges Dam. Among indicators of hydrology, the high-pulse duration, water level, and the rising rate of discharge were found to be the properties worst affected by dam, which in turn has affected the success rate of carp reproduction. Zhang et al. [33] simulated the runoff change of Dongjiang River and made a scientific evaluation of the impact of the reservoir on the Dongjiang River. Timbadiya et al. [34] employed one-dimensional hydrodynamic model to predict water level and submergence degree of the lower reaches of the Tapi River, which is useful for developing flood control measures for the Tapi River. Jimeno et al. [35] established the hydrological model to simulate runoff. The appropriate model was selected for different research cases and the watershed scale and daily time interval are used to evaluate the model.

At present, numerous research results on examines of time complementarity between hydropower and PV power [36, 37]. However, there are few studies on the hydrological regime under the influence of hydro-PV complementary operation.

This paper presents the analysis of pre- and post-dam hydrologic changes from dams that cover the independent and hydro-PV complementary operation of HSA. The hydrological changes were clearly quantified. Hydrological variables characterizing crucial elements of the hydrological regime, such as water level, flow velocity, and water surface width, were chosen for analysis. The daily variation of water level in spawning grounds was limited, which was used to formulate the operation scheme of HSA. The results in this paper serve as a technical reference for studying changes in the hydrological regime and lessening their impacts on aquatic organisms in hydro-photovoltaic complementary development.

2 Materials and Methods

2.1 Study Area

The River X is located in the inland basin of China, with rich hydraulic resources. The largest planned hydropower dam in the basin, Hydropower Station A (HSA), is currently being prepared for construction. HSA has multi-year regulation capacity. In addition to participating in hydropower generation and runoff regulation, HSA is also used to participate in peak shaving of hydro-PV complementary energy base in the planning.

Fish that live in the X river breed and spawn mostly from April to July. The 175 km downstream of HSA is a crucial area to protect river ecology and important spawning area. Two sections—a section downstream of the dam site (DB) and spawning ground (SG) were selected to study the changes of hydrological regime under different operation scenarios.

2.2 Model Equations

Continuity equation:

$$\frac{\partial Q}{\partial x} + \frac{\partial A}{\partial t} = q \tag{1}$$

Momentum equation:

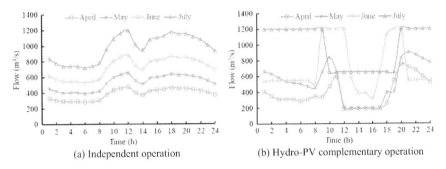

Fig. 1 Outflow process of power station A

$$\frac{\partial Q}{\partial t} + \frac{\partial \left(\frac{Q^2}{A}\right)}{\partial x} + gA\frac{\partial h}{\partial x} + \frac{gQ|Q|}{C^2AR} = 0 \tag{2}$$

where t is the coordinate of the calculation point time, s; q is the lateral inflow, m³/s; Q is the flow, m³/s; x is the coordinate of the calculation point space, m; A is the area of the cross section, m²; C is the Chezy coefficient; R is the hydraulic radius, m; g is the gravity acceleration, m/s²; and h is the water level, m.

2.3 Parameter Determination

The river roughness is a crucial parameter that must be taken into account. In this study, the roughness coefficient is 0.053, which was determined by inverse calculation of the observation data.

2.4 Boundary Conditions

The planned outflows of HSA were chosen as boundary conditions, as shown in Fig. 1.

3 Results and Discussion

3.1 Flow Variations

After the completion of HSA, the average flows of typical day in April and May were lower than the natural average flow, with a maximum decrease of 254 m³/s.

Fig. 2 Flow process of the representative sections in July

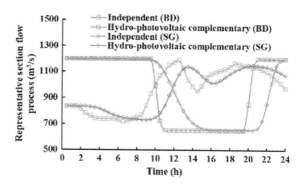

With a maximum rise of 334 m³/s, the average flow of typical day in June and July exceeded the natural average flow.

The hydrologic changes associated with the regulation of flow are reflected in the shape and characteristics of the hydrograph and/or flow duration curve. The daily flow procedures of DB and SG in July served as a representation, as shown in Fig. 2. Daily flow mostly had two peaks caused by the independent operation of HSA. In contrast, the hydro-PV complementary operations were high on both sides and low in the middle. In the process of the unsteady flow moving downstream, the peak flow will gradually decrease, the valley flow will gradually increase, and the flow process will gradually flatten,

The flow variabilities of the representative section are displayed in the Fig. 3. The daily and hourly flow variations significantly change when the results of HSA operation are compared to those of natural river. The daily variances of hydro-PV complimentary operation were greater than those of natural conditions, with the largest variances being seen in June. In June, the variances of DB and SG were increased by 485.3 m³/s and 350.7 m³/s respectively. For HSA's independent operation, a less pronounced pattern became apparent. The typical daily flow variances were lower than that under natural conditions in April, May, and June thanks to HSA's independent operation. In July, the typical daily flow variance was greater than that under natural condition.

The maximum hourly flow variation of natural river was 16 m³/s in July. The maximum hourly flow variations in the independent and hydro-PV complementary operations were substantially bigger than those found in the natural condition.

3.2 Water Level Variations

The water level variations of the representative section in July were displayed in Fig. 4. The hourly water level process of representative section shows that it is similar to the flow change characteristics, the variation from upstream to downstream is smaller.

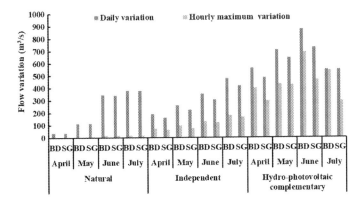

Fig. 3 Flow variation of the representative sections

Fig. 4 Water level process of the representative sections in July

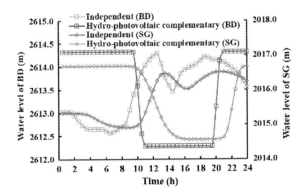

The maximum daily and hourly variations of the water level of the representative section were described in Fig. 5. According to the results, the maximum daily and hourly water level variations caused by HSA were greater than the natural conditions of the corresponding month. Under natural condition, the water level variations of DB and SG were 0.2–1.2 m and 0.3–1.2 m, respectively. When HSA operates independently, the daily water level variations of representative sections DB and SG were 1.0–1.7 m and 1.0–1.6 m respectively. When the hydro-PV complementary operation, the daily water level variations of DB and SG were increased to 2.0–3.6 and 2.1–3.5 m.

The maximum hourly variation of the natural water level was about 0.1 m. The maximum hourly variation of water level during independent operation was 0.6 m in July, while the maximum hourly variation during the hydro-PV complementary process was 2.4 m.

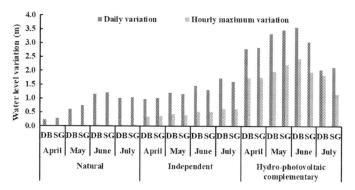

Fig. 5 Water level variation of the representative sections

3.3 Velocity Variations

The velocities of the representative sections were shown in Fig. 6. Before the construction of the dam, the velocity ranges of the DB and SG were 1.3–3.0 and 2.0–4.2 m/s, respectively. The maximum flow velocities of the DB and SG when HSA was operated independently were 2.3 and 3.5 m/s and the minimum flow velocities were 1.4 and 2.4 m/s. Under the hydro-PV complementary operation, the maximum and minimum flow velocities at DB were changed to 2.4 m/s and 1.1 m/s, respectively. Accordingly, the SG were changed to 3.6 and 2.1 m/s.

The flow velocities after the construction of the HSA were less than the maximum flow velocity and greater than the minimum flow velocity of natural condition. They were less than the highest flow velocities under natural conditions but more than the minimum flow velocities. The flow velocity variation range of SG was narrowed after the construction of HSA.

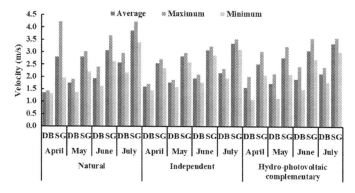

Fig. 6 Velocity of the representative sections

3.4 Water Surface Width Variations

Water surface width variations and water surface width change rates of the representative section were displayed in Figs. 7 and 8. From April to July, the average water surface width of DB and SG in natural conditions were 44–58 m and 40–63 m. The maximum daily variations of water surface width at DB and SG were 3.0 and 4.5 m, accounting for 5.9 and 7.2% of the average water surface width.

The maximum daily water surface width variations at DB and SG during independent operation were 4.4 and 4.8 m, accounting for 8.3 and 8.7% of the average water surface width under natural condition. The maximum daily variations of the water surface width of the hydro-PV complementary operation at DB and SG were 9.1 and 14.6 m, accounting for 18.1 and 30.1% of the average water surface width.

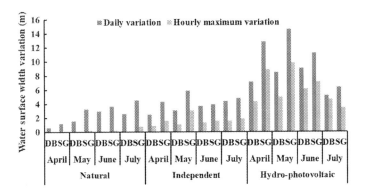

Fig. 7 Water surface width variation of the representative section

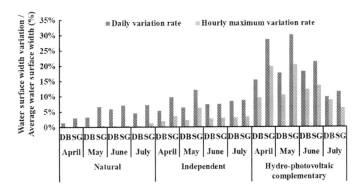

Fig. 8 Water surface width change rate of the representative section

3.5 *The Impacts on Fish*

Fish behavior is strongly influenced by the hydrodynamic environment of the river [38]. Abrupt change of water level can lead fish to strand and even die. A sudden increase in flow velocity may cause viscous spawning fish to lose their spawning factors [39]. A series of hydrological data was utilized to simulate the daily variation in water level and velocity at SG during spawning and feeding from April to July under natural conditions. The results are shown in Figs. 9 and 10.

The daily variations of the water level at SG on a typical day from April to July under independent operation ranged between 1.0 and 1.6 m. The maximum daily variation of the water level under complementary hydro-PV operation was increased to 2.1 m, which was significantly greater than the daily variations under natural condition. The maximum flow velocity at SG for typical day of hydro-PV complementary operation was 3.6 m/s, while it was less than 4.2 m/s during natural condition. During the spawning and breeding period, the daily variations of water level were larger than they would have been under nature.

The effect of the hydro-PV complementary operation was stronger than the independent operation of HSA. The variation amplitudes of different flow velocities and

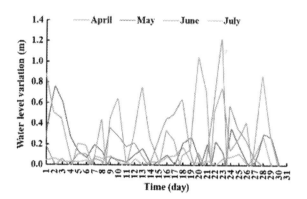

Fig. 9 Daily variation of water level at section SG under natural condition

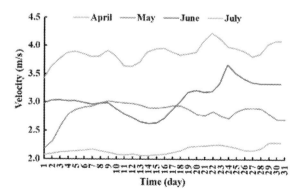

Fig. 10 Velocity of representative section SG under natural condition

Fig. 11 Distribution of the daily water level variation under natural condition

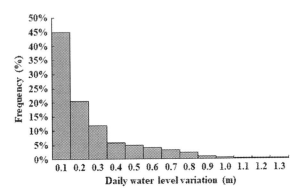

Fig. 12 Distribution of the daily velocity under natural condition

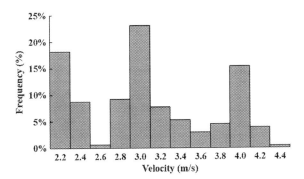

water levels were extracted in sections to determine the acceptable hydraulic parameter standards for fish. As shown in Figs. 11 and 12, under natural conditions, the highest flow velocity and water level variation ratios were 2.8–3.0 m/s and 0–0.1 m, respectively. Rivers should be naturalized in order to lessen the effects of a dam on the river as much as feasible [40, 41]. It can be seen from Figs. 11 and 12 that the water level variations were within 1.2 m, and the flow velocities were within 4.4 m/s. The daily variation in water level and velocity had a substantial influence on fish spawning and reproduction as the limit value.

3.6 Analysis of the Operational Scheduling Optimization

Under different operations for the HSA, the daily variations of water level at SG caused by daily scheduling during spawning and breeding were larger than variation levels under natural conditions. Therefore, the high variation in the water level may not meet the requirements of hydrological stability for fish spawning and breeding.

During the spawning and breeding season, which lasted from April to July, the flows of hydropower plant A were maintained at 186–2240 m³/s. As trial calculation circumstances, a number of typical flows, such as 200, 300, and 400 m³/s, were

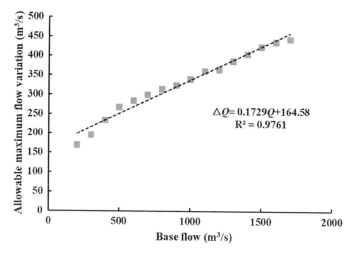

Fig. 13 The results of the allowable flow variation under different base flow conditions

employed. As a restriction for the trial computation, the daily variation of the water level at SG was less than 1.2 m. The allowable variation of the outflow of HSA was shown in Fig. 13.

$$\Delta Q = 0.1729Q + 164.58 \tag{3}$$

where ΔQ is the allowable flow variation of the HSA, and Q is the HSA base flow.

4 Conclusion

In this paper we assessed the hydrological impacts caused by HSA in China. A one-dimensional unsteady flow model was used to conduct the study. We found that the operation of HSA would significantly alter the hydrological regime of downstream. On the basis of the simulation findings, the following conclusions were made.

The average flow on a typical day in April or May after the construction of HSA was lower than the yearly natural flow, with a largest reduction of 254 m³/s. With a maximum rise of 334 m³/s, the average flow on a typical day in June and July exceeded the yearly natural flow.

Under the independent operation of HSA, the daily variation of water level at SG is 1.0–1.6 m during spawning period; the flow velocity is 2.4–3.5 m/s, the maximum daily variation of water surface width is 4.8 m, and the variation rate is 8.7%. Under the hydro-PV complementary operation of HSA, the daily variation of water level is 2.1–3.5 m and the flow velocity is 2.1–3.6 m/s at section SG from April to July. The maximum daily variation of water surface width is 14.6 m, and the variation rate is 30.1%.

The maximum daily variation in water level at the spawning ground during the spawning and breeding period under natural condition was 1.2 m. The flow velocities ranged from 2.0 to 4.2 m/s, and the daily water level variations of independent and hydro-PV complementary operations were far greater than that under natural conditions. The flow velocity variation was consistent with what would be seen in natural conditions.

The daily water level variation at the spawning ground was more affected by a power station's independent and hydro-PV complementary operation than it would have been under nature. To account for the maximum daily variation of outflow and various base flow relation, the trial calculations for the spawning ground used a daily water level variation limit of 1.2 m.

Research on the best scheduling is required to support the ecological protection of the rivers where hydropower and other energy generation projects are developed, as well as research on the effects of the combined operation of hydro-PV complementary and counter-regulating hydropower stations on hydrological changes.

References

1. Liu J, Zang C, Tian S et al (2013) Water conservancy projects in China: achievements, challenges and way forward. Glob Environ Chang 23(3):633–643
2. Talukdar S, Pal S (2019) Effects of damming on the hydrological regime of Punarbhaba river basin wetlands. Ecol Eng 135:61–74
3. Song X, Zhuang Y, Wang X et al (2020) Analysis of hydrologic regime changes caused by dams in China. J Hydrol Eng 25(4):05020003
4. Shi H, Chen J, Liu S et al (2019) The role of large dams in promoting economic development under the pressure of population growth. Sustainability 11(10):2965
5. Magilligan FJ, Nislow KH (2015) Changes in hydrologic regime by dams. Geomorphology 71(1–2):61–78
6. Wang Y, Rhoads BL, Wang D (2016) Assessment of the flow regime alterations in the middle reach of the Yangtze River associated with dam construction: potential ecological implications. Hydrol Process 30(21):3949–3966
7. Gierszewski PJ, Habel M, Szmańda J et al (2020) Evaluating effects of dam operation on flow regimes and riverbed adaptation to those changes. Sci Total Environ 710:136202
8. Kuriqi A, Pinheiro AN, Sordo-Ward A et al (2021) Ecological impacts of run-of-river hydropower plants—Current status and future prospects on the brink of energy transition. Renew Sustain Energy Rev 142:110833
9. Abdollahi K, Bazargan A, McKay G (2018) Water balance models in environmental modeling. Handb Environ Mater Manag 1–16
10. Somaweera R, Nifong J, Rosenblatt A et al (2020) The ecological importance of crocodylians: towards evidence-based justification for their conservation. Biol Rev 95(4):936–959
11. Han D, Lv G, He X (2022) A research on the ecological operation of reservoirs based on the indicators of hydrological alteration. Sustainability 14(11):6400
12. Wang Y, Guo C, Chen X et al (2021) Carbon peak and carbon neutrality in China: goals, implementation path and prospects. China Geology 4(4):720–746
13. Moreira M, Hayes DS, Boavida I et al (2019) Ecologically-based criteria for hydropeaking mitigation: a review. Sci Total Environ 657:1508–1522
14. Batalla RJ, Gibbins CN, Alcázar J et al (2021) Hydropeaked rivers need attention. Environ Res Lett 16(2):021001

15. An Y, Fang W, Ming B et al (2015) Theories and methodology of complementary hydro/photovoltaic operation: applications to short-term scheduling. J Renew Sustain Energy 7(6):063133
16. Zhang X, Ma G, Huang W et al (2018) Short-term optimal operation of a wind-PV-hydro complementary installation: Yalong River, Sichuan Province, China. Energies 11(4):868
17. Zhang Y, Ma C, Lian J et al (2019) Optimal photovoltaic capacity of large-scale hydro-photovoltaic complementary systems considering electricity delivery demand and reservoir characteristics. Energy Convers Manage 195:597–608
18. Zhang Y, Lian J, Ma C et al (2020) Optimal sizing of the grid-connected hybrid system integrating hydropower, photovoltaic, and wind considering cascade reservoir connection and photovoltaic-wind complementarity. J Clean Prod 274(24):123100
19. Poff NLR (2019) A river that flows free connects up in 4D. Nat Int Weekly J Sci 569(7755):201–202
20. Ali R, Kuriqi A, Abubaker S et al (2019) Hydrologic alteration at the upper and middle part of the Yangtze river, China: towards sustainable water resource management under increasing water exploitation. Sustainability 11(19):5176
21. Cui T, Tian F, Yang T et al (2020) Development of a comprehensive framework for assessing the impacts of climate change and dam construction on flow regimes. J Hydrol 590:125358
22. Bejarano MD, Sordo-Ward Á, Alonso C et al (2020) Hydropeaking affects germination and establishment of riverbank vegetation. Ecol Appl 30(4):e02076
23. Poff NL, Zimmerman JKH (2010) Ecological responses to altered flow regimes: a literature review to inform the science and management of environmental flows. Freshw Biol 55(1):194–205
24. Schülting L, Feld CK, Graf W (2016) Effects of hydro-and thermopeaking on benthic macroinvertebrate drift. Sci Total Environ 573:1472–1480
25. Virbickas T, Vezza P, Kriaučiūnienė J et al (2020) Impacts of low-head hydropower plants on cyprinid-dominated fish assemblages in Lithuanian rivers. Sci Rep 10(1):1–14
26. Bakken TH, Harby A, Forseth T et al (2021) Classification of hydropeaking impacts on Atlantic salmon populations in regulated rivers. River Res Appl
27. Braun-Cruz CC, Tritico HM, Beregula RL et al (2021) Evaluation of hydrological alterations at the sub-daily scale caused by a small hydroelectric facility. Water 13(2):206
28. Richter BD, Baumgartner JV, Powell J et al (1996) A method for assessing hydrologic alteration within ecosystems. Conserv Biol 10(4):1163–1174
29. Richter BD, Baumgartner JV, Braun DP et al (1998) A spatial assessment of hydrologic alteration within a river network. Regulated Rivers: Res Manag Int J Devoted River Res Manag 14(4):329–340
30. Richter B, Baumgartner J, Wigington R et al (1997) How much water does a river need? Freshw Biol 37(1):231–249
31. Singh RK, Jain MK (2021) Reappraisal of hydrologic alterations in the Roanoke River basin using extended data and improved RVA method. Int J Environ Sci Technol 18(2):417–440
32. Ban X, Diplas P, Shih WR et al (2019) Impact of Three Gorges Dam operation on the spawning success of four major Chinese carps. Ecol Eng 127:268–275
33. Zhang C, Shoemaker CA, Woodbury JD et al (2013) Impact of human activities on stream flow in the Biliu River basin, China. Hydrol Proc 27(17):2509–2523
34. Timbadiya PV, Patel PL, Porey PD (2014) One-dimensional hydrodynamic modelling of flooding and stage hydrographs in the lower Tapi River in India. Curr Sci 708–716
35. Jimeno-Sáez P, Senent-Aparicio J, Pérez-Sánchez J et al (2018) A comparison of SWAT and ANN models for daily runoff simulation in different climatic zones of peninsular Spain. Water 10(2):192
36. Kougias I, Szabo S, Monforti-Ferrario F et al (2016) A methodology for optimization of the complementarity between small-hydropower plants and solar PV systems. Renew Energy 87:1023–1030
37. Ming B, Liu P, Cheng L et al (2018) Optimal daily generation scheduling of large hydro–photovoltaic hybrid power plants. Energy Convers Manage 171:528–540

38. Silva AT, Bærum KM, Hedger RD et al (2020) The effects of hydrodynamics on the three-dimensional downstream migratory movement of Atlantic salmon. Sci Total Environ 705:135773
39. Zhou H, Yu C, Guo Q et al (2022) Spatial suitability evaluation of spawning reach revealing the location preference for fish producing drifting eggs. Front Marine Sci 1713
40. Wade RJ, Rhoads BL, Rodríguez J et al (2002) Integrating science and technology to support stream naturalization near Chicago, Illinois 1. JAWRA J Am Water Res Assoc 38(4):931–944
41. Peng F, Shi X, Li K et al (2022) How to comprehensively evaluate river discharge under the influence of a dam. Eco Inform 69:101637

Effect of Dry-Wet Cycling on Shear Strength of Phyllite-Weathered Soil in Longsheng, Guilin

Jianliang Yin, Zhikui Liu, Zhanfei Gu, Yan Yan, Yong Xiong Xie, and Bingyan Huang

Abstract The phyllite-weathered soil is a regional speciality. It is essential to study the changes in shear strength of phyllite-weathered soil under dry-wet cycles to understand the changes in mechanical properties of phyllite-weathered soil in the process of dry-wet climate and to manage the slope of phyllite-weathered soil. This paper simulated 12 dry-wet cycles on the specimens of remodelled phyllite-weathered soil. Direct shear and SEM tests were conducted on the specimens in the 0th, 3rd, 6th, 9th, and 12th drying paths. The effects of moisture content and the number of dry-wet cycles on the shear strength of phyllite-weathered soil were analysed macroscopically and microscopically. The following conclusions were obtained: (1) The cohesion of the weathered soil of phyllite will be reduced by increasing the number of cycles, and the more the number of dry-wet cycles, the more pronounced the reduction; the internal friction angle of the weathered soil of phyllite will be reduced by increasing the number of cycles, but the pattern of the decrease in the internal friction angle is not obvious. (2) The increase in the number of dry-wet cycles will increase the stiffness and brittleness of the phyllite-weathered soil specimen, and it will change from the weak hardening type of plastic damage to the solid softening type of brittle damage after a certain number of cycles. (3) The SEM test found that phyllite-weathered soil particles in Longsheng, Guilin are large, and most of the particles are in face-to-face and angle-to-face contact, which is easy to form a hollow structure, and the dry density value of the soil in the natural state is small. At the same time, the soil is reddish-brown in colour because of the leaching of Fe_2O_3. The shear strength index of the cemented phyllite-weathered soil with Fe_2O_3 is more significant than that of phyllite-weathered soil in other areas. The soil has a good shear strength index and a small dry density.

J. Yin · Z. Liu (✉) · Z. Gu · Y. Yan · Y. X. Xie · B. Huang
College of Civil and Architectural Engineering, Guilin University of Technology, Guilin 541004, China
e-mail: 1998009@glut.edu.cn

J. Yin · Z. Liu · Y. Yan · Y. X. Xie · B. Huang
Guangxi Key Laboratory of Geomechanics and Geotechnical Engineering, Guilin 541004, China

Z. Gu
College of Civil Engineering and Architecture, Zhengzhou University of Aeronautics, Zhengzhou 450046, China

© The Author(s) 2023
S. Wang et al. (eds.), *Proceedings of the 2nd International Conference on Innovative Solutions in Hydropower Engineering and Civil Engineering*, Lecture Notes in Civil Engineering 235, https://doi.org/10.1007/978-981-99-1748-8_42

Keywords Phyllite-weathered soil · Dry-wet cycles · Shear strength ·
Micro-scanning · Longsheng county

1 Introduction

Longsheng County is located in the mountainous area northwest of Guilin City, which
is a typical phyllite rock area in China. The "Longji Terraces" is a famous tourist
scenic spot in the area. Longsheng County has less land and more mountains, an
inch of land, and sizeable topographic relief, resulting in natural landslides and slope
instability caused by cutting slopes to build houses, which is one of the most developed areas of landslide disasters in northern Guilin. The main soil layer in Longsheng
County is the residual clayey soil of the Quaternary System weathered by the phyllite
rock. The lower rock layer is the phyllite rock of the Arch cave group (Pt3ng) of the
proterozoic danzhou group. As a particular regional soil, phyllite-weathered soil is
formed by metamorphic phyllite rock through long-time physicochemical and weathering transportation. Longsheng County of Guilin belongs to a typical subtropical
climate zone with a hot and rainy environment. Under the specific climate conditions
of the region, the soil of the phyllite-weathered soil slope of phyllite is often in a
saturated and unsaturated state and has experienced a long period of dry-wet cyclic
action. The shear strength of the soil will be affected by the damp and dry cyclic
action, thus affecting the stability of the landslide or slope.

Allam [1] found that the soil's stiffness and brittleness would increase with the
number of dry-wet cycles, and the earth's compression modulus and shear strength
would decrease with the number of dry-wet cycles. Zeng [2] obtained from the test
of swelling soil in the Nanning area that the shear strength of swelling soil will be
weakened continuously with the increase of the number of dry-wet cycles until it
tends to be stable. Liu [3] summarised the empirical formula of shear strength decay
of expansive soil by increasing the number of dry-wet cycles through the direct shear test
and also proposed that the shear strength decay is due to the development of fissures
during the dry-wet cycles. Chen [4] Through the test, it is concluded that the cohesive
force of red clay tends to decrease with increasing dry density, and the angle of internal
friction continues to grow under the same water content. After the remodelled red
clay cement bond is broken, the cementing force is weakened and cannot be recovered
in a short time. The increase in dry density causes the effective cementing area of
soil particles to decrease to a greater extent than it increases, leading to a decrease
in cohesion. The analysis of mechanical properties of swelling soil during dry-wet
cycles by Xu [5] found that its strength is closely related to the number of dry-wet
cycles and water content status. Zhang [6] argues that the mechanical properties of
unsaturated soils undergo irreversible changes after dry-wet cycles by finding that the
hygroscopic-dehygroscopic cycle process not only decreases the effective internal
friction angle ϕ' of unsaturated soils but also has a specific effect on the value of the
suction internal friction angle ϕb. Liang [7] measured soil microstructure parameters
rapidly and accurately using scanning electron microscopy and IPP image processing

techniques, and fractal theory contributed Ideas to studying the pore microstructure of red clay soils. Wan [8] conducted a systematic experimental study on the mechanical properties and microstructural characteristics of compacted clay under the action of dry-wet cycles (indoor simulated landfill climate) to reveal the intrinsic nature of deformation characteristics and strength decay of compacted clay under the act of dry-wet cycles from the microscopic level, in response to the problems such as damage of compacted clay impermeable structure of landfill closure cover system under the action of dry-wet cycles. The influence of factors such as threshold size, analysis area size, scan point location, and magnification on the quantitative study of soil microstructure was investigated by calculating its apparent porosity and soil particle morphology fractional dimensional number from a series of SEM photographs by Tang [9], and the influence mechanism of each factor was explored. In addition, about there are also changes in the shear strength of various types of soils under the action of dry-wet cycles. Liu [10–12] investigated the effects of dry density and dry-wet cyclic activity on the water-holding capacity of soils. Zhang [13–15] studied the law of fracture evolution of soils under the action of dry-wet cycles. Zhao [16–19] researched the shear strength test and the fracture evolution law of phyllite-weathered soil and red clay mixture.

There are few studies on the mechanical properties of phyllite-weathered soil, and the mechanisms affecting the mechanical properties of phyllite-weathered soil are not very clear, especially no studies on the soil–water interaction of phyllite-weathered soil (including dry-wet cycles). According to previous research [16–19], phyllite-weathered soils are powdered soils or powdered clay soils with a small plasticity index, low clay content, quiet strength, poor water stability and water retention, and poor soil adhesion. In contrast, the phyllite-weathered soil in the Longsheng region has a high Fe_2O_3 content after leaching, and following cementation, the soil's characteristics alter in numerous ways. This paper used the shear test to investigate the strength change of unsaturated phyllite-weathered soil during dry-wet cycles. The test results were analysed from unsaturated soil mechanics and soil structure perspectives. The research results are significant for understanding the change law of the mechanical properties of phyllite-weathered soil during dry-wet climatic processes. The significance of the research findings lies in their contribution to a better understanding of the variations in mechanical characteristics of phyllite-weathered soil in dry and wet climates and the management of phyllite-weathered soil slopes.

2 Test Material

The test soil was taken from the weathered soil of phyllite rock on the south side of the unstable slope of Heng Yi Road, Long Ji Xue Fu, Longsheng, Guilin City. The soil is reddish brown, with a depth of 1–2 m, a natural moisture content of 30–55%, and a natural dry density of 1.15–1.35 g/cm^3.

The main clay minerals were kaolinite and illite. Other primary physical property indexes are shown in Table 1.

Table 1 Basic physical property indexes of phyllite-weathered soil

Specific gravity (g/cm³)	Free expansion rate (%)	Sand mass fraction (%)	Powder particle mass fraction (%)	Clay mass fraction (%)	Liquid limit (%)	Plastic limit (%)	Plasticity index (%)
2.68	44	18.2	48.2	43.6	57	34.5	22.5

3 Test Method

3.1 Specimen Preparation

The air-dried moisture content of the soil samples was measured to be 3.7% after being air-dried, crushed, and sieved through a 2-mm mesh. The original moisture level of the crushed soil sample was adjusted to 28, 30, 32, 34, 36, 38 and 40% by adding the appropriate amount of water and mixing carefully. Soil samples were sealed in plastic bags, placed in a humidifying cylinder, and allowed to stand for 48 h to ensure that the moisture in the models was evenly distributed. Then, weigh the required amount of soil sample, pour it into the ring knife, and crush it with a hydraulic jack such that the specimen's diameter is 61.8 mm and its height is 20 mm. Referring to the density index of the in-situ soil, the initial dry density of the test design specimen is 1.25 g/cm³.

3.2 Dry-Wet Cycle Test Scheme

The experiment was structured to include 12 dry-wet cycles. Seven moisture content control points were established during the drying process: 28, 30, 32, 34, 36, 38 and 40%. The process of the dry-wet cycle is shown in Fig. 1. Firstly, some pressed heavy plastic soil specimens were sealed and stored with an initial moisture content of 40%. The remaining heavy plastic soil specimens were placed on a porous plate and dried at room temperature (temperature: (20 ± 3) °C). The water content was calculated by weighing the specimens and obtaining the change in mass. When the desired moisture content control point of the specimens was reached, the specimens were taken out and sealed with cling film for 48 h to allow for a more uniform distribution of moisture inside the specimens, thus reducing the experimental error. When the desired moisture content control point is reached, the specimens are taken out and stored in cling film for 48 h so that the internal moisture of the specimens can be distributed more evenly to reduce the experimental error. For the straight shear test, four parallel samples were prepared for each moisture content control point. The remaining specimens were dried until the residual moisture content; then immersed in water to saturate the specimens under side-limited conditions for 48 h. After that,

Fig. 1 Schematic diagram of the dry-wet cycle process

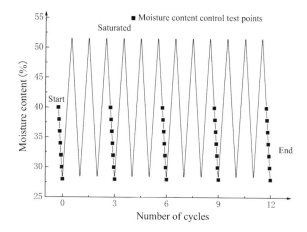

the specimens were removed, and the drying process was repeated. The straight shear test was performed at the corresponding moisture content control point at 0, 3, 6, 9, and 12 cycles.

3.3 Direct Shear Test

The test instrument is the ZJ-type strain-controlled straight shear instrument produced by Nanjing Ningxi Soil Instrument Co. To determine the effect of dry-wet cycles on the shear strength indexes (cohesion and internal friction angle), two sets of parallel direct shear tests were conducted for each group of specimens at a shear rate of 0.8 mm/min under normal stress conditions of 100, 200, 300 and 400 kPa when the specimens' moisture content reached the control point. A total of 280 samples were pressed to carry out the above dry-wet cycle tests, and the test protocol is shown in Table 2. Figure 1 depicts the process of the dry-wet cycle in schematic form.

3.4 Scanning Electron Microscope Experiments

The electron microscope scanning instrument used in this experiment is Field Emission Scanning Electron Microscope (FESEM), model S-4800. The technical parameters are the magnification range from $25\times$ to $800,000\times$, and the chemical element analysis range from 4Be to 99Es. In this experiment, the effect of dry-wet cycles on the weathered soil of Longsheng phyllite was mainly studied, so the test group with 34% moisture content was taken as an example, and tiny cubes of about 1 cm^3 were selected from the middle part after the dry-wet cycles for freeze-drying with liquid nitrogen. Since the phyllite-weathered soil of Longsheng is not a conductive

Table 2 Test scheme

Water content	Direct shear test options under different positive pressures				
	Dry-wet cycle 0 times	Dry-wet cycle 3 times	Dry-wet cycle 6 times	Dry-wet cycle 9 times	Dry-wet cycle 12 times
28	√	√	√	√	√
30	√	√	√	√	√
32	√	√	√	√	√
34	√	√	√	√	√
36	√	√	√	√	√
38	√	√	√	√	√
40	√	√	√	√	√

material, the sample was sprayed with gold to complete the SEM preparation. At $1000\times$, $5000\times$, $10,000\times$ and $20,000\times$ magnification, soil samples of phyllite rock subjected to 0, 3, 6, 9, and 12 cycles of dry-wet conditions were photographed.

4   Test Results and Discussion

4.1   Direct Shear Test Results and Analysis

The specimens meet the test requirements after completing the fabrication and dry-wet cycle process according to the scheme in Sect. 3.2. By geotechnical criteria [20], specimens of phyllite-weathered soil were submitted to direct shear testing, and the test results are presented in Tables 3 and 4.

Table 3 Direct shear test of phyllite-weathered soil with dry-wet cycles under different water content conditions (cohesion)

Water content (%)	The cohesion of phyllite-weathered soils (KPa)				
	Dry-wet cycle 0 times	Dry-wet cycle 3 times	Dry-wet cycle 6 times	Dry-wet cycle 9 times	Dry-wet cycle 12 times
28	32	41.3	39.6	35.9	31.1
30	39	50.7	47.2	45.3	39.6
32	46	54.2	52.7	51.3	43.2
34	53.2	58	56.8	54.9	48.2
36	48	53.3	51.7	50.2	46.1
38	42	47.1	46.3	44.7	38.2
40	38	38.2	36.7	35.1	29.2

Table 4 Direct shear test of phyllite-weathered soil with dry-wet cycles under different water content conditions (internal friction angle)

Water content	Internal friction angle of phyllite-weathered soil (°)				
	Dry-wet cycle 0 times	Dry-wet cycle 3 times	Dry-wet cycle 6 times	Dry-wet cycle 9 times	Dry-wet cycle 12 times
28	26.7	23.2	24	26.2	27.5
30	24.8	21.7	22.3	23.1	24.5
32	22	17.9	18.6	19.7	20.3
34	18.7	16.2	17.1	18.5	19.2
36	20.1	18.2	17.8	18.9	19.2
38	20.5	16.2	16.7	18.9	20.1
40	18.7	14.3	15.2	16.4	17.1

The graphs of the variation of cohesion and internal friction angle with moisture content under the controlled number of dry-wet cycles (Fig. 2a and b) and the graphs of the variation of cohesion and internal friction angle with moisture content under the controlled number of dry-wet cycles (Fig. 3a and b) are plotted according to the contents of Tables 3 and 4.

From Fig. 2a, it can be observed that the cohesive forces of the remodelled phyllite-weathered soil specimens increased significantly from 0 times dry-wet cycles to 3 times dry-wet cycles due to the action of dry-wet cycles in the case of controlled water content, except in the case of 40% water content, and that the cohesive forces of the remaining six controlled water content specimens underwent varying degrees of cohesive reduction from 3 times dry-wet cycles. Among these, the extent of the drop in force of cohesion is shown to be at its greatest between 9 and 12 cycles. The cohesive strength of the specimen group with 34% water content was the largest

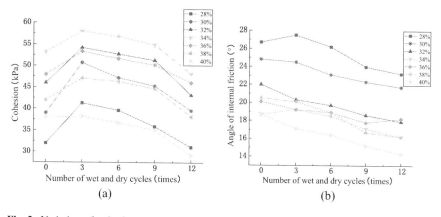

Fig. 2 Variation of cohesiveness and the angle of internal friction with the number of dry-wet cycles under-regulated moisture content

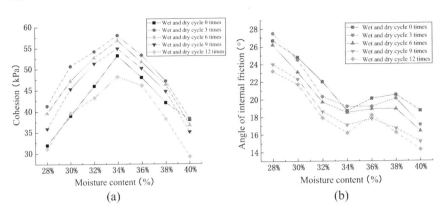

Fig. 3 Variation of cohesion and internal friction angle with moisture content

group in this test, and the maximum cohesive force of 54.2 kPa was obtained after the third cycle of the test. The magnitude and direction of the change in cohesion with the number of dry-wet cycles for these two test groups with a control moisture content of 32 versus 36% were comparable. The cohesive forces of the two groups of tests with 28 and 40% control moisture content had lower values in this test. Under the condition of constant water content, the cohesive force of the weathered soil specimens of phyllite showed an overall trend of increasing and gradually decreasing after the dry-wet cycles.

From Fig. 2b, it can be seen that the angle of internal friction shows a general trend of gradually decreasing after the action of dry-wet cycles with constant water content. The value of the angle of internal conflict for the group with low water content is more significant than that for the group with high water content.

From Fig. 3a, it can be observed that the cohesive force of the phyllite-weathered soil tends to increase and then decrease as the water content increases over the same number of cycles. The phyllite-weathered soil with three dry-wet cycles had the highest cohesion among the group. In this test, the cohesive force of the phyllite-weathered soil sample after 12 wet–dry cycles has the lowest value. And it can be observed that: the cohesive force after 3 dry-wet cycles > the coherent force after 6 dry-wet cycles > the cohesive force after 9 dry-wet cycles > the coherent force after 0 dry-wet cycles > the cohesive force after 12 dry-wet cycles.

From Fig. 3b, it can be observed that the internal friction angle of the phyllite-weathered soil decreases, then increases by a certain amount with the increase of water content over the same number of cycles, and then decreases again. The values of internal friction angle are more significant for the two groups of phyllite-weathered soil after 0 times cycles and after 3 times cycles and smaller for the group after 12 times cycles.

4.2 Comparative Microscopic Analysis of Phyllite-Weathered Soil

Figure 4 shows the SEM images of the weathered soil specimens with water content controlled at 34% phyllite-weathered soil after 0 cycles. The electron microscope scan can be used to obtain the microstructure of the phyllite-weathered soil, observe the particle morphology of the soil particles, and estimate the division size of the soil particles from the scale in the picture. It can be clearly marked from Fig. 4d that the grains of the phyllite-weathered soil without circulating remodelled maintain a lamellar structure with flaky, long flat lamellar grains with a grain length of 10–40 μm and a thickness of 0.5–3 μm. Most soil particles are in face-to-face and edge-to-face contact with each other. Therefore, the interior of the phyllite-weathered soil will form a fly-over structure. The natural dry density is small. At the same time, it can be seen that the arrangement of the particles of the phyllite-remoulded soil is basically disordered, and the directional arrangement trend of the particles cannot be obviously found. The overall structure of this combination of soil aggregates is unstable in geotechnical mechanics. A laminar structure dominates the microstructure of phyllite-weathered soil. The soil comprises varying-sized lamellar units, primarily in face-to-face contact but partially in edge-to-face and edge-to-edge connections to pile up and gather, forming erected lamellar agglomerates. The particles of cemented oxidised cemented clay are dispersed on the surface of the soil or fill the pores above. Under the contact bonding of these cement, the flake aggregates are connected into the laminated structure. The contact between the particle units is not close, the distribution is irregular, the pores between the particles are large, and the structural compactness is poor. The soil samples were damaged under pressure, which is mainly due to the damage to the body of the contact unit, that is, the damage to the bonding force between the contact soil particles, the change of the soil structure, the transformation of the contact mode, the arrangement direction between the soil particles and other tissues.

Figure 5 is the SEM image of the phyllite-weathered soil sample with a water content of 34% after different cycles of 20,000 times. Figure 5a demonstrates that the particles of remodelled soil without a dry-wet cycle are relatively complete, with more large particles, a greater pore depth, and a large number of small particles on the surface of large particles. After three dry-wet cycles, as depicted in Fig. 5b, the pore depth becomes shallow, the large particles are shattered, and the small particles on the surface of the large particles are drastically diminished. After 6 and 9 cycles, the particles were further broken, and the fine particles increased significantly. The arrangement and combination of particles cannot find clear rules and show anisotropy.

The SEM image scanning is binarised, and MATLAB calculates the pores and particles to obtain the proportion of pores and particles, and then the line chart is drawn.

It can be seen from Fig. 6 that the proportion of pores in phyllite-weathered soil samples under four different magnifications decreases first when the dry-wet cycle is 0 to 3 times. In the case of dry-wet cycles of 3–9 times, the pore ratio of samples

(a) 1000X   (b) 5000X

(c) 10,000X   (d) 20,000X

Fig. 4 SEM images of specimens of phyllite-weathered soil at different magnifications after 0 times cycles

(a) 0 cycles   (b) 3 cycles

(c) 6 cycles   (d) 9 cycles   (e) 12 cycles

Fig. 5 Phyllite weathered soil sample magnified 20,000 times SEM image after different cycles

Fig. 6 Variation of pore proportion with cycle times under different magnifications

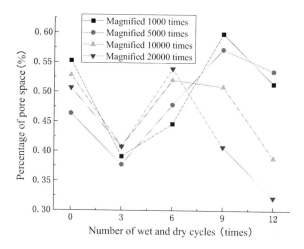

amplification 1000 and 5000 times continue to increase. The proportion of pores in the samples magnified by 10,000 times and 20,000 times increased in the case of 3–6 times dry-wet cycles and decreased in the case of 6–9 cycles. This differs from the law of 1000 and 5000 times enlarged samples. Since the image of the sample magnified 10,000 and 20,000 times have larger soil particles and pores, the entire image is occupied by fewer soil particles and pores, and the data obtained by the proportion of pores in the image is more one-sided. So magnified 1000 and 5000 times, the response of the image results is more convincing. In the case of 9–12 wetting–drying cycles, the proportion of pores in phyllite-weathered soil continues to decrease. The proportion of pores in phyllite-weathered soil decreases when experiencing 0–3 wetting–drying cycles, increases when experiencing 3–9 cycles, and decreases again when experiencing 9–12 cycles. After 0, 3, 6, 9, 12 dry-wet cycles, the proportion of pores in SEM images is about 55%, 40%, 44%, 59% and 51%, respectively.

5   Discussion and Mechanism Analysis

In the direct shear test results, the corresponding strength indexes (c and φ) of the remoulded soil samples after 0 cycles are lower than those of the remoulded soil samples after 3 cycles. This is because when the cycle is not experienced, the particles of the phyllite-weathered soil are large, the degree of compaction is not very high (the dry density of the remoulded soil sample is designed to be 1.25 g/cm^3), the overall structure of the remoulded soil sample is loose, and there are many aggregates. This leads to large pores and cracks in the remoulded soil sample. There will be bubbles and a large amount of free water in the pores, so the obtained shear strength index is low. After three dry-wet cycles, these pores and cracks will gradually be filled by the solid phase components of clay, non-clay minerals, organic matter, and precipitated salts. Therefore, the shear strength index of phyllite-weathered soil samples after 3 times of dry-wet cycle is higher than that of phyllite-weathered soil samples after 0 times the dry-wet cycle. This falls in line with the findings of other cohesive soils.

In most instances, the soil is an integrated body of soil particles, and the displacement between soil particles determines the deformation process. In the process of dry-wet cycles, the content of clay minerals and soil organic matter will be lost more and more with the increase of cycle times, which is one of the reasons why the shear strength index of phyllite weathered soil decreases with the increase of cycle times.

The 0 dry-wet cycle test with a relatively large test error was temporarily taken out. The shear strength indexes of phyllite-weathered soil samples after 3, 6, 9, and 12 wetting–drying cycles were analysed. In the same drying process, in the optimal moisture content (33.7%) dry side, with the decrease of moisture content, the cohesion of the sample decreases the internal friction angle increases. On the side higher than the optimal moisture content, as the moisture content decreases, the cohesion of the sample increases, and the internal friction angle increases [21–23]. This is because the sample suction gradually increases during drying, contributing to cohesion and shear strength. In addition, the volume of the specimen shrinks during drying, the pore ratio decreases, and the structure become denser, which also contributes to the increase in the shear strength index. In addition, during the drying process, the sample volume shrinks, the void ratio decreases, and the structure becomes denser, which is also an important reason for the increase in the shear strength index.

As shown in Fig. 7, after 3, 6, 9, and 12 wetting–drying cycles, the shear stress-displacement curve (normal stress 300 kPa) of phyllite soil with 34% moisture content are taken as an example. It can be seen from the figure that after 9 and 12 dry-wet cycles, the phyllite samples with 34% water content showed apparent brittle failure. The peak strength loss of the sample with 12 dry-wet cycles is more significant than that of the sample with 9 dry-wet cycles. Combined with Fig. 8, phyllite weathered soil samples after 3 times, 6 times, 9 times, and 12 times dry-wet cycles after a direct shear test. It can be seen that after 3 and 6 dry-wet cycles, the sample has apparent displacement but no shear fracture, and the shear stress does not decrease. Combined with Fig. 7, it is inferred that the failure mode of the sample is weak hardening plastic deformation. After the 9th and 12th wetting–drying cycles, the sample was sheared off. But it is worth mentioning that after the 9th wetting–drying cycle, the specimen still has a weak connection after being cut. The sample after the 12th dry-wet cycle was utterly cut off. Combined with the shear stress-displacement curve of Fig. 7, it can also be seen that the residual shear strength value after the 9th dry-wet cycle is more significant than that after the 12th dry-wet cycle.

The stiffness and brittleness of phyllite-weathered soil samples showed an increasing trend during the dry-wet cycle. The failure mode of the direct shear test changed from a weak hardening type of plastic failure to a strong softening brittle failure. This is because, during the drying process of the remoulded soil sample, the suction of the soil leads to the shrinkage deformation and structural adjustment of the soil and the crystal replacement of the clay minerals in the dry-wet cycle, which is irreversible. The principal clay mineral of phyllite-weathered soil is kaolinite. Isomorphous replacement and exchange capacity occur in the process of the dry-wet

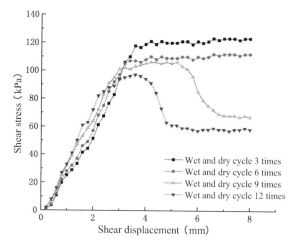

Fig. 7 Effect of wetting–drying cycles on shear behaviour of phyllite weathered soil with 34% water content (normal stress 300 kPa)

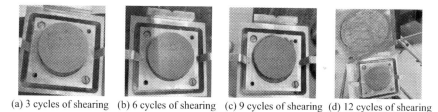

(a) 3 cycles of shearing (b) 6 cycles of shearing (c) 9 cycles of shearing (d) 12 cycles of shearing

Fig. 8 Pictures of 34% water content phyllite weathered soil after direct shear test fter the dry-wet cycle

cycles. These factors are deep and need to be analysed with geotechnical engineering soil properties.

In addition, the bound water film between soil agglomerates is an essential factor affecting the shear strength. During the wetting process of the sample, the thickness of the surface-bound water film between the soil particles is thickened, and the water film force is weakened. In Table 3, cohesion decreases in the optimal moisture content wet test, and internal friction angle decreases in the same dry-wet cycle with increased moisture content. During dry-wet cycles, the water film of soil particles becomes thinner with the increase of cycles. The soil develops cracks after experiencing multiple dry-wet cycles, and the number of cracks increases with the rise of dry-wet cycles, which is one of the reasons why the shear strength index decreases with the increase of dry-wet cycles.

The microscopic particles of the phyllite-weathered soil are large-sized particles with substantial occlusion, which may increase the dilatancy of the soil. Thus the shear strength is more significant. After the dry-wet cycles, the large-sized particles are broken, thus reducing the dilatancy and decreasing the strength of the soil. Figure 5 shows that the particles of the phyllite-weathered soil specimen after 12 cycles are significantly smaller than those of the phyllite-weathered soil specimen after 0 and 3 cycles. The clay minerals of the phyllite-weathered soil are mainly illite and kaolinite. Kaolinite is a hexagonal flake and correlates with its degree of crystallisation, while illite is lamellar. The water-holding capacity of the mineral particles of the clay is weakened after multiple dry-wet cycles, the particles are reduced, the needle-like particles become smooth after movement, and the cohesion and internal friction angle are reduced. Due to the dry-wet cycles, the clay minerals are leached. The conversion of secondary minerals and the weakening of Fe_2O_3 cementation are also reasons for weakening the shear strength of phyllite-weathered soil. The strength of phyllite-weathered soil is also related to the properties of adsorbed cations, the exchange of ions affects the stability of the soil, and the exchange of cations affects the shear strength.

6   Conclusion

(1)   In the case of the same water content, after the dry-wet cycling action, with the increase in the number of cycles, the cohesion of the phyllite-weathered soil will decrease. The more the number of dry-wet cycles experienced, the more pronounced the decrease is. Under the same water content, the internal friction angle of phyllite-weathered soil tends to increase and then decrease with the number of cycles in general, but the regularity is not obvious.

(2)   Under the same water content, the increase in dry-wet cycles increases the stiffness and brittleness of the phyllite-weathered soil specimens. When the 9th cycle is experienced, the failure mode of the specimen will change from weak hardening failure of plastic failure to muscular softening failure of brittle failure.

(3)   After 0, 3, 6, 9, and 12 dry-wet cycles, the proportion of pores occupied by pores in SEM images is about 55%, 40%, 44%, 59%, and 51%, respectively. During the cycling process, the small clay particles first fill part of the larger pores through the dry-wet cycling motion, and the proportion of the area occupied by the pores decreases. Then, after the subsequent cycles, the large particles are broken, increasing the area occupied by the pores.

(4)   As the number of cycles increases, the larger particles and agglomerates in the phyllite-weathered soil decrease, the fine particles increase, and the cementation strength decreases, thus reducing the shear strength of the phyllite-weathered soil.

(5)   The SEM test revealed that the phyllite-weathered soil in the Longsheng area has larger particles, and most of the particles are in surface-surface and corner-surface contact, which can easily form a hollow structure. The dry density value of the soil is small in its natural state. At the same time, the reddish-brown colour of the soil is due to the leaching of Fe_2O_3. The shear strength index (cohesion and internal friction angle) of the phyllite-weathered soil cemented

by Fe_2O_3 is larger than that of the previously studied phyllite-weathered soil. At the same time, the soil has a better shear strength index while having a smaller dry density.

Acknowledgements This research was funded by the National Natural Science Foundation of China (41867039), the Foundation of Technical Innovation Center of Mine Geological Environmental Restoration Engineering in Southern Area (CXZX 2020002), project funded by Guangxi Key Laboratory of Geotechnical Engineering (20-Y-XT-03).

References

1. Allam MF (1981) Effect of wetting and drying on shear strength. J Geotech Eng 107(4):421–438
2. Zeng ZT (2012) Dry-wet cycle effect of expansive soil and its influence on slope stability. J Eng Geol 20(06):934–939
3. Liu HQ (2010) Experimental study on the effect of cracks on shear strength index of expansive soils. Geotechnics 31(03):727–731
4. Chen JY (2019) Effect of Water content and dry density on shear strength of guilin red clay. Karst China 38(06):930–936
5. Xu D (2018) Experimental study on the influence of wetting-drying cycles on shear strength of expansive unsaturated soil. Earth Sci Front 25(01):286–296
6. Zhang FZ (2010) Study on the influence of repeated dry-wet cycles on the mechanical properties of unsaturated soils. Chin J Geotech Eng 32(01):41–46
7. Liang SH (2020) Study on Guizhou Qiannan red clay microstructure based on SEM. Hydropower Energy Sci 38(02):151–154
8. Wan Y (2015) Study mechanical properties and microscopic mechanism of compacted clay under dry-wet cycle. Geotechnics 36(10):28152824
9. Tang CS (2008) Analysis of influencing factors in soil microstructure based on SEM. J Geotech Eng 04:560–565
10. Liu L (2022) Study on water holding characteristics of phyllite fully weathered soil. Henan Univer Sci Technol (Nat Sci Ed) 43(6):53–58
11. Cai GQ (2020) Experimental study of the soil-water characteristic curve of sandy loess. Chin J Geotech Eng 42(Z1):11–15
12. Liu QQ (2021) Experimental study on water retention characteristics of saline soil in the full suction range. Geotechnics 42(3):713–722
13. Zhang JJ (2011) Experimental study on crack evolution law of expansive soil under dry-wet cycle. Geotechnics 32(09):2729–2734
14. Yang HP (2012) Crack development rule of rolled expansive soil under drying - wetting cycles. Traffic Sci Eng 28(01):1–5
15. Li Y (2018) The influence of multiple dry-wet cycles on the fracture and mechanical properties of laterite. J Nanchang Univer (Engineering Edition) 40(03):253–256+261
16. Zhao XS (2022) Experimental study on shear strength of phyllite-red clay mixed soil. J Chongqing Jiaotong Univer (Nat Sci Ed) 41(8):120–126
17. Zhao XS (2021) Road performance of fully weathered phyllite composite improved soil. J Transp Eng 21(6):147–159
18. Zhao XS (2021) Fissure evolution law of fully weathered phyllite, red clay and its improved soil. China J Highway Transp 34(12):323–334
19. Zhao XS (2020) Experimental study on microstructure and compression characteristics of phyllite soil mixed with red clay. Sci Technol Eng 20(21):8732–8738
20. Earthwork Testing Method Standard (GB/T50123-2019) (2019) China Plan Publishing House

21. Li KP (2021) Experimental study on improved expansive soil of sandstone. Hydroelectric Energy 39(12):147–150
22. Liu XR (2017) Study on particle breakage characteristics of soil - rock mixture based on large direct shear test. Chin J Geotech Eng 39(08):1425–1434
23. Yang HP (2006) Variation of total strength index of expansive unsaturated soil with saturation. Chin Civ Eng J 04:58–62

Quantitative Evaluation of Four Kinds of Site Seismic Response Analysis Methods Using DTW

Rui Sun and Wanwan Qi

Abstract In order to quantitatively evaluate the one-dimensional site seismic response analysis methods, this article selected 2418 ground motion records of Japan KiK-net strong-motion seismograph network and 2418 groups of acceleration response spectra calculated by DEEPSOIL, SHAKE2000, SOILQUAKE and SOILRESPONSE, and then' used the dynamic time warping (DTW) algorithm to calculate the DTW distance between the measured acceleration response spectrum and the calculated acceleration response spectrum. The average DTW distance and change trend in different PGA ranges were compared and analyzed. The average DTW distance of the four methods in weak ground motion were similar, and in the strong ground motion, the average DTW distance of SOILRESPONSE was smaller than the other three methods. The DTW distance of the four methods increased with the increase of PGA, the growth rate of SOILRESPONSE was significantly lower than the other three methods. DTW distance can accurately and effectively reflect the difference between response spectrum, which provides a new method for quantitative evaluation of one-dimensional site seismic response analysis method.

Keywords DTW · Site seismic response analysis · Quantitative evaluation

1 Introduction

Site seismic response analysis is an important part of geotechnical seismic engineering. At present, one-dimensional numerical simulation method is mainly used to site seismic response analysis. The frequency domain equivalent linearization method and the nonlinear time domain method are two main methods for one-dimensional site

R. Sun (✉) · W. Qi
Key Laboratory of Earthquake Engineering and Engineering Vibration, Institute of Engineering Mechanics, China Earthquake Administration, 29 Xuefu Road, Nangang District, Harbin, Heilongjiang, China
e-mail: iemsr@163.com

Key Laboratory of Earthquake Disaster Mitigation, Ministry of Emergency Management, 29 Xuefu Road, Nangang District, Harbin, Heilongjiang, China

S. Wang et al. (eds.), *Proceedings of the 2nd International Conference on Innovative Solutions in Hydropower Engineering and Civil Engineering*, Lecture Notes in Civil Engineering 235, https://doi.org/10.1007/978-981-99-1748-8_43

seismic response analysis. According to these methods, programs such as LSSRLI-1 [1], SHAKE [2], DEEPSOIL [3], etc. have been developed and widely used in the domestic and international engineering community. However, the reliability and applicability of these methods need to be verified by comparing with the measured records.

Dynamic Time Warping (DTW) algorithm is a method to calculate the similarity between two time series by finding the optimal mapping, which has been widely used in many fields [4, 5]. In earthquake engineering, the DTW algorithm has also been preliminarily applied. For example, Joswig et al. [6] used the DTW distance to judge the main events in earthquake, Li et al. [7] showed that DTW distance can quantitatively reflect the similarity of simulated seismic field.

This paper based on 2418 ground motion records of Japan KiK-net strong-motion seismograph network and 2418 groups of acceleration response spectra calculated by DEEPSOIL, SHAKE2000, SOILQUAKE and SOILRESPONSE. We calculated the DTW distance between the measured acceleration response spectrum and the calculated acceleration response spectrum, comprehensively evaluated the DTW distance of different PGA ranges and the overall change trend, and quantitatively analyzed the application scope of the four numerical simulation methods.

2   DTW Algorithm

2.1   Computing Method of DTW Distance

The core idea of DTW algorithm is to find the optimal mapping between two given time series through dynamic adjustment, and judge the similarity of two time series by calculating the distance between the optimal mapping. The smaller the distance, the higher the similarity [8]. This distance is called DTW distance. The DTW distance between two spectra is calculated as follows:

(1) Assuming the measured acceleration response spectrum P and the calculated acceleration response spectrum Q, the number of period is m and n respectively, where $P = \{p_1, p_2, ..., p_3, p_m\}$, $Q = \{q_1, q_2, ..., q_n\}$;

(2) The Euclidean distance is used to calculate the corresponding distance between any two points of the two sequences, which is recorded as $d(p_i, q_j)$;

(3) Calculate the cumulative distance to get the matrix D, the calculation is shown in Eq. 1 and the $D(m, n)$ is the DTW distance.

$$D = \begin{cases} d(p_1, q_j) + D(1, j-1), i = 1 \\ d(p_i, q_1) + D(i-1, 1), j = 1 \\ d(p_i, q_j) + \min(D(i-1, j), D(i, j-1), D(i-1, j-1)), i \neq 1, j \neq 1 \end{cases} \tag{1}$$

The path formed by the optimal mapping is called the regular path w, as shown by the red solid line in Fig. 1. There are three conditions to be satisfied by the regular

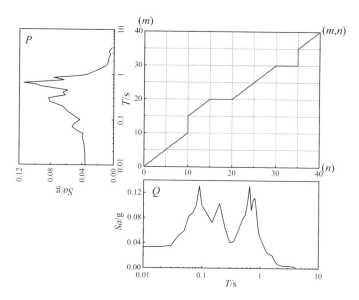

Fig. 1 DTW algorithm regularization path

path w. Firstly, the regular path must start from the lower left and end at the upper right, that is, from $w_1 = (1, 1)$ to $w_k = (m, n)$, because the period of the response spectrum has a sequence. Secondly, an element $w_{k-1} = (i', j')$ in the path must satisfy $I - i' \leq 1$ and $j - j' \leq 1$, so that each element in P and Q appears in the path without jumping, and ensure the continuity of the regular path. Finally, an element $w_{k-1} = (i', j')$ in the path and the next point $w_{k-1} = (i, j)$ must meet the requirements of $i - i' \geq 0$ and $j - j' \geq 0$, so that the points on the path are monotonous with the time axis, and ensure that the mapping lines do not cross.

2.2 DTW Distance Verification

In this paper, the original response spectrum was enlarged by 2 and 3 times respectively, and the DTW distance between the original response spectrum and the enlarged response spectrum was calculated. As shown in Fig. 2, the DTW distance between the original response spectrum and the enlarged 2 times response spectrum is 1.781, and the DTW distance which enlarged 3 times is 3.563. It can be seen that the greater the difference between the response spectra, the greater the DTW distance.

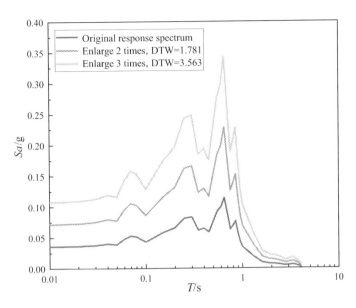

Fig. 2 DTW algorithm verification

Table 1 Number of ground motion records in different PGA ranges

PGA range/g	≤ 0.05	0.05–0.1	0.1–0.2	> 0.2	Total
Number	993	1045	302	78	2418

3 Data Sources

Based on the Japan KiK-net strong-motion seismograph network [9], this paper selected 2418 horizontal ground motion records, and collected the acceleration response spectra calculated by four numerical simulation methods [10], which named DEEPSOIL5.0 (DP), SHAKE2000 (SHAKE), SOILQUAKE16 (SQ), and SOILRE-SPONSE (SR). Table 1 shows the number of ground motion records in different PGA ranges.

4 Calculation Results

4.1 Comparison of DTW Distance in Different PGA Ranges

In this paper, 38 period points were selected uniformly in the 0–4 s spectral period according to the logarithmic distance, and calculated the DTW distance between the measured acceleration response spectrum and the calculated acceleration response spectrum.

The average DTW distance of each method in different PGA ranges is shown in Fig. 3, and the error line is 0.5 times of the standard deviation. It can be seen from Fig. 3 that the average DTW distance of the four methods has little difference for ranges with PGA less than 0.2 g, which the difference is about 0.1, the average DTW distance and standard deviation of SQ method are relatively large. When PGA is greater than 0.2 g, the average DTW distance and standard deviation of SR method are the smallest, and the average DTW distance of the other three methods are similar. It can be seen that the average DTW distance and standard deviation of the four methods increase with the increase of PGA ranges.

Figure 4 is the typical working condition comparison between the calculated acceleration response spectra of the four methods and the measured response spectrum. The measured PGA is 0.049 g in Fig. 4a, the calculated response spectra of the four methods are basically consistent, and the corresponding DTW distances are all about 0.5. The measured PGA is 0.088 g in Fig. 4b, the difference between the calculated response spectrum of DP method and the measured response spectrum is slightly greater than that of the other three methods, and the DTW distance of DP method is correspondingly greater than the other three methods. The measured PGA is 0.104 g in Fig. 4c, the calculated response spectrum of SR method is the closest to the measured response spectrum, the DTW distance is 1.477, which is the smallest, and the DTW distance of the other three methods is about 2. The measured PGA is 0.436 g in Fig. 4d, the difference between the calculated response spectrum of SR

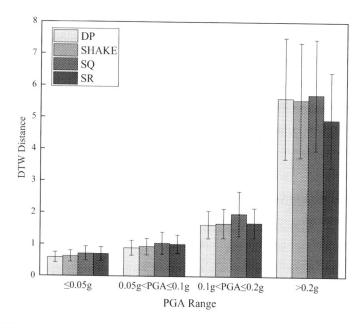

Fig. 3 DTW distance for different PGA ranges

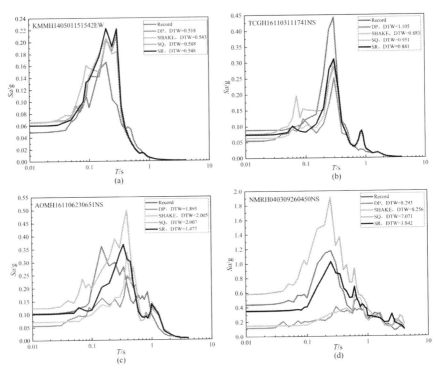

Fig. 4 Comparison between calculated and measured acceleration response spectra

method and the measured response spectrum is the smallest, and the DTW distance is 3.842, which is the smallest, and the DTW distance of the other three methods is greater than 7. It can be seen that the greater the difference between response spectra, the greater the DTW distance. DTW distance can truly reflect the difference more realistically.

4.2 General Trend Analysis

The change trend of DTW distance with PGA is shown in Fig. 5. It can be seen from Fig. 5 that there is an obvious linear relationship between DTW distance and PGA, and the fitting curve is shown by the red solid line. In terms of discreteness, DP and SHAKE methods have less discreteness, while SQ method has the largest discreteness. In terms of the general trend, DP and SHAKE methods have little difference, and the growth rate is relatively fast, and SQ method takes the second place. SR method has the smallest growth rate, and the DTW distance in strong ground motion is smaller than the other three methods, with the maximum value of 14.2.

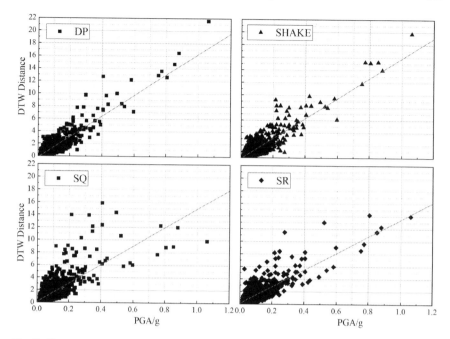

Fig. 5 Change trend of DTW distance with PGA

5 Calculation Results

(1) DTW distance is used to calculate the similarity between two response spectra by finding the optimal mapping, which can better reflect the difference between two response spectra, and can be used to quantitatively evaluate the difference of response spectra.

(2) When PGA is less than 0.2 g, the calculation errors of the four numerical simulation methods are similar. When PGA is greater than 0.2 g, the calculation errors of SR method are less than those of the other three methods.

(3) The calculation errors of the four methods increase with the increase of PGA, while the growth rate of SR method is relatively small. Especially in strong ground motions, the DTW distance of SR method does not show an obvious growth trend, and the calculation error is relatively small.

Acknowledgements This research was funded by the Scientific Research Fund of Institute of Engineering Mechanics, China Earthquake Administration (Grant No. 2020C04), and the Heilongjiang Provincial Natural Science Foundation of China (Grant No. LH2020E019).

References

1. Li XJ (1989) A computer program for calculating earthquake response of ground layered soil. Seismological Press, Beijing (in Chinese)
2. Schnabel PB, Lysmer J, Seed HB (1972) SHAKE: a computer program for earthquake response analysis of horizontal layer sites. University of California, Berkeley
3. Hashash YMA, Park D (2001) Non-linear one-dimensional seismic ground motion propagation in the mississippi embayment. Eng Geol 62:185–206
4. Shanker AP, Rajagopalan AN (2007) Off-line signature verification using DTW. Pattern Recogn Lett 28:1407–1414
5. Huang SF, Lu HP (2020) Classification of temporal data using dynamic time warping and compressed learning. Biomed Sig Proc Control 57:1–14
6. Joswig M, Theis HS (1993) Master-event correlations of weak local earthquakes by dynamic waveform matching. Geophys J Int 113(3):562–574
7. Li YM, Chen HG, Wu ZQ (2010) Dynamic time warping distance method for similarity test of multipoint ground motion field. Math Prob Eng
8. Sakoe H, Chiba S (1978) Dynamic programming algorithm optimization for spoken word recognition. IEEE Trans Acoust Speech Sig Proc 26(1):43–49
9. National Research Institute for Earth Science and Disaster Resilience of Japan (KiK-net) https:// www.kyoshin.bosai.go.jp/
10. Sun R, Yuan XM (2021) Holistic equivalent linearization approach for seismic response analysis of soil layers. Chin J Geotech Eng 43(4):603–612 (in Chinese)

Study on Adsorption Model and Influencing Factors of Heavy Metal Cu^{2+} Adsorbed by Magnetic Filler Biofilm

Zhaoxu Li, Xiaoping Zhu, Min Zhang, Wei Guo, Qian Wu, and Jianguo Wang

Abstract The use of biofilm to repair heavy metal pollution in rivers has become a research hotspot in various countries and has attracted more and more attention. The adsorption of heavy metal ions by biofilm depends on many physical and chemical factors. In this paper, the model and its influencing factors of the adsorption of heavy metals by biofilm attached onto the magnetic fillers containing 10% strontium ferrite were studied. The study found that pH is the most important factor that interferes with the adsorption of heavy metal Cu^{2+} in biofilm. When the pH is 6.2, the adsorption capacity of heavy metal Cu^{2+} reaches maximum value. In comparison, temperature has no significant effect on the adsorption of Cu^{2+} by biofilm. When the temperature increases from 5 to 30°C, the adsorption rate of Cu^{2+} increased by 11.2% accordingly. This study has important theoretical reference value for heavy metal Cu^{2+} in-situ repairation in rivers or lakes using magnetic filler biofilm method.

Keywords Heavy metal Cu^{2+} · Magnetic filler biofilm · Adsorption model · Influencing factors

1 Introduction

In recent years, with the rapid development of economy, the population has increased dramatically, and a large number of industrial wastewater and domestic sewage containing heavy metals have been discharged, resulting in serious pollution of heavy metals in rivers [1]. It is increasingly urgent to repair heavy metal pollution in rivers.

Z. Li · X. Zhu · M. Zhang · W. Guo · Q. Wu · J. Wang (✉)
Pearl River Water Resources Commission, Pearl River Water Resources Research Institute, Guangzhou 510610, China
e-mail: lzhxu@mail3.sysu.edu.cn

Guangdong Provincial Engineering Technology Research Center for Life and Health of River & Lake, Guangzhou 510610, China

J. Wang
Key Laboratory of the Pearl River Estuary Regulation and Protection of Ministry of Water Resources, Guangzhou 510610, China

S. Wang et al. (eds.), *Proceedings of the 2nd International Conference on Innovative Solutions in Hydropower Engineering and Civil Engineering*, Lecture Notes in Civil Engineering 235, https://doi.org/10.1007/978-981-99-1748-8_44

At present, in-situ remediation of polluted rivers by biofilm has become a hot research topic in many countries. While people pay attention to the removal effect of biofilm on ammonia nitrogen, nitrogen, phosphorus and COD in river and lake water, they also pay much more attention to the heavy metal adsorption by biofilm [2, 3]. It is of great practical significance to study the adsorption performance of biofilm for heavy metals in river and lake environment [4].

With the increasing attention paid to water environment treatment, magnetic induced microbial effect has been introduced into the field of water environment treatment, and its application has attracted the attention of more and more researchers at home and abroad [5–8]. The related research group of this paper found that the magnetic filler containing 10% strontium ferrite has strong adsorption efficiency for heavy metals [9]. In this paper, the newly developed magnetic filler was used to culture biofilm, and the adsorption model of heavy metals by magnetic filler biofilm and its influencing factors were studied, which provided basic theory and data support for the study of in-situ remediation of heavy metal Cu^{2+} polluted rivers or lakes with the magnetic fillers.

2 Materials and Methods

2.1 Cultivation System Installation and Operation

In this study, a biofilm culture device was made to carry out the experiment for testing the biofilm adsorption capacity of heavy metal Cu^{2+} (Fig. 1a). For the culture device, its main component is the glass container, which size was $1200 \times 400 \times 650$ mm (L $\times$ W $\times$ H). The glass container is divided into 5 water flow channels by 4 glass baffles (Fig. 1b). There were 5 water flow channels are sequentially connected by four flow holes, which are vertically arranged on the staggered edges of the glass diaphragm to facilitate the water flow. The water pump was fixed to pump the culture liquid from the plastic tank into the culture device through the water inlet. The nutrient solution in the glass container overflows back to the plastic tank from the water outlet. In order to maintain a certain concentration of DO in the culture medium, an air pump was designed to pump air directly into each water channel.

When the biofilm was cultivated with the biofilm culture device, the surface sediment of the river course was collected from the river and evenly spread on the bottom of the glass container, with a thickness of 5 cm. The surface sediment of the river contained abundant microorganisms, which were used to inoculate microorganisms when cultivating biofilm.

The biofilm culture solution contained a variety of nutrients (all in mg/L), mainly including: 4.60 of $Na_3PO_4 \cdot 12H_2O$; 1.89 of $NaH_2PO_4 \cdot 2H_2O$; 17.7 of $(NH_4)_2SO_4$; 2.05 of $MgSO_4 \cdot 7H_2O$; 2.54 of NaCl; 0.74 of $CaCl_2 \cdot 2H_2O$; and 1.91 of KCl. The biofilm culture solution was continuously pumped into the glass container from the plastic tank, and then flowed back to the plastic tank through the glass container outlet

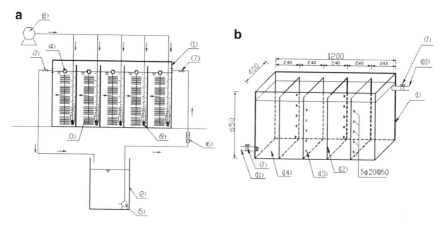

Fig. 1 Schematic diagram of biofilm culture device: **a** Biofilm culture device. **b** Elevation of glass container for cultivating biofilm; (1) glass container; (2) plastic tank; (3)three-dimensional fillers; (4) cylindrical buoy; (5) water pump; (6)flow meter; (7) water switch; (8) air pump; (9) cylindrical aeration head; (10) water flow inlet; (11) water flow outlet; (12) glass baffles; (13) water flow hole; (14) flow channel

again by overflow. In order to accelerate the growth of biofilm, glucose was added to the biofilm culture medium to maintain the concentration of chemical oxygen demand (COD) at about 120 mg/L. The flow rate of biofilm nutrient solution in the reactor was set as approximate 0.3 $m^3/(m^2 \ s)$.

The filler used for cultivating biofilm was three-dimensional magnetic elastic filler. The main components of the magnetic filler were shown in Table 1. Each magnetic filler had a length of 10 cm and a diameter of 10 cm. Five pre weighed magnetic fillers were strung together and assembled into a bunch. These bunches of magnetic filler were strung onto PVC pontoons, and then they were put into the water channels of the biofilm culture device, respectively.

During the biofilm cultured, the DO concentration in the culture solution was controlled at about 4.0 mg/L. With the aid of a time control switch, the biofilm culture solution was aerated for about 10 min every half an hour. The pH of the biofilm culture solution is controlled at 6–8 with the aid of 1 mol/L hydrochloric acid and 1 mol/L sodium hydroxide. The SC100e instrument (American Hach Company) was used to monitor the DO and pH parameters simultaneously. In order to ensure the normal growth of microorganisms in the biofilms, 40% of the culture medium in the plastic tank was replaced every two days to maintain the nutrient contents at a certain level.

Table 1 Percentages of filler components (%)

Polyethylene terephthalate (PET)	Strontium ferrite powder	Dispersant	Others (such as surface active agent)
83	10	2	5

By domesticating and culturing, the microorganisms in the culture solution, the biofilms adhered onto the magnetic fillers in the culture device and grew cumulatively. After continuous culturing for 60 days, the biofilms adhered onto the fillers accumulated to a certain amount, when the adsorption and influencing factors experiments of heavy metal Cu^{2+} could be carried out.

2.2   Sampling Method of the Biofilm

On the 60th day of biofilm culture, the magnetic fillers adhered with biofilms were carefully taken out from the culture device, placed them in a 2000 mL beaker, add 1000 mL phosphate buffer (PBS, pH 7.2), which contained 0.092 g L^{-1} KH_2PO_4, 0.036 g L^{-1} K_2HPO_4 and 0.493 g L^{-1} NaCl. Then, the magnetic fillers were stirred vigorously with a glass rod to remove the biofilm from the magnetic fillers. The obtained biofilm suspension was transferred into a 2000 mL volumetric flask. 200 ml PBS was added into the beaker, and the elution operation was repeated again, and the operation was repeated three times, then all the obtained biofilm suspension was transferred into the volumetric flask. Finally, the PBS was fixed into the volumetric flask to 2000 ml for the following experiment.

The total biomass of biofilm in unit volume suspension was determined by measuring the dry weight (DW) of biofilm. The specific operation method was as the following. Firstly put 50 ml uniform biofilm suspension into a crucible [10]. After weighed, the crucible was placed into the electric blast drying oven at 105 °C until it reaches constant weight, and then placed the crucible in a desiccator to cool to room temperature and weighed it again. The total biomass of biofilm in the unit volume suspension was determined by using the mass difference of the two weight. Made three parallel determinations, and took the average of the three mass differences as the final determination result.

2.3   Determination of Adsorption Isotherm

Adsorption Model of Heavy Metal Ions by Biofilm. Heavy metal adsorption by biofilm is a complex process, mainly including transmembrane diffusion and transport. For the single component adsorption system of biofilm, there are two classical equilibrium adsorption models. For multicomponent adsorption, an extended adsorption model had been developed based on the single component adsorption system model [11].

Langmuir adsorption mode:

$$q_e = \frac{QbC_e}{1 + bC_e} \tag{1}$$

Frenndlich adsorption mode:

$$q_e = K_F C_e^{\frac{1}{n}} \tag{2}$$

In the formula:

q_e—the amount of heavy metals adsorbed by the adsorbent per unit mass when the adsorption equilibrium is reached, mg/g; Q—the amount of metal adsorbed by the adsorbent per unit mass, mg/g; b—the adsorption constant, L/g;

C_e—the mass concentration of heavy metals in solution at adsorption equilibrium, mg/L; K_F, $\frac{1}{n}$—Frenndlich constant.

Determination of Adsorption Isotherm. In the first step, took six 150 ml conical flasks and added 150 ml of Cu^{2+} solutions with the mass concentration (C0) of 10, 20, 30, 40, 50 and 60 mg/L respectively. In the second step, adjusted the pH of the solution to 5.0, and added 10 g biofilm (wet weight) to the conical flask. Put the conical flask into the constant temperature freezer and shaked it for 1 h. The speed of the freezer is 120 r/s. And then, the suspension in every conical flask was filtered using 0.45 μm microporous membrane and fixed the volume to 250 ml. At last, the mass concentrations of Cu element in the treated samples were determined by ICP-AES.

2.4 Experiment on the Influencing Factors

The pH Influencing Factor. Six 250 ml conical flasks numbered from P1 to P6 were taken and added 150 ml of Cu^{2+} solution with a mass concentration of 40 mg/L, then 20 g biofilm (wet weight) were added to every conical flask, and adjusted the pH of the solution to 2.0, 3.0, 4.0, 5.0, 6.0. 7.0 and 8.0, respectively. Then, put the conical flask into the constant temperature freezer at 20°C and shaked them for 1 h. The rotating speed of the freezer is 120 r/s. And then, the suspension in every conical flask was filtered using 0.45 μm microporous membrane and fixed the volume to 250 ml. At last, the mass concentrations of Cu element in the treated samples were determined by ICP-AES.

The Temperature Influencing Factor. Similar to above, six 250 ml conical flasks numbered from W1 to W6 were taken and added 150 ml of Cu^{2+} solution with a mass concentration of 40 mg/L, then 20 g biofilm (wet weight) were added to every conical flask, and adjusted the pH of the solution to 5.0. Secondly, put the conical flask into the constant temperature freezer at a series of temperatures (5, 10, 15, 20, 25, 30 °C) and shaked them for 1 h, respectively. The rotation speed and subsequent filtering operation and detection were the same as those in subsection.

Table 2 Working parameters of the ICP-AES instrument

ICP-AES working parameters	Setting value
RF power (KW)	1.3
Flow rate of cooling gas (L/min)	15.0
Flow rate of spray gas (L/min)	0.8
Flow rate of auxiliary gas (L/min)	0.2
Solution uptake rate (mL/min)	15.0
Analytical wavelength of Cu element (nm)	327.393

2.5 Determination of Heavy Metal Cu^{2+}

The concentration of heavy metal Cu^{2+} in the above samples were determined by ICP-AES method. The main working parameters of ICP-AES instrument were as follows: the radio frequency (RF) power is 1.3 kW; the cooling gas flow rate, the spray gas flow rate and the auxiliary gas flow rate were 15.0, 0.8 and 0.2 L/min, respectively; the observation mode was radial or axial, and the solution uptake rate was 15.0 m L/min; the analytical wavelength of heavy metal Cu determined was 327.393 nm (Table 2). The concentration of heavy metal Cu^{2+} in the samples were determined three times in parallel, and the average values were taken. The amounts of heavy metal Cu^{2+} adsorbed were calculated according to the concentrations of Cu^{2+} before and after the adsorption and the corresponding volumes. The adsorption amounts of heavy metal Cu^{2+} in the experiment were calculated according to the concentration values of Cu^{2+} before and after the adsorption and their corresponding volumes.

3 Results and Discussion

3.1 Determination of Adsorption Model of Heavy Metal Cu^{2+}

The adsorption model of heavy metal Cu^{2+} adsorbed by magnetic filler biofilm was studied as an example showed in Fig. 2.

The adsorption isotherm of Cu^{2+} adsorbed by magnetic filler biofilm was performanced as the Fig. 2a. It can be seen from Fig. 2b that the saturation adsorption capacity of the magnetic filler biofilm containing 10% strontium ferrite component for Cu^{2+} is about 0.37 mg/g. The experimental results of the saturated adsorption capacity of the magnetic filler biofilm containing 10% strontium ferrite on heavy metal Cu^{2+} were processed to perform linear fitting of Langmuir model and Freundlich model, and the equilibrium model of the biofilm adsorption of heavy metal Cu^{2+} was discussed, as was shown in Fig. 2b and c. The results showed that the adsorption of heavy metal Cu^{2+} by biofilm is in better agreement with Langmuir model than Freundlich model. The correlation coefficient R^2 of Langmuir model is

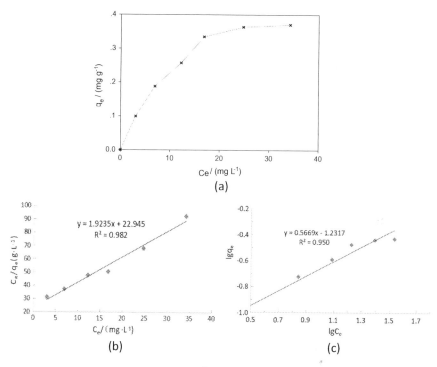

Fig. 2 Adsorption model of heavy metal Cu^{2+} adsorbed by magnetic filler biofilm: **a** Adsorption isotherm; **b** Linear fitting of Langmuir model; c Linear fitting of Freundlich model

0.982, and the correlation coefficient ruler of Freundlich model is 0.950. Therefore, the adsorption of heavy metal Cu^{2+} can be fitted by Langmuir model.

3.2 Study on Influence FACTORS of Heavy Metal Adsorption by Magnetic Filler Biofilm

The pH Influencing Factor. As shown in Fig. 3, the adsorption capacity of the magnetic filler biofilms for heavy metal Cu^{2+} showed a parabolic relationship with the pH value. There is an optimal pH range for adsorbing heavy metal Cu^{2+}, and the optimal adsorption pH range is 5–7. When the pH value reached 6.2, the adsorption capacity was the largest. It can be inferred that the adsorption capacity of the magnetic filler biofilm for heavy metal Cu^{2+} decreases when it deviates from the most appropriate pH range. As is the reason that when the acidity increases, the degree of amino protonation in biofilm proteins, which exist in biofilm microorganisms or extracellular polymers, etc., increases, and its coordination ability with metals weakens. Moreover, a large amount of H^+ and H_3O^+ in the system will compete for adsorption sites with heavy metal Cu^{2+}, which results in a decrease in the adsorption

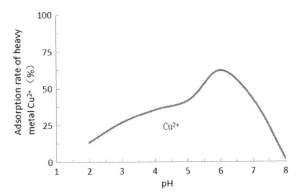

Fig. 3 Effect of pH value on Cu^{2+} adsorption by magnetic filler biofilm

ability of magnetic filler biofilm for heavy metal Cu^{2+}. When the pH value is greater than the optimal range, heavy metal Cu^{2+} will hydrolyze to generate corresponding hydroxides and deposit on the surface of the biofilm, which will affect its biosorption active sites to a certain extent, and also lead to the decline of the ability for the magnetic filler biofilm to adsorb heavy metal Cu^{2+}.

The pH value can not only have a significant impact on the active sites of heavy metals bound by biofilms, but also have a significant impact on the chemical reactions of heavy metal solutions (such as inorganic coordination, organic complexation, redox, hydrolysis, precipitation, etc.) [12]. The adsorption capacity of biofilm for heavy metal Cu^{2+} increases with the increase of pH value, but there is no linear correlation between the two. At the same time, the heavy metal Cu^{2+} will slightly precipitate when the pH value is 5.5. Thus it can be seen, too high pH value will not be conducive to the heavy metal Cu^{2+} biological adsorption [13].

The Temperature Influencing Factor. It can be seen from Fig. 4 that temperature has a certain effect on Cu^{2+} adsorption by magnetic filler biofilms, but it is not very significant. At 5°C, the Cu^{2+} adsorption rate was 46.3%, and at 30 °C, it was 57.5%. When the temperature increased by 25 °C, the Cu^{2+} adsorption rate only increased by 11.2%. Compared with pH value, the effect of temperature on biosorption is relatively small. Generally speaking, biosorption is an exothermic reaction process. Therefore, the adsorption capacity of microorganisms to heavy metal Cu^{2+} increases with the decrease of temperature. However, sometimes biosorption is also an endothermic reaction process. Han et al. [14] found that S Cerevisiae's adsorption capacity for Cu increases with the increase of temperature (at 293 k, the adsorption capacity for Cu is 0.00809 mmol g^{-1}DW., and when the temperature rises to 323 K, the adsorption capacity continuously increases to 0.0206 mmol g^{-1}DW.), which is an endothermic reaction. However, too high a temperature will not only destroy the active site of the biosorbent, resulting in a decrease in the amount of heavy metal Cu^{2+} adsorbed, but also increase the operating cost [15, 16].

The adsorption of heavy metal Cu^{2+} by magnetic filler biofilm depends on many physical and chemical factors. In addition to pH value and the temperature, other factors such as the adsorption time, the amount of magnetic filler biofilm, microbial

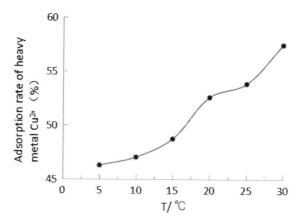

Fig. 4 Effect of temperature on Cu^{2+} adsorption by magnetic filler biofilm

activity in the biofilm, the initial concentration of heavy metal Cu^{2+}, and other coexisting heavy metal ions will affect the adsorption of heavy metal Cu^{2+} by magnetic filler biofilm. For example, the adsorption capacity of cells with biosorbents for heavy metal ions at the early and late growth stages is stronger than that at the stable growth stage (plateau stage) [17]. Some studies have shown that the magnetotactic bacteria preferentially adsorb Zn^{2+} in the binary competitive adsorption system for the coexistence system of Cu^{2+} and Zn^{2+} (the concentration of both ions is the same), and the adsorption rate is reduced by 36.76 and 24.5% compared with the single system of them, respectively [18].

4 Conclusions

In this paper, the adsorption model and influencing factors of heavy metal Cu^{2+} by magnetic filler biofilm were studied. According to the research, the adsorption of heavy metal Cu^{2+} by magnetic filler biofilm can better fit Langmuir model linearly than Freundlich model. The adsorption of heavy metal Cu^{2+} by magnetic filler biofilm is affected by many factors. Among the many factors affecting the adsorption of heavy metal Cu^{2+}, pH is the most important factor affecting the adsorption of heavy metal Cu^{2+} by biofilm. The temperature have certain effects on the adsorption of heavy metal Cu^{2+} by biofilm. In addition, the microbial activity in the biofilm and the initial concentration of heavy metal Cu^{2+} in the solution will also affect the biosorption of heavy metal Cu^{2+}. This study on heavy metal Cu^{2+} absorbed in magnetic filler biofilm has a very broad application prospect in water environment treatment, which will play a far-reaching influence and role in the treatment and improvement of aquatic ecological environment quality in China.

Acknowledgements This work was supported by the Special Foundation for National Science and Technology Basic Research Program of China (2019FY101900), the National Natural Science

Foundation of China (Grant No. 5170929) and the Guangdong Foundation for Program of Science and Technology Research (2020B1111530001).

References

1. Zheng PM, Xu KF, Dong XB (2022) Analysis on stabilization and solidification technology of heavy metal polluted river sediment. Leather Mak Environ Protect Technol 3(10):189–191
2. Liao HY, Sun L, Li JB, He X (2021) Research progress of bacteria algae symbiotic biofilm wastewater treatment. J Civ Environ Eng (Chin Eng) 43(04):141–153
3. Miao J, Mao MJ, Li J, Shen Z, Zhang YL (2022) Research progress of microbial immobilization technology in sewage treatment. J Fujian Normal Univer (Nat Sci Ed) 38(01):117–124
4. Li ZX, Yuan LH, Lei HY, Wang YJ, Tang HL (2011) Study on adsorption and migration regularity of heavy metals by biofilm in Tidal River. J Anhui Agric Sci 39(34):21217–21220
5. Lei N (2012) Preparation of magnetic bio-carriers and their characteristics on the Cd^{2+}/phenol complex wastewater treatment. Xiangtan University
6. Yang F, Yao C, Luo H, Li ZX, Xu LQ (2014) Magnetically induced microbial effect and its application in wastewater treatment. Pearl River 35(05):76–79
7. Wang DW (2016) Preparation of new magnetic carrier filler and its application in domestic sewage treatment. Kunming University of Science and Technology
8. Liao LY (2017) Preparation and properties of magnetic material/waterborne polyurethane composite. South China University of Technology
9. Li ZX, Yang F, Jiang R, Xu LQ (2016) Study on adsorption capacities of heavy metals in biofilms grown on magnetic carriers. In: The 16th international conference on environment pollution and human health (EPHH), 2016/12/10–2016/10/12. Wuhan, pp 256–263
10. Figueira MM, Volesky B, Ciminelli VST (1997) Assessment of interference in biosorption of a heavy metal. Biotechnol Bioeng 54(2):344–350
11. Patricia A, Terry WS (2002) Biosorption of cadmium and copper contaminated water by scenedesmus abundans. Chemosphere 47(3):249–255
12. Volesky B (2001) Detoxification of metal- bearing effluents: biosorption for the next century. Hydrometallurgy 59:203–216
13. Kratochvil D, Volesky B (1998) Advances in the biosorption of heavy metals. Trends Biotechnol 16:291–300
14. Han RP, Li HK, Li YH et al (2006) Biosorption of copper and lead ions by waste beer yeast. J Hazard Mater B 137:1569–1576
15. Kapoor A, Viraraghavan T (1995) Fungi biosorption- an alternative treatment option for heavy metal bearing wastewaters: a review. Biores Technol 53:195–206
16. Özer A, Özer D (2003) Comparative study of the biosorption of Pb (II), Ni (II) and Cr (VI) ions onto S. cerevisiae: determination of biosorption heats. J Hazard Mater B 100:219–229
17. Yang Q (2018) Research Progress on treatment of heavy metal ions in water by biosorption. Yunnan Chem Ind 45(07):76–78 + 81
18. Liu J, Zhou PG, Zhang QS (2013) Treating wastewater including Cu^{2+}, Zn^{2+} with magnetotactic bacteria. Environ Sci Technol 26(04):20–24

Reliability Evaluation Methods of Accelerated Degradation Test for Fiber-Optic Gyroscope Under Temperature Environment

Renqing Li, Jin Li, Yan Song, and Kun Wang

Abstract Fiber Optic Gyroscope (FOG) is a kind of rotation sensor with high reliability and long lifetime. Accelerated life test is hard to obtain enough failure data for reliability evaluation. By analyzing failure mechanism, the sensitive environment stress and performance degradation parameters of FOG are determined. Then an accelerated degradation test (ADT) is undertaken to obtain reliability information. Wiener process is introduced to describe performance degradation path and the Arrhenius model is selected as accelerated model. Considering epistemic uncertainty in monitoring, an interval linear regression model for ADT is introduced. Then the reliability of FOG at operating environment stress is evaluated. By discussing the reliability assessment results, the methods in this paper are feasible for FOG.

Keywords Fiber optic gyroscope · ADT · Wiener process · Reliability · Interval regression model

1 Introduction

Fiber Optic Gyroscope (FOG) is a kind of rotation sensor, which has the merits of high reliability, long work lifetime, large dynamic range, simple architecture, light weight, and so on [1]. As an important navigation equipment, FOG has been widely used in aircraft, ships and satellite. It's necessary to evaluate the reliability of FOG accurately to manage products scientifically and effectively.

With high reliability, accelerated test is usually introduced to obtain more reliability information of FOG. Ma et al. [2] summarized the accelerated test technology

R. Li (✉) · J. Li · Y. Song
CEPREI Laboratory, Guangzhou, China
e-mail: lrqhnxn@126.com

J. Li
Guangdong Provincial Key Laboratory of Electronic Information Products Reliability
Technology, Guangzhou, China

K. Wang
China Ship Development and Design Center, Wuhan, China

© The Author(s) 2023
S. Wang et al. (eds.), *Proceedings of the 2nd International Conference on Innovative Solutions in Hydropower Engineering and Civil Engineering*, Lecture Notes in Civil Engineering 235, https://doi.org/10.1007/978-981-99-1748-8_45

of mechanical gyroscopes and optic gyroscopes. And the conclusion is that ALTs and ADTs of critical component are valid way to evaluate the overall life of the gyroscopes. Chao et al. [1] developed research on the reliability of superfluorescent light diode (SLD) through accelerated life test. By analyzing the failure mechanism and the configuration of SLD, an Arrhenius life-stress relationship was obtained. By utilizing Weibull distribution as its life distribution function, reliability assessment was achieved. Zhang et al. [3] developed research on accelerated life test of the key component of FOG and assessed the reliability.

As FOG is high reliable, it's hard to obtain failure data by ALT. ADT is also utilized to evaluate reliability of FOG. Chen et al. [4] studied feasibility of accelerated degradation test for FOG. By analyzing the performance data under temperature-accelerated stress, the accelerated ability of performance degradation was demonstrated. In paper [5], step-stress accelerated degradation test with two accelerating stresses was proposed to evaluate FOG's reliability and lifetime. And a degradation path model including residual standardized coefficients was presented to describe degradation path.

The degradation data is usually processed as precise data. But during tests, there exists measurement tolerance and different cognition from testers, which results in epistemic uncertainty of performance degradation data [6]. For imprecise data, Wang et al. [7] developed reliability analysis on competitive failure processes under fuzzy degradation data. David et al. [8] researched a non-linear fuzzy regression for estimating reliability in a degradation process. Liu et al. developed an evaluation method for accelerated degradation testing with interval analysis based on Wiener process [6].

The main purpose of this paper is to investigate reliability evaluation methods of ADT for fiber-optic gyroscope under temperature environment. The sensitive stress and performance degradation parameters are obtained through failure mechanism analysis. An ADT is undertaken to obtain performance degradation data of FOG. Then a reliability evaluation method based on Wiener process and interval regression model is introduced to assess the reliability of FOG. The remaining parts are organized as follows. Section 2 is devoted to describe the ADT undertaken for FOG and the performance degradation data. In Sect. 3, the reliability of FOG is evaluated based on Wiener process. In Sect. 4, an assessment method based on Wiener process and interval regression analysis is utilized to evaluate the reliability. Conclusions are drawn in Sect. 5.

2 An ADT Testing for Fiber Optic Gyroscope

2.1 Failure Mechanism and Sensitive Stress

Fiber Optic Gyroscope is a kind of rotation sensor, which has the merits of long work lifetime, large dynamic range, small size, and so on. FOGs usually contain optical components, electronic components and architectural components. As key components, optical components consist of superfluorescent light diode (SLD), photodetector, phase modulators and optic coupler. Compared with well-developed electrical components, optical components are less reliable [2]. Ma et al. [9] and Zhang et al. [3] conducted failure mode, effect and crisis analysis to FOGs. From the crisis matrix, the drop of light power of the SLD is the most critical problem, which means SLD is the most fragile part of FOGs. SLDs usually contain lighting tube, semiconductor, heat radiator structure, thermal resistor, fiber pigtail and so on [1]. Research on the failure mechanism of SLD is conducted to obtain cause of failure and sensitive stress. From literature [3], the failure of lighting tube itself and the coupling failure between fiber pigtail and lighting tube is the frequent failure modes of SLD. And abrupt failure, quick failure and degradation failure are three primary failure modes of lighting tube. The degradation failure is prevalent. Abrupt and quick failure modes usually result from technologic defects, which can be carried out by stress screening test. But degradation failure is nonreversible and always occurs with the accumulation of working stress. The performance degradation is affected by environment temperature and temperature is the sensitive stress of FOG [10]. So, temperature is selected as accelerated stress for ADT in this paper. The main performance parameters of FOG are zero bias, zero-bias stability and scale factor.

2.2 An Undertaken ADT

As FOGs have long life and high reliability, an ADT is undertaken to obtain reliability information of FOGs efficiently. In this paper, test samples and test stresses are shown in Table 1.

The performance parameters, monitoring period and corresponding failure thresholds are selected in Table 2.

Table 1 Test samples and test stresses

Num	Test samples No.	Number of test samples	Accelerated test stresses ($^\circ$C)
1	001#, 002#	2	$T_1 = 75$
2	003#, 004#	2	$T_2 = 80$

Table 2 The performance parameters

Num	Performance parameter	Failure threshold	Monitoring period (h)
1	Zero bias	$-5 \sim 5^\circ$/h	96
2	Zero-bias stability	≤ 0.2	96
3	Scale factor	≤ 600 ppm	96

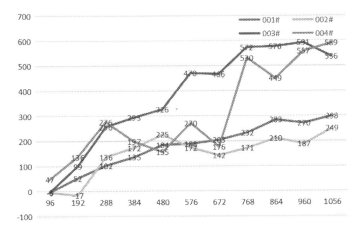

Fig. 1 The processed degradation data of scale factor parameter

An ADT with censoring time $t = 1056$ h is conducted to obtain performance degradation data. From testing results, only scale factor has an obvious degradation trend and the processed data are shown in Fig. 1.

3   Reliability Evaluation Method for ADT Based on Wiener Process

With good statistical prosperity, stochastic modeling and analysis is usually introduced for degradation modeling for highly reliable products [11]. And Wiener process is widely used in degradation modeling [12–15]. The performance parameter $X(t)$ is assumed to follow a general Wiener process as following [16].

$$X(t) = \mu \Lambda(t) + \sigma B(\Lambda(t)) \tag{1}$$

If $X(0) \neq 0$, $X(t) = X(t) - X(0)$. Here, σ is diffusion coefficient. $B(.)$ is standard brown motion. $\Lambda(t)$ is a monotone increasing function and $\Lambda(t = 0) = 0$. With a goodness-of-fit, let $\Lambda(t) = t$. And μ is drift coefficient. For Weiner process, increments of degradation $\triangle X_i$ is expressed as follow.

$$\Delta X_i \sim N(\mu \Delta t_i, \sigma^2 \Delta t_i) \tag{2}$$

Here, $\triangle X_i = X(t_i) - X(t_{i-1})$ and $\triangle t_i = \Lambda(t_i) - \Lambda(t_{i-1})$.

As temperature is the accelerated stress, Arrhenius is selected as accelerated model shown as [17, 18].

$$\frac{\partial M}{\partial t} = A_0 \exp\left(-\frac{Ea}{kT}\right) \tag{3}$$

Here, $\partial M / \partial t$ is degradation rate. Ea is activation energy. T indicates temperature. And k is Boltzmann constant. A_0 is constant.

The relationship of drift coefficient μ and T can be expressed as following [2].

$$\ln \mu = \alpha + \beta / T$$

Here, $\alpha = \ln(A_0)$ and $\beta = -Ea/k$.

When $X(t)$ crosses a pre-specified failure threshold C, the product is failure and the first-passage-time T_C satisfies Inverse Gaussian distribution as following.

$$\Lambda(Tc) \sim IG(C/\mu, C^2/\sigma^2) \tag{4}$$

And the reliability function $R(t)$ is expressed as following.

$$R(t|C, \sigma, \mu) = \Phi\left(\frac{C - \mu\Lambda(t)}{\sigma\sqrt{\Lambda(t)}}\right) - \exp\left(\frac{2\mu C}{\sigma^2}\right)\Phi\left(-\frac{C + \mu\Lambda(t)}{\sigma\sqrt{\Lambda(t)}}\right) \tag{5}$$

The log-likelihood function of Eq. (2) can be maximized to obtain the maximum likelihood estimators of parameters in Eq. (5). When the operating stress is 25 °C, the $R(t)$ curve is shown in Fig. 2.

From Fig. 2, $R(t) = 0.80662$ when $t = 8000$ h.

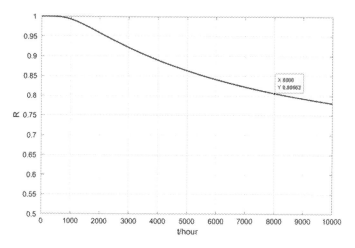

Fig. 2 The $R(t)$ curve based on Wiener process

4 Reliability Evaluation Method for ADT Based on Wiener Process and Interval Regression Analysis

There is epistemic uncertainty in ADT modeling [3]. In order to describe the epistemic uncertainty in ADT, interval analysis is introduced to evaluate reliability of ADT. An interval linear regression model can be written as following [4].

$$F(Y) = A_0 + A_1 y_1 + \cdots + A_q y_q = AY \tag{6}$$

Here, $y = (1, y_1 \ldots, y_q)$ is an input vector. $F(Y)$ is the corresponding estimated interval. $A = (A_0 \ldots, A_q)$ is an interval coefficient vector and $A_i = (a_i, c_i)$, where a_i is a center and c_i is a radius. By interval arithmetic, $F(Y_i)$ is expressed as.

$$F(Y_i) = \langle a_0, c_0 \rangle + \langle a_1, c_1 \rangle y_{i1} + \cdots + \langle a_q, c_q \rangle y_{iq}$$
$$= \langle a^{\mathrm{T}} y_i, c^{\mathrm{T}} |y_i| \rangle$$

Here, $a = [a_0, a_1 \ldots, a_q]^{\mathrm{T}}$, $c = [c_0, c_1, \ldots, c_q]^{\mathrm{T}}$ and $y_i = [1, y_{i1}, \ldots, y_{iq}]^{\mathrm{T}}$.

By interval arithmetic, the possibility and necessity estimation models are expressed as [4].

$$F^*(Y) = A_0^* + A_1^* y_1 + \cdots + A_q^* y_q = A^{*T} Y \tag{7}$$

$$F_*(Y) = A_{*0} + A_{*1} y_1 + \cdots + A_{*q} y_q = A_*^T Y \tag{8}$$

It's assumed that $A_i^* = \langle a_i, d_i \rangle$ and $A_{*i} = \langle a_i, c_i \rangle$. Quadratic programming approach is introduced to solve Eqs. (7) and (8) as following.

$$\min_{a,d} d^T \left(\sum_{i=1}^n |y_i| |y_i|^T \right) d + \xi a^T a$$
$$a^T y_i + d^T |y_i| \geq \overline{F}(Y_i)$$
$$subject\ a^T y_i - d^T |y_i| \leq \underline{F}(Y_i)$$
$$to\quad d_j \geq 0 \quad j = 0, 1, \ldots, q$$
$$i = 1, 2, \ldots, n \tag{9}$$

$$\min_{a,c} \; c^T \left(\sum_{i=1}^{n} |y_i| |y_i|^T \right) c + \xi c^T c$$

$$a^T y_i + c^T |y_i| \le \overline{F}(Y_i)$$

$$subject \quad a^T y_i - c^T |y_i| \ge \underline{F}(Y_i)$$

$$to \quad c_j \ge 0 \quad j = 0, 1, \ldots, q$$

$$i = 1, 2, \ldots, n \tag{10}$$

The monitored degradation interval vector $F(Y)$ satisfies the following constrains.

$$F_*(Y) \subseteq F(Y) \subseteq F^*(Y) \tag{11}$$

According to Eqs. (9–11) can be expressed as following quadratic programming problem.

$$\min_{a,d} \; d^T \left(\sum_{i=1}^{n} |y_i| |y_i|^T \right) d + \xi(a^T a + c^T c)$$

$$a^T y_i + c^T |y_i| + d^T |y_i| \ge \overline{F}(Y_i)$$

$$a^T y_i - c^T |y_i| + d^T |y_i| \le \underline{F}(Y_i)$$

$$subject \quad a^T y_i + c^T |y_i| \le \overline{F}(Y_i)$$

$$to \quad a^T y_i - c^T |y_i| \le \underline{F}(Y_i)$$

$$c_j \ge 0 \; d_j \ge 0 \; j = 0, 1, \ldots, q$$

$$i = 1, 2, \ldots, n \tag{12}$$

For Wiener process satisfied Eq. (1), the expectation $E(X(t)) = \mu \Lambda(t)$. For a data set with crisp inputs $\Lambda(t)$ and interval outputs $[\underline{X}, \overline{X}]$, μ can be abstained.

For the processed data are shown in Fig. 1, let $\Delta x \sim N(2, 0.04)$ and let $X \pm |\Delta x|$, where the higher degree of epistemic uncertainty with large Δx. Then interval degradation data $[x_{ijk} \overline{x_{ijk}}]$ is obtained. Here, i indicates the number of accelerated stress levels and $i = 1, 2$. The j is the number of testing samples in ith stress level and $j = 1, 2$. The k indicates the monitoring time and $k = 1, 2 \ldots, 11$. For interval degradation data $(t_{ijk}, [\underline{x_{ijk}}, \overline{x_{ijk}}])$, model (9) is selected to obtain $[\underline{\mu_{ij}}, \overline{\mu_{ij}}]$. Then $[\underline{\mu_i}, \overline{\mu_i}]$ is expressed as.

$$\begin{cases} \underline{\mu}_i = \min \left\{ \overset{2}{\underset{j=1}{U}} [\underline{\mu}_{ij}, \overline{u}_{ij}] \right\} \\ \overline{u}_i = \max \left\{ \overset{2}{\underset{j=1}{U}} [\underline{\mu}_{ij}, \overline{u}_{ij}] \right\} \end{cases}$$

The diffusion coefficient σ is expressed as following.

$$\sigma^2 = \frac{\sum_{i=1}^{2} \sum_{j=1}^{2} \sum_{k=2}^{11} \left(\frac{\sqrt{(\overline{x}_{ijk}^* - \overline{x}_{ijk})^2 + (x_{ijk}^* - x_{ijk})} - }{\sqrt{(\overline{x}_{ij(k-1)}^* - \overline{x}_{ij(k-1)})^2 + (x_{ij(k-1)}^* - x_{ij(k-1)})}} \right)^2}{\sum_{i=1}^{2} \sum_{j=1}^{2} \sum_{k=2}^{11} \Delta t_{ijk}}$$

For $(T_i, \left[\underline{\mu_i}, \overline{\mu_i}\right])$, Eq. (6) is introduced to develop interval regression analysis. By solving model (12), a, c and d can be obtained. With known operating stress, $[\underline{\mu_0}, \overline{\mu_0}]$ is obtained. And $R(t)$ is expressed as.

$$\left[\underline{R}, \overline{R}\right]\left(t | C, \sigma, \mu_0 \in \left[\underline{\mu_0}, \overline{\mu_0}\right]\right) = \left[R(t | C, \sigma, \overline{\mu_0}), R\left(t | C, \sigma, \underline{\mu_0}\right)\right]$$

When $T_0 = 25°C$, the reliability $R(t)$ is shown is Fig. 3.

From Fig. 3, $\left[\underline{R}, \overline{R}\right] = [0.80137, 0.83118]$ when $t = 8000$ h.

From Figs. 2 and 3, the epistemic uncertainty can be considered in the reliability evaluation model based on Wiener and interval regression analysis. When Δx varied, $\left[\underline{R}, \overline{R}\right] = [0.80747, 0.83601]$ when $t = 8000$ h with $\Delta x \sim N(1, 0.02)$ and $\left[\underline{R}, \overline{R}\right] = [0.79545, 0.82648]$ when $t = 8000$ h with $\Delta x \sim N(3, 0.06)$. When mean value of Δx increases, the value of $\Delta R = \overline{R} - \underline{R}$ increases, which means it's necessary to reduce epistemic uncertainty.

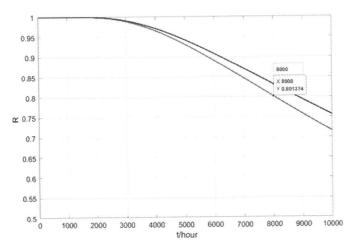

Fig. 3 The $R(t)$ curve based on Wiener process and interval regression analysis

5   Conclusions

In this paper, failure mechanism analysis is introduced to obtain sensitive environmental stress and performance degradation parameter of FOG. An ADT is undertaken to obtain performance degradation data. Then a reliability evaluation method based on Wiener process is introduced. Considering epistemic uncertainty, a reliability assessment method based on Winer process and interval regression model is introduced. The reliability assessment results illustrate the feasibility of ADT for FOG.

References

1. Chao DH, Ma J, Li XY (2009) Research on the reliability of SLD through accelerated life testing. In: 2009 8th international conference on reliability, maintainability and safety, pp 1263–1267
2. Ma X, Li M, Zhang SN et al (2011) The application status of accelerated test technology in gyroscopes life assessment. In: 2011 prognostics and system health management conference, pp 1–4
3. Zhang B (2008) Research on accelerated life test of the key component of fiber optical gyroscope. Beijing, Beijing University of 24
4. Chen SY, Zhao JJ, Ma J (2008) Feasibility of accelerated degradation test for FOG. J Chin Inertial Technol 16(5):623–626
5. Wang W, Zheng X, Meng XT (2012) Step-stress accelerated degradation technique with two accelerating stresses for fiber-optic gyroscope. Infrared Laser Eng 41(6):1007–2276
6. Liu L, Li XY, Jiang TM (2015) Evaluation method for accelerated degradation testing with interval analysis. J Beijing Univer Aeronaut Astronaut 41(12):1001–5965
7. Wang ZL, Huang HZ, Du L (2011) Reliability analysis on competitive failure processes under fuzzy degradation data. Appl Soft Comput 11(3):2964–2973
8. David GG, Rolando JPA, Mario CS et al (2014) A non-linear fuzzy regression for estimating reliability in a degradation process. Appl Soft Comput 16:1568–4946
9. Ma J, Wang DH, Chao DH et al (2009) Reliability evaluation of FOG based on key apparatus. J Chin Inertial Technol 17(5)
10. Wei W (2004) Relationship between performance of a 1.3μm double heterojunction superluminescent diode and its operation current and temperature. Infrared
11. Ye ZS, Xie M (2015) Stochastic modelling and analysis of degradation for highly reliable products. Appl Stochastic Mod Bus Ind 31(1):1524–1904
12. Park C, William P (2006) Stochastic degradation models with several accelerating variables 55(2):0018–9529
13. Ye ZS, Chen N, Shen Y et al (2015) A new class of Wiener process models for degradation analysis. Reliab Eng Syst Saf 139:0951–8320
14. Zhang ZX, Si XS, Hu CH et al (2018) Degradation data analysis and remaining useful life estimation: a review on Wiener-process-based methods. Euro J Oper Res 271(3):0377–2217
15. Si XS, Wang WB, Hu CH et al (2013) A Wiener-process-based degradation model with a recursive filter algorithm for remaining useful life estimation. Mech Syst Sig Proc 35(1–2):0888–3270
16. Wang XL, Jiang P, Guo B et al (2014) Real-time reliability evaluation with a general Wiener process-based degradation model. Q Reliab Eng Int 30(2):0748–8017
17. Escobar LA, Meeker WQ (2006) A review of accelerated test models. Statist Sci 0883–4237
18. Yu IT, Chang CL (2012) Applying Bayesian model averaging for quantile estimation in accelerated life tests. IEEE Trans Reliab 61(1):0018–9529

Printed in the United States
by Baker & Taylor Publisher Services